Lecture Notes in Computer Science 15483

Founding Editors

Gerhard Goos
Juris Hartmanis

Editorial Board Members

Elisa Bertino, *Purdue University, West Lafayette, IN, USA*
Wen Gao, *Peking University, Beijing, China*
Bernhard Steffen , *TU Dortmund University, Dortmund, Germany*
Moti Yung , *Columbia University, New York, NY, USA*

The series Lecture Notes in Computer Science (LNCS), including its subseries Lecture Notes in Artificial Intelligence (LNAI) and Lecture Notes in Bioinformatics (LNBI), has established itself as a medium for the publication of new developments in computer science and information technology research, teaching, and education.

LNCS enjoys close cooperation with the computer science R & D community, the series counts many renowned academics among its volume editors and paper authors, and collaborates with prestigious societies. Its mission is to serve this international community by providing an invaluable service, mainly focused on the publication of conference and workshop proceedings and postproceedings. LNCS commenced publication in 1973.

Minsu Cho · Ivan Laptev · Du Tran · Angela Yao ·
Hong-Bin Zha
Editors

Computer Vision – ACCV 2024 Workshops

17th Asian Conference on Computer Vision
Hanoi, Vietnam, December 8–12, 2024
Revised Selected Papers, Part II

Springer

Editors
Minsu Cho ⓘ
Pohang University of Science
and Technology (POSTECH)
Pohang-si, Korea (Republic of)

Ivan Laptev
Mohamed bin Zayed University of Artificial
Intelligence
Abu Dhabi, United Arab Emirates

Du Tran
Google
Mountain View, CA, USA

Angela Yao ⓘ
National University of Singapore
Singapore, Singapore

Hong-Bin Zha
Peking University
Beijing, China

ISSN 0302-9743   ISSN 1611-3349 (electronic)
Lecture Notes in Computer Science
ISBN 978-981-96-2643-4   ISBN 978-981-96-2644-1 (eBook)
https://doi.org/10.1007/978-981-96-2644-1

© The Editor(s) (if applicable) and The Author(s), under exclusive license
to Springer Nature Singapore Pte Ltd. 2025

This work is subject to copyright. All rights are solely and exclusively licensed by the Publisher, whether the whole or part of the material is concerned, specifically the rights of translation, reprinting, reuse of illustrations, recitation, broadcasting, reproduction on microfilms or in any other physical way, and transmission or information storage and retrieval, electronic adaptation, computer software, or by similar or dissimilar methodology now known or hereafter developed.
The use of general descriptive names, registered names, trademarks, service marks, etc. in this publication does not imply, even in the absence of a specific statement, that such names are exempt from the relevant protective laws and regulations and therefore free for general use.
The publisher, the authors and the editors are safe to assume that the advice and information in this book are believed to be true and accurate at the date of publication. Neither the publisher nor the authors or the editors give a warranty, expressed or implied, with respect to the material contained herein or for any errors or omissions that may have been made. The publisher remains neutral with regard to jurisdictional claims in published maps and institutional affiliations.

This Springer imprint is published by the registered company Springer Nature Singapore Pte Ltd.
The registered company address is: 152 Beach Road, #21-01/04 Gateway East, Singapore 189721, Singapore

If disposing of this product, please recycle the paper.

Foreword

Welcome to the 17th Asian Conference on Computer Vision (ACCV) 2024. This is the first time the conference is being hosted in Vietnam. The idea of bringing this prestigious event to Vietnam began in 2021, when two general and program chairs envisioned promoting the growth of computer vision in this region. By hosting the conference in a developing country, we aim to encourage other nations with less representation in the field to participate in this rapidly evolving discipline that is transforming lives globally. With Vietnam's affordable food and accommodation, we also hope that more participants will be able to attend and benefit from the conference.

ACCV 2024 is the first fully in-person ACCV event since the onset of the COVID-19 pandemic, with the last in-person edition held in 2018. This six-year gap meant that we couldn't rely on data from the past two conferences to accurately forecast submission and attendance numbers, making planning, budgeting, and logistics more challenging. As a result, we had to revise our estimates, plans, and budgets multiple times. We deeply appreciate the support of our Professional Conference Organizer, Nicole Finn, who enthusiastically (and courageously) agreed to help organize ACCV in a new region, outside the traditional North American and European venues she typically works with. Nicole worked tirelessly to navigate these changes, coordinating with hotels and convention centers to secure a venue that met both our needs and budget. We are also deeply grateful to our general chair, Richard Hartley, who took on the additional role of finance chair. Richard worked closely with Nicole, meticulously overseeing budget projections to maintain low registration fees while ensuring the financial sustainability of the conference. His dedication and sense of responsibility to the community, given his level of seniority, is a true testament to his commitment.

While ACCV 2024 aims for greater diversity and inclusion, our primary goal remains the exchange of scientific ideas while maintaining integrity and quality. This commitment is reflected in the program curation, led by our Program Chairs: Minsu Cho, Ivan Laptev, Du Tran, Angela Yao, and Hong-Bin Zha. They oversaw a rigorous process involving a thousand submissions and reviewers, ensuring that each paper received thorough and fair evaluation. Although smaller than conferences like CVPR, ACCV posed its own logistical challenges, including a tight timeline and budget constraints. Despite these difficulties, the Program Chairs were determined to hold an in-person Area Chair meeting to ensure high-quality decisions, even though this required additional effort and coordination—a clear reflection of their dedication. The area chairs who participated also valued the opportunity to connect and receive mentorship from more senior area chairs and the Program Chairs. As in-person meetings have become impractical for larger conferences, we hope that smaller conferences like ACCV can continue to offer this valuable in-person training for future area chairs.

This year's main conference is complemented by eleven workshops and two tutorials. We extend our thanks to the Workshop and Tutorial Chairs, Vineeth N. Balasubramanian, Li Liu, Anh Tran, and Kota Yamaguchi, who organized this crucial part of the program and coordinated with the ACML conference for a shared workshop day. This year, registered attendees for either ACCV or ACML can participate in workshops and tutorials across both conferences, fostering greater collaboration.

Our Publication Chairs, Yu-Lun Liu and Phong Nguyen, deserve special recognition for their outstanding work in assembling the camera-ready papers for the Springer volumes, ensuring their timely distribution. They courageously took on this challenging role without prior experience, fully aware of the limited time they would have to meet Springer's requirements once they received the camera-ready papers and final program. Their dedication and effort were crucial to the success of this process.

We are also grateful to our Local Chairs, Thi-Lan Le, Cuong Pham, and Minh-Triet Tran, who handled various local logistics, from working with vendors to organizing the area chair meeting and liaising with government officials for endorsements and approvals. VinAI, as a registered business entity in Vietnam, also played a crucial role in ensuring compliance with local regulations, coordinating with authorities, and handling endorsements. We thank the Vietnam's Ministry of Information and Communication for their support of this scientific event.

We are thankful to our Diversity and Inclusion Chairs, Supasorn Suwajanakorn and Miaomiao Liu, for their efforts in ensuring a diverse and inclusive selection process, even amidst budget uncertainties. Their work allowed us to support the participation of many who might not otherwise have attended. Financial support was made possible through sponsorships from several organizations: Asian Federation of Computer Vision, Sapien, Google, Springer, and the Australian Institute for Machine Learning. We also acknowledge the help from the Microsoft Research team, who provided the Conference Management Toolkit and technical support for ACCV 2024.

We hope you enjoy ACCV 2024 and take full advantage of the rich scientific program and networking opportunities this conference offers.

December 2024

Richard Hartley
C. V. Jawahar
Minh Hoai Nguyen
Dimitris Samaras
ACCV 2024 General Chairs

Preface

The 17th Asian Conference on Computer Vision (ACCV) 2024 was held in-person in Hanoi, Vietnam during December 8–12, 2024. The conference featured novel research contributions from almost all sub-areas of computer vision.

The ACCV 2024 main conference received 839 valid submissions (after desk rejections), which then entered the review stage. Fifty Area Chairs (ACs) and 1,020 reviewers made great efforts to ensure that every submission received thorough and high-quality reviews. Each submission received at least 3 reviews and was discussed by a panel of three expert ACs. Following previous editions of ACCV, this conference adopted a double-blind review process in which the identities of authors were not visible to the reviewers or area chairs, and the identities of the assigned reviewers and area chairs were not visible to the authors. The program chairs did not submit papers to the conference.

After receiving initial reviews, the authors had the option of submitting a rebuttal. In the post-rebuttal period reviewers reconsidered all paper materials including rebuttals and other reviews and provided their final recommendations. Area chairs were grouped into 17 conflict-free AC triplets. The area chairs led the discussions and made final recommendations based on reviews, author rebuttals, and discussions between reviewers. With the confirmation of three area chairs for each paper, 269 papers were accepted, making the acceptance rate of 32%. Among 269 accepted papers, 47 papers were selected for oral presentations.

ACCV 2024 also included eleven workshops and two tutorials. The ACCV 2024 workshops and tutorials were held on two days of December 8 and 9, 2024, with the first day (December 8th) being shared with ACML 2024 workshops. The proceedings of ACCV 2024 are open access at the Computer Vision Foundation website, by courtesy of Springer. The quality of the papers presented at ACCV 2024 demonstrates the research excellence of the international computer vision communities.

This year, ACCV 2024 also featured three keynote speeches given by Professor Michal Irani (Weizmann Institute of Science, Israel), Dr. Gérard Medioni (Amazon, USA), and Professor Deepak Pathak (Carnegie Mellon University, USA).

We would like to thank all the organizers, keynote speakers, area chairs, reviewers, and authors who made great contributions to ensure a successful ACCV 2024. Last but not least, we would also like to thank the attendees of ACCV 2024. Their presence showed strong commitment and appreciation towards this conference.

December 2024

Minsu Cho
Ivan Laptev
Du Tran
Angela Yao
Hong-Bin Zha
ACCV 2024 Program Chairs

Organization

General Chairs

Richard Hartley — Australian National University, Australia
C. V. Jawahar — IIIT Hyderabad, India
Minh Hoai Nguyen — VinAI and The University of Adelaide, Australia
Dimitris Samaras — Stony Brook University, USA

Program Chairs

Minsu Cho — Pohang University of Science and Technology, South Korea
Ivan Laptev — Mohamed bin Zayed University of Artificial Intelligence, United Arab Emirates
Du Tran — Google, USA
Angela Yao — National University of Singapore, Singapore
Hongbin Zha — Peking University, China

Workshop Chairs

Anh Tran — VinAI, Vietnam
Vineeth N. Balasubramanian — IIIT Hyderabad, India

Tutorial Chairs

Kota Yamaguchi — CyberAgent, Japan
Liu Liu — National University of Defense Technology, China

Publication Chairs

Yu-Lun Liu — National Yang Ming Chiao Tung University, Taiwan (R.O.C.)
Phong Nguyen — VinAI, Vietnam

Social Media Chair

Victor Escorcia   Samsung AI Center Cambridge, UK

Diversity and Inclusion Chairs

Miaomiao Liu	Australian National University, Australia
Supasorn Suwajanakorn	VISTEC

Technical Chairs

Seungwook Kim	Pohang University of Science and Technology, South Korea
Dayoung Gong	Pohang University of Science and Technology, South Korea

Local Chairs

Cuong Pham	PTIT and VinAI Research, Vietnam
Thi-Lan Le	Hanoi University of Science and Technology, Vietnam
Minh-Triet Tran	Ho Chi Minh City University of Science, Vietnam

Area Chairs

Abhinav Shrivastava	University of Maryland, USA
Anh Tran	VinAI, Vietnam
Asako Kanezaki	Tokyo Institute of Technology, Japan
Bohyung Han	Seoul National University, South Korea
Boxin Shi	Peking University, China
Chen Chen	University of Central Florida, USA
Fahad Shahbaz Khan	MBZUAI, United Arab Emirates
Gianni Franchi	ENSTA Paris, France
Go Irie	Tokyo University of Science, Japan
Guofeng Zhang	Zhejiang University, China

Hazel Doughty	Leiden University, Netherlands
Heng Wang	TikTok, USA
Hieu Pham	VinUniversity (VinUni), Vietnam
Hyung Jin Chang	University of Birmingham, UK
Jian Sun	Xi'an Jiaotong University, China
Jiwen Lu	Tsinghua University, China
Junsong Yuan	State University of New York at Buffalo, USA
Karteek Alahari	Inria, France
Khoi Nguyen	VinAI Research, Vietnam
Leonid Sigal	University of British Columbia, Canada
Makarand Tapaswi	IIIT Hyderabad, Wadhwani AI, India
Manmohan Chandraker	UC San Diego, USA
Ming-Hsuan Yang	University of California at Merced, USA
Ngan Le	University of Arkansas, USA
Rei Kawakami	Tokyo Institute of Technology, Japan
Saeed Anwar	Australian National University, Australia
Salman Khan	MBZUAI, United Arab Emirates
Seon Joo Kim	Yonsei University, South Korea
Shiguang Shan	Institute of Computing Technology, Chinese Academy of Sciences, China
Shijian Lu	Nanyang Technological University, Singapore
Shizhe Chen	Inria, France
Simon Lucey	University of Adelaide, Australia
Suha Kwak	POSTECH, South Korea
Tae-Hyun Oh	POSTECH, South Korea
Tae-Kyun Kim	KAIST/Imperial College London, South Korea
Tat-Jun Chin	The University of Adelaide, Australia
Triet Tran	Ho Chi Minh University of Science, VNU, Vietnam
Venkatesh Babu Radhakrishnan	Indian Institute of Science, India
Vicky Kalogeiton	Ecole Polytechnique, IP Paris, France
Vincent Lepetit	Ecole des Ponts ParisTech, France
Wen-Sheng Chu	Google, USA
Xiang Bai	Huazhong University of Science and Technology, China
Xiaodan Liang	MBZUAI, United Arab Emirates
Xiaojun Chang	Mohamed bin Zayed University of Artificial Intelligence, United Arab Emirates
Yasuyuki Matsushita	Osaka University, Japan
Yoichi Sato	University of Tokyo, Japan
Young Min Kim	Seoul National University, South Korea
Yusuke Matsui	The University of Tokyo, Japan

Zhanyu Ma — Beijing University of Posts and Telecommunications, China
Zuzana Kukelova — Center for Machine Perception, CTU in Prague, Czech Republic

Reviewers

Abbas Anwar
Abdallah Dib
Abdul Jabbar Siddiqui
Abhinav Kumar
Abulikemu Abuduweili
Aditya Arun
Aditya Sahdev
Aditya Singh
Adriano Fragomeni
Adrien Lafage
Ahmed Abdelkader
Ahyun Seo
Aishik Konwer
Akash Awasthi
Akin Caliskan
Akisato Kimura
Akos Godo
Akshay Kulkarni
Akshita Gupta
Alberto Marchisio
Alex Jinpeng Wang
Alexander Binder
Ali Athar
Ali Zia
Alper Yilmaz
Ameya Joshi
Amir Atapour-Abarghouei
Amitangshu Mukherjee
Anders Dahl
Andong Tan
Andrew Gilbert
Andrew Beng Jin Teoh
Aneeshan Sain
Anh Tran
Anh Vu Nguyen
Anh-Dzung Doan
Anh-Quan Cao
Aniket Roy
Anil Batra
Anoop Cherian
Anshul Shah
Anthony Hu
Anyi Rao
Ao Luo
Aoran Xiao
Arjan Kuijper
Asm Iftekhar
Atsushi Hashimoto
Avideep Mukherjee
Avinash Sharma
B. V. K. Vijaya Kumar
Bang Liu
Baoteng Li
Benjamin Killeen
Berthy Feng
Bharadwaj Ravichandran
Bin Chen
Bin Chen
Bingfeng Zhang
Binod Bhattarai
Bo Li
Bo Liu
Bo Miao
Bo Sun
Bo Wang
Bo Yang
Bo Zhang
Boeun Kim
Bogdan Raducanu
Bohong Chen
Bojian Wu
Boseung Jeong
Boshen Xu
Bowen Cai

Bowen Wen
Boyu Yang
Brian Clipp
Bruno Korbar
Buzhen Huang
Byung-Kwan Lee
Cameron Gordon
Carlos Rodriguez-Pardo
C. Kumar Mummadi
C. Hewa Koneputugodage
Chandra Kambhamettu
Chang Liu
Changmin Lee
Changtao Miao
Chanyong Jung
Chao Qu
Chao Wen
Chaojian Li
Chaoqin Huang
Chaowei Fang
Chau Pham
Che Sun
Chee Kheng Chng
Chen Feng
Cheng Chen
Cheng Long
Cheng Luo
Cheng Perng Phoo
Chengxin Liu
Chengxu Liu
Cheng-Yen Yang
Chengyuan Zhuang
Chengzhi Mao
Chengzhou Tang
Chenhongyi Yang
Chenwei Tang
Chien-Yao Wang
Chi-Han Peng
Ching Lam Choi
Chongjian Ge
Chongruo Wu
Cho-Ying Wu
Christos Kyrkou
Chuanguang Yang
Chuhua Xian

Chul Lee
Chun-Hsiao Yeh
Chunlei Peng
Chunxia Xiao
Chuong Huynh
Cong Wu
Congli Wang
Congpei Qiu
Daan de Geus
Daekyu Kwon
Da-Han Wang
Dahyun Kang
Daisuke Miyazaki
Dan Zeng
Danfeng Hong
Danpeng Chen
Darshan Singh S.
Dasong Li
David Chan
David Hart
Da-Wei Zhou
Decheng Liu
Deepika Bablani
Denis Baručić
Denys Rozumnyi
Deyi Ji
Di Yuan
Diego Thomas
Difei Gao
Dimitrios Sakkos
Ding-Jie Chen
Dingyi Yang
Dingyuan Zhang
Divy Kala
Divya Choudhary
Dong Wang
Dongdong Wang
Donggon Jang
Donghao Zhou
Donghwan Kim
Donghyeon Kwon
Dong-Jin Kim
Dongkai Wang
Dongkeun Kim
Dongliang Cao

Dongliang Chang
Dongqing Zou
Dongyoung Kim
Dongze Lian
Driton Salihu
Duc Vu
Duc Anh Nguyen
Duc Minh Vo
Duc-Tien Dang-Nguyen
Duolikun Danier
Duy Le
Duy Minh Ho Nguyen
Eldar Insafutdinov
En Yu
Erickson Nascimento
Eshika Khandelwal
Ethan Elms
Ethan Tseng
Fabio Pizzati
Fahad Shamshad
Faisal Qureshi
Fan Lu
Fan Yang
Fan Yang
Fangjinhua Wang
Fangyi Chen
Federico Stella
Fei Pan
Fei Xie
Fei Xue
Feiran Li
Fenggen Yu
Fengting Yang
Filippo Maggioli
Fiora Pirri
Florian Kleber
Fu-En Yang
Fu-Jen Chu
Fu-Jen Tsai
Fumihiko Sakaue
Fumio Okura
Furkan Kınlı
Gaku Nakano
Gangming Zhao
Gangwei Xu
Gianfranco Doretto
Gihyun Kwon
Giulio Rossolini
Gonçalo Dias Pais
Gongjie Zhang
Gu Wang
Guanglei Yang
Guangming Lu
Guangyu Sun
Guangzhi Wang
Guile Wu
Guo Chen
Guoqing Wang
Guorong Li
Gustavo Perez
Gwangtak Bae
Gyeongsik Moon
Gyuseong Lee
Hai D. Pham
Hai X. Pham
Haiming Xu
Haithem Turki
Haiyang Jiang
Han Hu
Han Qiu
Hanbyel Cho
Hangil Park
Hanjiang Hu
Hanjung Kim
Hanqing Sun
Hanxiao Jiang
Hao Wang
Hao Wang
Hao Zhao
Haobo Yuan
Haokun Lin
Haotong Lin
Haowei Tai
Haoyue Bai
Harry Cheng
Hashmat Shadab Malik
Heeseung Yun
Helder Araujo
Helena Maia
Heng Guo

Heng Li
Hengli Wang
Heran Yang
Heydi Mendez-Vazquez
Hideaki Uchiyama
Hieu Le
Himangi Mittal
Hiroaki Santo
Hirokatsu Kataoka
Hoàng-Ân Lê
Hoang-Quan Nguyen
Hojun Jang
Hong Wang
Hongchen Luo
Hong-Han Shuai
Hongje Seong
Hongji Guo
Hongjia Zhai
Hongyi Fan
Hsin-Ping Huang
Hsuan-I Ho
Hsu-Kuang Chiu
Hu Wang
Huafeng Wang
Huaiwen Zhang
Huan Wang
Huan Zheng
Huan Zheng
Huangying Zhan
Huanrui Yang
Huayi Zhou
Huei-Fang Yang
Hui Lin
Huijing Zhan
Huimin Ma
Huiming Sun
Huiyuan Yang
Hung Nguyen
Hung Tran
Hyeokjun Kweon
Hyung-gun Chi
Iago Suárez
Inho Kim
Inhwan Bae
Iuliia Kotseruba

Jaegul Choo
Jaeho Lee
Jaeseong Lee
Jaewon Lee
Jaime Cardoso
Janghoon Choi
Jean-Philippe Tarel
Jeonghun Baek
Jhih-Ciang Wu
Ji Hou
Ji Liu
Jia Wan
Jiachen Li
Jiachen Sun
Jiacheng Li
Jiafan Zhuang
Jiahao Nie
Jiahong Ouyang
Jiajun Tang
Jiali Duan
Jiaman Li
Jiaming Zhang
Jiande Sun
Jiang Liu
Jiange Yang
Jianglong Ye
Jiangpeng He
Jianing Li
Jianing Xi
Jianjia Wang
Jianqiao Zheng
Jianqiu Chen
Jianyang Gu
Jianzhong He
Jiaqi Li
Jiashuo Yu
Jia-Wang Bian
Jiawei He
Jiayi Ji
Jiaze Wang
Jiazhen Wang
Jichang Li
Jicheol Park
Jie Guo
Jie Hong

Jie Min
Jie Tang
Jie Yang
Jie Zhang
Jie Zhao
Jierui Lin
Jierun Chen
Jieyu Li
Jihyun Lee
Jin Fang
Jingchun Cheng
Jinghao Shi
Jinghua Hou
Jinghuan Shang
Jingjing Deng
Jingjing Xiong
Jingyi Zhang
Jinho Jeong
Jinhong Deng
Jinjian Wu
Jinsu Yoo
Jinsung Lee
Jinwoo Kim
Jinyoung Choi
Jinyu Cai
Jiyuan Liu
Jonathan Donnelly
Jongbin Ryu
Jongmin Lee
Jongwon Choi
JoonKyu Park
Joya Chen
Juan C. Sanmiguel
Jue Wang
Julian Tanke
Julie Mordacq
Junbin Xiao
Junbo Zhang
Junfei Yi
Jungchan Cho
Junha Lee
Junhan Chen
Junhao Dong
Junhyeong Cho
Junhyug Noh
Junhyun Lee
Junke Wang
Junlin Hu
Junsong Fan
Junwei Liang
Kai Chen
Kai Katsumata
Kai Wang
Kai Zhu
Kaihong Wang
Kailun Yang
Kamal Nasrollahi
K. Vaishnavi Gandikota
Kashu Yamazaki
Kavisha Vidanapathirana
Kazuhiro Hotta
Ke Xu
Keita Takahashi
Keke Tang
Keyan Wang
Khoi Pham
Kibok Lee
Kim Jun-Seong
Konstantinos Alexandridis
Konstantinos M. Dafnis
Koutilya Pnvr
Krishna Kanth Nakka
Kuan-Chih Huang
Kumar Ashutosh
Kun Fang
Kun Li
Kun Su
Kun Xia
Kun Xiang
Kun Zhou
Kunming Luo
Kwang Moo Yi
Kyong Hwan Jin
Kyoungkook Kang
Lang Nie
Latha Pemula
Lei Jin
Lei Tan
Lei Wang
Lei Zhu

L. Sampaio Ferraz Ribeiro
Leonardo Iurada
Leonardo Nunes
Li Ding
Li Niu
Li Song
Liang An
Liang Chen
Liang-Jian Deng
Liangke Gui
Lidong Yu
Ligong Han
Lile Cai
Lin Liu
Lin Geng Foo
Lingdong Kong
Linghao Chen
Linlin Shen
Liwei Yang
Li-Wei Kang
Lixiang Ru
Long Ma
Long Pham
Longfei Han
Longkun Zou
Lujun Li
Lumin Xu
Luwei Yang
Lv Tang
M. Yashwanth
Mahmoud Afifi
Manogna Sreenivas
Maoxun Yuan
Marc A. Kastner
Marco Piccirilli
M.-Luliana Georgescu
Marios Loizou
Martin Eisemann
Martin Mundt
Martin Weinmann
Matej Grcić
Matteo Dunnhofer
Matthew Beveridge
Matthew Gwilliam
Max Ehrlich

Maxwell Collins
Meghshyam Prasad
Mei Wang
Meirui Jiang
Meng Liu
Miaohui Wang
Miaomiao Liu
Michael Greenspan
Michael Wray
Mikhail Kennerley
Min Je Kim
Minesh Mathew
Mingde Yao
Mingfei Han
Ming-Feng Li
Mingfu Liang
Mingfu Xue
Minggui Teng
Minghui Hu
MingKun Yang
Ming-Kun Xie
Mingon Kang
Mingyuan Liu
Minh Luu
Minh Tran
Minkwan Kim
Mizuki Kojima
Mohammadreza Babaee
Mohammed Mahmoud
Mohit Goyal
Momin Abbas
Moon Ye-Bin
Mosam Dabhi
Mouin Ben Ammar
Mouxing Yang
Muhammad Maaz
Mutian Xu
Myungsub Choi
Nacim Belkhir
Nagabhushan Somraj
Nakamasa Inoue
Nakul Agarwal
Namyup Kim
Nanqing Dong
Nanyang Ye

Nayeong Kim
Necati Cihan Camgoz
Nhat Chung
Ni Zhang
Nico Messikommer
Nicola Garau
Niki Foteinopoulou
Nikita Durasov
Nikolaos Gkanatsios
Nikolaos Zioulis
Ningli Xu
Nirat Saini
Noranart Vesdapunt
Oh Hyun-Bin
Olivier Laurent
Omkar Thawakar
P. J. Narayanan
Pablo Garrido
P. Shivakumara
Paola Cascante-Bonilla
Parikshit Sakurikar
Paritosh Parmar
P. Kumar Anasosalu Vasu
Pedro Castro
Peiqi Duan
Peng Dai
Peng Wu
Peng-Hao Hsu
Pengliang Ji
Pengpeng Li
Pengpeng Zeng
Peter Kulits
Petra Bevandić
Pha Nguyen
Pingping Zhang
Pratik Vaishnavi
Pravin Nagar
Priyam Dey
Pulkit Kumar
Qi Bi
Qi Yu
Qiang Nie
Qiangmin Chen
Qiaole Dong
Qichen Fu

Qihao Liu
Qing Yu
Qing Zhang
Qingji Guan
Qingsong Zhao
Qingtian Zhu
Quan Dao
Quan Tang
Quang Nguyen
Quang Nguyen
Rahul Sajnani
Rakib Hyder
Ran Xu
Ratnesh Kumar
Rémi Kazmierczak
Renato Martins
Renjiao Yi
Renjie Wan
Reyer Zwiggelaar
Ricardo Garcia Pinel
Ridouane Ghermi
Robert Sablatnig
Robin Courant
Rohit Bharadwaj
Rohit Jena
Rohit Keshari
Rohit Kundu
Ronglai Zuo
Ronny Haensch
Rui Li
Rui Qian
Rui Xu
Rui Zhao
Rui Zhu
Rui Zhu
Ruikang Xu
Ruili Feng
Ruilong Li
Ruiqi Zhao
Ruixuan Yu
Ruizhi Shao
Runpei Dong
Runtian Zhai
Runze Li
Ruoshi Liu

Ruoteng Li
Ryo Furukawa
Ryo Hachiuma
Ryo Kawahara
Ryosuke Furuta
Ryota Yoshihashi
Ryuhei Hamaguchi
Salil Tambe
Salma Abdel Magid
Sampath Chanda
Sandesh Kamath
Sang Min Kim
Saptarshi Sinha
Sara Elkerdawy
Sara Pieri
Sateesh Kumar
Scott McCloskey
Sehyun Hwang
Sejong Yang
Seogkyu Jeon
Seonguk Seo
Seung Hyun Lee
Seungheon Kim
Seungho Lee
Seungjun Nah
Seungryul Baek
Seungwook Kim
Shady Abu-Hussein
Shailaja Keyur Sampat
Shaobing Gao
Shaolin Su
Shaoxiong Zhang
Shao-Yuan Lo
Shengeng Tang
Shengsheng Qian
Shenqi Lai
Sherry Chen
Shikun Li
Shin-Fang Chng
Shin'ichi Satoh
Shintaro Yamamoto
Shiqi Tian
Shiqiang Ma
Shivanand Venkanna Sheshappanavar
Shiyu Li
Shiyu Zhao
Shiyue Zhang
Shuaicheng Liu
Shuangrui Ding
Shuhong Zheng
Shuo Cheng
Shuvendu Roy
Shuwei Huo
Shuxiao Ding
Shuzhe Wu
Sicheng Zhao
Sihui Luo
Simon Reiß
Simon Woo
Sinisa Segvic
Sixun Dong
Siyuan Yang
Sonia Raychaudhuri
Soumya Banerjee
Srijan Das
Srinivas Rana
Sua Choi
Sucheng Ren
Sukrit Shankar
Sunny Bhati
S. Narasimhaswamy
Syed Talal Wasim
Sze Jue Yang
Taeyun Woo
Taihong Xiao
Taiki Sekii
Taisong Jin
Takahiro Kushida
Takahiro Okabe
Takashi Shibata
Takayuki Okatani
Takeshi Saitoh
Takuma Yagi
Takumi Karasawa
Tam Nguyen
Tao Hu
Tao Sun
Tao Wang
Tao Wu
Tao Wu

Tao Yu
Tao Zhang
Tao Zhou
Taoyue Wang
Tarun Kalluri
Tat-Jun Chin
Tengfei Liu
Thanh Le
Thao Nguyen
Thinh Phan
Thi-Thu-Huong Le
Thomas Leimkuehler
Thomas Westfechtel
Tiange Xiang
Tianpei Gu
Tianrui Chai
Tianyun Zhang
Tian-Zhu Xiang
Tiesong Zhao
Tingting Xie
Toby Breckon
Tomas F Yago Vicente
Tongda Xu
Tongyu Yang
Tooba Imtiaz
Toshihiko Yamasaki
Trong Thang Pham
Trong-Thuan Nguyen
Trung Pham
Trung-Nghia Le
Truong Vu
Tsai-Shien Chen
Tu Van Ninh
Tuan-Anh Vu
Tu-Khiem Le
Tung Do
Tung Kieu
Uy Tran
Van Nguyen Nguyen
Vasileios Mezaris
Vatsal Agarwal
Viet Nguyen
Vincent Cartillier
Vincent Gaudilliere
Vincent Tao Hu

Vinh-Tiep Nguyen
Viraj Shah
Vu Truong
Wataru Shimoda
Wei Liao
Wei Mao
Wei Wan
Weicheng Zhu
Weidong Cai
Weifeng Liu
Weihao Li
Weihao Xia
Weijie Lyu
Weijun Mai
Weilian Song
Weiwei Cai
Weiwei Xu
Weixiao Liu
Weixuan Tang
Wei-Yi Chang
Wenbin Li
Wenhan Yang
Wenhao Wang
Wenhui Zhou
Wenjia Wang
Wenke Huang
Wenqian Wang
Williem Williem
Wolfgang Fuhl
Woo Jae Kim
Woobin Im
Wooseok Lee
Xi Cheng
Xi Wang
Xi Yu
Xiang Chen
Xiang Gu
Xiang Wen
Xianghui Xie
Xiangwei Kong
Xiangyang Li
Xiangyu Xu
Xiankai Lu
Xiao Zhang
Xiaoguang Li

Xiaohan Chen
Xiaohua Huang
Xiaole Tang
Xiaopeng Ji
Xiaosong Jia
Xiaotao Hu
Xiaotong Luo
Xiaoyun Yuan
Xiaoyun Zhang
Xibin Song
Xijun Wang
Xin Feng
Xin Li
Xin Li
Xin Liao
Xin Wei
Xin Yang
Xin Yuan
Xin Zhou
Xindi Wu
Xingjiao Wu
xingkui Zhu
Xingtong Liu
Xingyu Liu
Xinhang Liu
Xinxin Zhu
Xiu Su
Xiujun Li
Xiujun Shu
Xiuwei Chen
Xu Cao
Xu Cao
Xu Yao
Xu Zhao
Xuan Ju
Xudong Liu
Xuepeng Shi
Xueqian Li
Xueting Liu
Xueying Jiang
Xugong Qin
Xuhua Huang
Xuqian Ren
Xuxin Lin
Yajie Wang
Yajing Zheng
Yan Wang
Yan Xia
Yan Yang
Yanan Li
Yanan Zhang
Yang Yang
Yangguang Zhu
Yanjing Li
Yannan Pu
Yanqing Shen
Yanru Xiao
Yansong Tang
Yao-Chih Lee
Yaojie Liu
Yaosi Hu
Yaqing Ding
Yash Bhalgat
Yash Mukund Kant
Yasunori Ishii
Yawen Zeng
Yazeed Alharbi
Ye Du
Ye Liu
Yeying Jin
Yi Chang
Yi Zhang
Yichang Shih
Yichao Cao
Yicong Li
Yifan Xing
Yifei Huang
Yihao Huang
Yihua Cheng
Yijie Lin
Yijie Zhong
Yilin Wen
Yingcheng Liu
Yingliang Zhang
Yinglin Zheng
Yingqian Wang
Yining Jiao
Yinjie Lei
Yinuo Jing
Yinyu Nie

Yipeng Qin
Yiqi Lin
Yiqi Zhong
Yiqing Shen
Yiqun Lin
Yiqun Wang
Yiran Guan
Yiran Xu
Yisi Luo
Yixuan Ren
Yizhak Ben-Shabat
Yizhen Lao
Yizhou Wang
Yizhou Wang
Yongtuo Liu
Yoonwoo Jeong
You Xie
YoungBin Kim
Youngseok Yoon
Youwei Lyu
Yu Liu
Yu Yin
Yu Zhang
Yuan Shen
Yuan Tian
Yuanhao Zhai
Yuchen Pei
Yucheng Zhao
Yuchong Sun
Yucong Shen
Yue Fan
Yue Xu
Yuecong Min
Yuedong Chen
Yueying Gao
Yueying Kao
Yufei Xie
Yuhe Jin
Yujia Chen
Yu-Jie Yuan
Yujun Cai
Yujun Tong
Yukang Cao
Yuki Fujimura
Yukun Huang

Yuliang Liu
Yu-Lun Liu
Yuming Gu
Yun Liu
Yun Xing
Yunfan Li
Yunhao Zou
Yunhua Zhang
Yunhui Guo
Yuning Cui
Yunqiu Xu
Yuqi Yang
Yuru Pei
Yusuke Hirota
Yusuke Kurose
Yusuke Sekikawa
Yusuke Sugano
Yuta Nakashima
Zakaria Laskar
Ze Yang
Zeeshan Khan
Zehua Sheng
Zerui Chen
Zeyu Xiao
Zezeng Li
Zhang Chen
Zhanzhan Cheng
Zhaodong Sun
Zhao-Min Chen
Zhaopei Huang
Zhaowen Li
Zhaoxin Fan
Zhaoyi An
Zhaoyu Chen
Zhen Chen
Zhen Chen
Zhen Liu
Zheng Chen
Zheng Chong
Zheng Qin
Zhengyi Luo
Zhennan Wang
Zhenwei Shi
Zhenyu Zhang
Zhenyue Qin

Zhepeng Wang
Zhewei Huang
Zhexiong Wan
Zhi Chen
Zhi Gao
Zhicheng Sun
Zhigang Tu
Zhihua Liu
Zhijian Huang
Zhijie Deng
Zhijie Wang
Zhikang Wang
Zhiliang Wu
Zhiming Zou
Zhineng Chen
Zhipeng Fan
Zhiqi Huang
Zhiqi Kang
Zhi-Qi Cheng
Zhixuan Yu
Zhiyong Wang
Zhiyuan Mao
Zhong Li

Zhongqun Zhang
Zhongyun Hu
Zhuangzhuang Chen
Zhuo Su
Zhuoran Yu
Ziang Cao
Zicheng Zhang
Zichun Zhong
Zikai Song
Zikui Cai
Zikun Zhou
Ziqi Huang
Ziqi Zhang
Ziqiang Li
Ziteng Cui
Ziteng Gao
Zitong Yu
Zixiang Zhao
Zixuan Jiang
Ziyao Zeng
Zi-Yi Dou
Zizheng Yan
Zutao Jiang

Contents – Part II

MLCSA2024

Adaptive Dual Attention into Diffusion for 3D Medical Image Segmentation 3
  Nhu-Tai Do, Van-Hung Bui, and Quoc-Huy Nguyen

The Application of Graph Attention Mechanism in the Automation of Analog Circuit Design 18
  Xinpeng Li, Minglei Tong, and Yongqing Sun

CROCODILE: Crop-Based Contrastive Discriminative Learning for Enhancing Explainability of End-to-End Driving Models 31
  Chenkai Zhang, Daisuke Deguchi, Jialei Chen, Zhenzhen Quan, and Hiroshi Murase

Deterministic Guided Progressive Medical Image Cross-Modal Generation Based on Deep Learning 47
  Chujie Zhang, Lanfen Lin, and Yen-Wei Chen

Separate Guided Denoising Training for Human-Object Interaction Detection 58
  Yuki Isoda and Daisuke Kobayashi

GTA: Global Tracklet Association for Multi-object Tracking in Sports 74
  Jiacheng Sun, Hsiang-Wei Huang, Cheng-Yen Yang, Zhongyu Jiang, and Jenq-Neng Hwang

IsoTGAN: Spatial and Geometrical Constraints at GAN and Transformer for 3D Contour Generation 88
  Thao Nguyen Phuong, Vinh Nguyen Duy, and Hidetomo Sakaino

Hierarchical Feature Aggregation Network Based on Swin Transformer for Medical Image Segmentation 105
  Hayato Iyoda, Yongqing Sun, and Xian-Hua Han

E2CANet: An Efficient and Effective Convolutional Attention Network for Semantic Segmentation 118
  Yuerong Mu and Qiang Guo

3-D Reconstruction from Consecutive Endoscopic Images Using Gaussian
Splatting 134
    Hung-Le Minh, Duy-Van Truong, Huy-Xuan Manh, Viet-Hang Dao,
    Phuc-Binh Nguyen, Thanh-Tung Nguyen, and Hai Vu

EMMA: EMotion Mixing Algorithm for Compound Expression
Recognition Using Angle-Based Metric Learning 147
    Riku Yamamoto and Noriko Takemura

LViTES: Leveraging Vision and Text for Enhancing Segmentation
of Endoscopic Images 163
    Thang La, Minh-Hanh Tran, Viet-Hang Dao, and Thanh-Hai Tran

RichMediaGAI

GameIR: A Large-Scale Synthesized Ground-Truth Dataset for Image
Restoration over Gaming Content 179
    Lebin Zhou, Kun Han, Nam Ling, Wei Wang, and Wei Jiang

Vector Logo Image Synthesis Using Differentiable Renderer 197
    Ryuta Yamakura and Keiji Yanai

HYPNOS: Highly Precise Foreground-Focused Diffusion Finetuning
for Inanimate Objects 211
    Oliverio Theophilus Nathanael, Jonathan Samuel Lumentut,
    Nicholas Hans Muliawan, Edbert Valencio Angky,
    Felix Indra Kurniadi, Alfi Yusrotis Zakiyyah, and Jeklin Harefa

GraVITON: Graph Based Garment Warping with Attention Guided
Inversion for Virtual-Tryon 228
    Sanhita Pathak, Vinay Kaushik, and Brejesh Lall

Image and Video Compression Using Generative Sparse Representation
with Fidelity Controls 245
    Lebin Zhou, Wei Wang, and Wei Jiang

Enhancing Continuous Skeleton-Based Human Gesture Recognition
by Incorporating Text Descriptions 261
    Thi-Lan Le, Viet-Duc Le, and Thuy-Binh Nguyen

Towards Robust Video Frame Interpolation with Long-Term Propagation ... 276
    Ziqi Huang, Kelvin C. K. Chan, Bihan Wen, and Ziwei Liu

WiCV

Enhanced Survival Prediction in Head and Neck Cancer Using
Convolutional Block Attention and Multimodal Data Fusion 295
  Aiman Farooq, Utkarsh Sharma, and Deepak Mishra

Spatial Clustering and Machine Learning for Crime Prediction: A Case
Study on Women Safety in Bhopal 306
  Yamini Sahu and Vaibhav Kumar

Building Usage Classification in Indian Cities: Utilizing Street View
Images and Object Detection Models 322
  Yamini Sahu, Vasu Dhull, Satyajeet Shashwat, and Vaibhav Kumar

3D-CmT: 3D-CNN Meets Transformer for Hyperspectral Image
Classification 337
  Sunita Arya, Shiv Ram Dubey, S. Manthira Moorthi, Debajyoti Dhar, and Satish Kumar Singh

Author Index 355

Contents – Part I

AWSS

BgSub: A Background Subtraction Model for Effective Moving Object Detection 3
Islam Osman and Mohamed S. Shehata

Physically Interpretable Probabilistic Domain Characterization 17
Anaïs Halin, Sébastien Piérard, Renaud Vandeghen, Benoît Gérin, Maxime Zanella, Martin Colot, Jan Held, Anthony Cioppa, Emmanuel Jean, Gianluca Bontempi, Saïd Mahmoudi, Benoît Macq, and Marc Van Droogenbroeck

Analysis of Adapter in Attention of Change Detection Vision Transformer 36
Ryunosuke Hamada, Tsubasa Minematsu, Cheng Tang, and Atsushi Shimada

Supervised Domain Adaptation with Disjoint Label Spaces for Fine-Grained Classification 52
Enrico Krohmer, Stefan Wolf, and Jürgen Beyerer

Leveraging Thermal Imaging for Robust Human Pose Estimation in Low-Light Vision 69
Mickael Cormier, Caleb Ng Zhi Yi, Andreas Specker, Benjamin Blaß, Michael Heizmann, and Jürgen Beyerer

U-ENHANCE: Underwater Image Enhancement Using Wavelet Triple Self-attention 87
Priyanka Mishra, Santosh Kumar Vipparthi, and Subrahmanyam Murala

WARMOS: Enhancing Weather-Affected Referred Moving Object Segmentation 105
Prafulla Saxena, Dinesh Kumar Tyagi, Santosh Kumar Vipparthi, and Subrahmanyam Murala

Robust Anomaly Detection Through Transformer-Encoded Feature Diversity Learning 118
Kuldeep Biradar, Dinesh Kumar Tyagi, Ramesh Babu Battula, and P. Subbarao

Adversarial Weather-Resilient Image Retrieval: Enhancing Restoration
Using Captioning for Robust Visual Search 132
  Prem Shanker Yadav, Kushall Singh, Dinesh Kumar Tyagi,
  and Ramesh Babu Battula

RW-SVD: A Surround View Rough Weather Video Anomaly Dataset
and a Brief Overview of Existing Datasets 146
  Sachin Dube, Dinesh Kumar Tyagi, and Ramesh Babu Battula

GAISynMeD

Unsupervised Skull Segmentation via Contrastive MR-to-CT Modality
Translation 165
  Kamil Kwarciak, Mateusz Daniol, Daria Hemmerling,
  and Marek Wodzinski

CleftLipGAN : Interactive GAN-Inpainting for Post-Operative Cleft Lip
Reconstruction 180
  Daniel Anojan Atputharuban, Christoph Theopold, and Aonghus Lawlor

A Comparative Study on Diffusion Sampling Methods Across Diverse
Medical Imaging Modalities 197
  Muhammad Ali Farooq, Ayman Abaid, Ihsan Ullah, and Peter Corcoran

Medical Imaging Complexity and Its Effects on GAN Performance 211
  William Cagas, Chan Ko, Blake Hsiao, Shryuk Grandhi,
  Rishi Bhattacharya, Kevin Zhu, and Michael Lam

LAMM

RSSep: Sequence-to-Sequence Model for Simultaneous Referring Remote
Sensing Segmentation and Detection 223
  Ngoc-Vuong Ho, Thinh Phan, Meredith Adkins, Chase Rainwater,
  Jackson Cothren, and Ngan Le

Smart Camera Parking System with Auto Parking Spot Detection 237
  Tuan T. Nguyen and Mina Sartipi

LAVA

Questioning, Answering, and Captioning for Zero-Shot Detailed Image
Caption 249
  Duc-Tuan Luu, Viet-Tuan Le, and Duc Minh Vo

Exploring Cross-Attention Maps in Multi-modal Diffusion Transformers
for Training-Free Semantic Segmentation 267
   Rento Yamaguchi and Keiji Yanai

Enhancing Visual Question Answering with Pre-trained Vision-Language
Models: An Ensemble Approach at the LAVA Challenge 2024 281
   *Trong-Hieu Nguyen-Mau, Nhu-Binh Nguyen Truc, Nhu-Vinh Hoang,
Minh-Triet Tran, and Hai-Dang Nguyen*

DermAI: A Chatbot Assistant for Skin Lesion Diagnosis Using Vision
and Large Language Models 293
   *Viet-Tham Huynh, Trong-Thuan Nguyen, Thao Thi-Phuong Dao,
Tam V. Nguyen, and Minh-Triet Tran*

Mitigating Backdoor Attacks Using Activation-Guided Model Editing 308
   *Felix Hsieh, Huy H. Nguyen, AprilPyone MaungMaung,
Dmitrii Usynin, and Isao Echizen*

Exploring Visual Multiple-Choice Question Answering with Pre-trained
Vision-Language Models 324
   Gia-Nghia Tran, Duc-Tuan Luu, and Dang-Van Thin

An Approach to Complex Visual Data Interpretation with Vision-Language
Models 338
   *Thanh-Son Nguyen, Viet-Tham Huynh, Van-Loc Nguyen,
and Minh-Triet Tran*

Author Index 355

MLCSA2024

Adaptive Dual Attention into Diffusion for 3D Medical Image Segmentation

Nhu-Tai Do[1], Van-Hung Bui[2], and Quoc-Huy Nguyen[1](✉)

[1] Saigon University, Ho Chi Minh City, Vietnam
{dntai,nqhuy}@sgu.edu.vn
[2] Ho Chi Minh University of Science, Ho Chi Minh City, Vietnam
19127414@student.hcmus.edu.vn

Abstract. Denoising diffusion models have recently demonstrated great success in generating detailed pixel-wise representations for image generation. Applications like Dall-E, Stable Diffusion, and Midjourney have showcased impressive image-generation capabilities, sparking significant discussion within the community. Recent studies have also highlighted the utility of these models in various other vision tasks, including image deblurring, super-resolution, and image segmentation. This work introduces a novel Adaptive Dual Attention into Diffusion model for 3D medical image segmentation. Applying diffusion models to 3D medical image segmentation presents significant challenges. The alignment of semantic features necessary for conditioning the diffusion process with noise embedding is often inadequate. Additionally, traditional U-Net backbones in diffusion models are not sufficiently sensitive to the contextual information required for accurate pixel-level segmentation during reverse diffusion. Our method, which integrates Adaptive Dual Attention into Diffusion, addresses these issues by capturing local and global contextual information, enhancing the precision and robustness of 3D image segmentation. Our approach surpasses current state-of-the-art methods on the BraTS2020 dataset, achieving higher segmentation accuracy. This improved performance can significantly aid in diagnosing and treating medical conditions by enabling highly accurate segmentation of anatomical structures in 3D medical images.

Keywords: Diffusion Probabilistic Model · Medical Image Segmentation · Dual Domain Attention · Volumetric Data

1 Introduction

Medical volumetric segmentation is a critical task in medical image analysis, enabling precise delineation of anatomical structures and lesions within high-dimensional datasets on a voxel-by-voxel basis [10]. Accurate segmentation provides essential diagnostic information to clinicians, aiding in disease diagnosis, treatment planning, and patient management. Traditional methods, typically utilizing encoder-decoder architectures with skip connections, often fall short in capturing global contextual information due to their reliance on convolutional neural networks (CNNs) with localized receptive fields [18].

Recent advancements in transformer-based models have shown promise in addressing these limitations by leveraging self-attention mechanisms to capture global dependencies. Models such as SwinUNETR [5] and UNETR [4] have demonstrated improved segmentation accuracy by integrating transformers with CNNs. However, these models still encounter challenges in efficiently extracting multi-scale features and handling the high computational complexity associated with volumetric data [9].

Denoising diffusion models have emerged as potent tools for generating semantically meaningful pixel-wise representations [8]. These models have been successfully applied to various generative tasks, including image synthesis and restoration. In medical image segmentation, diffusion models offer the advantage of progressively refining image representations, making them well-suited for generating detailed segmentations. However, existing diffusion-based approaches are often constrained to 2D and binary segmentation tasks, limiting their applicability in more complex medical imaging scenarios [1, 19].

1.1 Difference Between 2D and 3D Medical Image Segmentation

Medical image segmentation can be approached in two-dimensional (2D) and three-dimensional (3D) contexts, each presenting unique challenges and considerations.

2D Medical Image Segmentation: In 2D segmentation, images are processed slice by slice, treating each slice independently. This approach simplifies computational complexity and reduces memory requirements. However, it often fails to capture spatial continuity and contextual information across slices, leading to inconsistencies in the segmentation of adjacent slices. This limitation is particularly problematic in medical imaging, where anatomical structures extend across multiple slices.

3D Medical Image Segmentation: Conversely, 3D segmentation methods process volumetric data, simultaneously considering the entire stack of slices. This approach leverages the spatial relationships and continuity inherent in volumetric data, leading to more accurate and coherent segmentation results. However, 3D segmentation poses significant computational challenges due to the increased data dimensionality and memory requirements. Models designed for 3D segmentation must efficiently handle high-resolution volumetric data while capturing local and global contextual information.

This work focuses on 3D medical image segmentation to fully exploit the rich spatial information available in volumetric scans, such as MRI and CT images. By designing our framework to operate on 3D data, we aim to achieve higher segmentation accuracy and robustness than 2D approaches.

1.2 Contributions

The primary motivation for developing the Adaptive Dual Attention Diffusion (ADAD) framework stems from the limitations of traditional U-Net architectures and the challenges associated with aligning semantic features with noise embeddings in diffusion models. Traditional U-Nets, while effective for many segmentation tasks, cannot often capture global context due to their localized receptive fields. This limitation becomes pronounced in 3D medical image segmentation, where understanding the broader spatial relationships within the volume is crucial for accurately delineating anatomical structures.

Diffusion models, with their iterative denoising processes, offer a unique advantage in progressively refining image representations. However, effectively integrating semantic information with noise embeddings throughout the diffusion process remains challenging. Our approach addresses this by introducing dual-domain attention mechanisms that enhance the model's sensitivity to both local details and global context.

Our main contributions are as follows:

- **We introduce an end-to-end framework called Adaptive Dual Attention Diffusion (ADAD):** Our ADAD framework is designed to enhance 3D medical image segmentation by leveraging a 3D image diffusion model to improve segmentation accuracy.
- **We embed a 3D Dual Domain Attention (3D-DDA) block:** To augment the robustness and accuracy of our model, we incorporate a 3D-DDA block [2] within the feature encoder section of the network. This block captures local and global features simultaneously, enhancing the network's capability to handle the variability and complexity of medical images.
- **We introduce Dynamic Conditional Encoding:** To address the challenge of ambiguous and low-contrast regions in medical images, we integrate a dynamic conditional encoding strategy. This approach incorporates current-step segmentation information into the raw image encoding, dynamically enhancing segmentation accuracy at each step.
- **We implement a Step-Uncertainty based Fusion (SUF) module:** During the inference phase, we utilize an SUF module to aggregate predictions from multiple diffusion steps. This module leverages uncertainty information to produce more accurate segmentation results, improving the model's robustness.

We evaluated our ADAD framework on the BraTS2020 benchmark dataset, demonstrating its superior performance compared to state-of-the-art methods. Our results underscore the potential of the ADAD framework to significantly advance medical volumetric segmentation significantly, providing more precise and reliable tools for clinical diagnostics and treatment planning. This innovative approach holds promise for improving patient outcomes by enabling more accurate and detailed segmentation of anatomical structures.

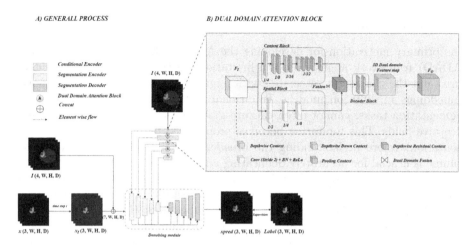

Fig. 1. The proposed method: A) General process of diffusion for 3D medical image segmentation, B) Detailed structure of the proposed 3D Dual Domain Attention (3D-DDA) block.

2 Proposed Method

2.1 Overview

The overall process is shown in Fig. 1. ADAD captures local and global features by combining advanced diffusion models with attention mechanisms, improving segmentation accuracy. The key components include the conditional encoder, segmentation encoder, segmentation decoder, and the 3D dual-domain attention (3D-DDA) block, each playing a critical role in the workflow.

The process begins with the input of 3D medical images I, which have dimensions $(4, W, H, D)$, where W, H, and D represent the width, height, and depth of the image, respectively. These images undergo initial preprocessing, resulting in a noisy version x used for the diffusion process. Unlike traditional U-net methods that use volume data to predict segmentation labels directly, the diffusion model learns by removing noise. It takes a noisy volumetric image and segmentation labels as input and gradually removes the noise to create clear segmentation results.

The conditional encoder processes the input image I, generating essential feature representations that condition the diffusion process at each time step t. These feature representations are crucial for guiding the segmentation process, ensuring the model retains important contextual information from the original image. The 3D-DDA block is strategically placed within the conditional encoder to enhance the feature extraction process. The 3D-DDA block employs dual-domain attention mechanisms to capture local and global features, significantly improving the network's ability to handle the complexity and variability inherent in medical images.

At each time step t, the segmentation encoder integrates the features from the conditional encoder and the 3D-DDA block, producing an intermediate representation x_t. These concatenated features (I, x_t) are passed through the denoising module. This module consists of a series of convolutional layers designed to denoise the input and refine the segmentation progressively.

The segmentation labels x_0 are first converted into one-hot encoded labels. To generate the noisy labels x_t used in the diffusion process, we apply the forward diffusion equation:

$$x_t = \sqrt{\bar{\alpha}_t} x_0 + \sqrt{1 - \bar{\alpha}_t} \epsilon \qquad (1)$$

where $\epsilon \sim \mathcal{N}(0, I)$ is Gaussian noise, and $\bar{\alpha}_t$ is a predefined variance schedule controlling the amount of noise added at each time step t.

In Fig. 1, this process is depicted where the segmentation labels x_0 are transformed into x_t through the addition of Gaussian noise, preparing them for the reverse diffusion process during training.

The final predicted segmentation x_{pred} is supervised using the ground truth labels Label, which have dimensions $(3, W, H, D)$. The supervision process involves comparing the predicted segmentation with the ground truth labels and calculating the loss. This loss is then backpropagated through the network to optimize the parameters, enhancing the model's accuracy and robustness over successive training iterations.

2.2 Adaptive Dual Domain Attention Encoder

As shown in Fig. 1, the conditional encoder is designed to process the input image and extract rich features necessary for accurate segmentation. The conditional encoder initially processes the input image I to generate feature representations crucial for conditioning the diffusion process at each time step t. These features retain essential contextual information from the original image, ensuring the subsequent segmentation steps are well-informed.

3D Dual Domain Attention (3D-DDA) block: The 3D Dual Domain Attention (3D-DDA) block is a crucial component of the ADAD framework, designed to capture both local spatial details and global contextual information. The architecture of the 3D-DDA block integrates spatial-domain and context-domain attentions, leveraging residual learning to refine feature maps at each stage of the encoder. The 3D-DDA block operates on the encoding feature map $\mathbf{F}_I$ and produces a refined feature map $\mathbf{F}_O$ by combining spatial and contextual information. Mathematically, this can be expressed as:

$$\mathbf{F}_O = \mathbf{F}_I + \text{DDA3D}(\mathbf{F}_I) \qquad (2)$$

As shown in Eq. 2, the output of the 3D-DDA block is added to the input feature map $\mathbf{F}_I$ to produce the refined feature map $\mathbf{F}_O$. This residual connection helps preserve the original features while enhancing them with attention mechanisms.

The dual-domain attention mechanism within the 3D-DDA block consists of two main components: the Spatial-Domain Block ($\mathcal{S}$) and the Context-Domain

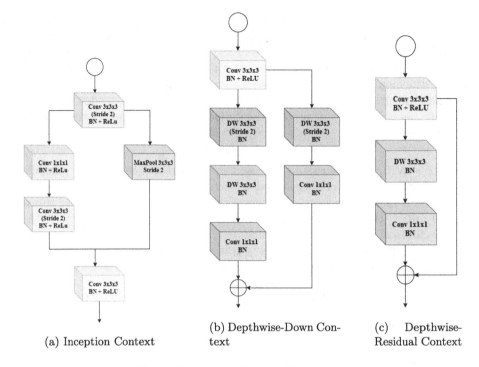

Fig. 2. 3D Context-Domain Block details

Block ($\mathcal{C}$). These components work together to enhance the feature map by focusing on different aspects of the data.

Spatial-Domain Block: The spatial-domain block ($\mathcal{S}$) stacks k convolutional layers to capture fine-grained spatial details. Each block f_i is defined:

$$\mathcal{S}(\mathbf{F}_I) = f_k \circ f_{k-1} \circ \cdots \circ f_1(\mathbf{F}_I) \tag{3}$$

where

$$f_i(\mathbf{x}) = \sigma(\mathrm{BN}(\mathrm{Conv3D}_{s=2}(\mathbf{x}))) \tag{4}$$

Here, k denotes the number of convolutional layers, σ is the ReLU activation function, and $\circ$ represents the composition of layers.

Context-Domain Block: The context-domain block ($\mathcal{C}$) consists of t sequential components designed to capture global contextual information:

$$\mathcal{C}(\mathbf{F}_I) = g_{\mathrm{pool}}\left(f_t \circ f_{t-1} \circ \cdots \circ f_1(\mathbf{F}_I)\right) \tag{5}$$

Each component f_i is defined as:

$$f_i(x) = g_{\mathrm{res}_i} \circ g_{\mathrm{down}_i}(x) \tag{6}$$

where g_{res_i} represents the i-th residual block, and g_{down_i} denotes the downsampling operation (Fig. 2).

Dual-Domain Fusion: Before combining the spatial-domain feature map $\mathcal{S}(\mathbf{F}_I)$ and the context-domain feature map $\mathcal{C}(\mathbf{F}_I)$, we apply resizing operations to ensure they have the same spatial dimensions. Specifically, we upsample the context-domain feature map using trilinear interpolation:

$$\tilde{\mathcal{C}}(\mathbf{F}_I) = \text{Upsample}(\mathcal{C}(\mathbf{F}_I)) \tag{7}$$

With both feature maps aligned in size, we perform element-wise addition:

$$\text{DDA3D}(\mathbf{F}_I) = \varphi_s \left(\mathcal{D} \left(\mathcal{S}(\mathbf{F}_I) + \tilde{\mathcal{C}}(\mathbf{F}_I) \right) \right) \tag{8}$$

This fusion effectively combines local and global information, enhancing the feature representation for segmentation tasks (Fig. 3).

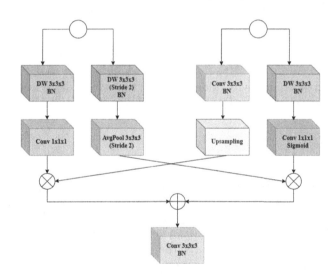

Fig. 3. Dual Domain Fusion Block

3D Decoder Block ($\mathcal{D}$): The 3D decoder block returns the fused feature map to the original resolution while preserving the learned features. this block consists of three transposed convolution layers, each paired with batch normalization and a ReLU activation unit, followed by a Conv3D layer with a kernel size of $1 \times 1 \times 1$. This setup decodes and normalizes the dual-domain feature map, resizing it to match the dimensions of the input feature map. The operation of the decoder block can be represented as:

$$\mathcal{D}(\mathbf{F}_{\text{fused}}) = \text{ConvTranspose3D}_{\text{BN, ReLU}}(\mathbf{F}_{\text{fused}}) \tag{9}$$

where $\mathbf{F}_{\text{fused}}$ is the fused feature map from the dual-domain fusion block.

The decoder block utilizes skip connections from corresponding encoder layers to retain high-resolution features and combine them with the upsampled feature maps. This approach helps preserve spatial details and improves overall segmentation accuracy. The final output of the decoder block is a refined feature map that has been upsampled to the original resolution of the input image, ready for generating the final segmentation map.

3 3D Diffusion

3.1 Label Embedding

This study introduces a Label Embedding operation to overcome the limitation of traditional diffusion models that can only handle binary segmentation. This operation converts the segmentation label map into one-hot encoded labels, enabling the model to segment multiple targets simultaneously [7]. The one-hot encoding process converts a label x_0 with dimensions $D \times W \times H$ into a multi-channel label $x_0 \in \mathbb{R}^{N \times D \times W \times H}$, where N is the number of labels.

$$x_t = \sqrt{\bar{\alpha}_t} x_0 + \sqrt{1 - \bar{\alpha}_t} \epsilon \tag{10}$$

where x_t is the label map with added noise at time step t, $\bar{\alpha}_t$ is a scaling factor, and ϵ is the Gaussian noise. The goal is to predict the clear label map x_0 from the noisy input x_t.

3.2 Denoising Module

The Denoising Module in the framework consists of an Adaptive Dual Domain Attention Encoder (ADDA Encoder) and a Denoising-UNet (DU). The ADDA Encoder extracts multi-scale features from the input volume data I, which are then combined with the noisy one-hot encoded label x_t and input into the encoder part of the DU. Subsequently, the DU processes these features through its encoder-decoder architecture to generate the denoised output [20]. This process can be described as:

$$\hat{x}_0 = \text{DU}(\text{cat}(I, x_t), t, \tilde{I}_f) \tag{11}$$

where $\hat{x}_0$ is the predicted clear label map, $\tilde{I}_f$ represents the multi-scale features from the ADDA Encoder, and cat denotes channel-wise concatenation. The integration of the ADDA Encoder and Denoising-UNet (DU) in the Denoising Module represents an advanced approach to image denoising. By utilizing the multi-scale features extracted by the ADDA Encoder and incorporating the noisy label information, the DU effectively denoises the input data, demonstrating the potential of deep learning models in addressing noise reduction challenges.

The Denoising-UNet (DU) is trained using a combination of loss functions to optimize its performance. Specifically, the training process incorporates the Dice Loss, Binary Cross-Entropy (BCE) Loss, and Mean Squared Error (MSE)

Loss. The Dice Loss function evaluates the overlap between predicted and ground truth segmentation masks, providing a pixel-wise measure of accuracy. The BCE Loss function focuses on the classification aspect, aiding in the pixel-wise segmentation of the input data. Additionally, the MSE Loss function contributes to minimizing the differences between the generated denoised output and the ground truth data, ensuring the preservation of essential details and structures in the denoised images [17].

$$\mathcal{L}_{\text{total}} = \mathcal{L}_{\text{dice}}(\hat{x}_0, x_0) + \mathcal{L}_{\text{bce}}(\hat{x}_0, x_0) + \mathcal{L}_{\text{mse}}(\hat{x}_0, x_0) \tag{12}$$

3.3 Step-Uncertainty Based Fusion

The testing phase of the diffusion model involves iteratively refining the segmentation prediction through multiple steps using the Denoising Diffusion Implicit Models (DDIM) method. In addition to the final prediction from the last step, the Step-Uncertainty based Fusion (SUF) module is incorporated to amalgamate predictions from each iterative step. The uncertainty for each prediction step is estimated similarly to Monte Carlo Dropout [3], where the average prediction $\bar{p}_i$ and the uncertainty u_i are computed as:

$$u_i = -\bar{p}_i \log(\bar{p}_i), \quad \text{where} \quad \bar{p}_i = \frac{1}{S} \sum_{s=1}^{S} p_{s,i} \tag{13}$$

and S is the number of forward passes. The fusion weights w_i for each prediction step are determined by combining the step index and uncertainty, expressed as:

$$w_i = \exp(\sigma(\frac{i}{\text{scale}}) \times (1 - u_i)) \tag{14}$$

where σ is the sigmoid function, and scale is a normalization factor. The final segmentation result Y is obtained by weighting the predictions from all steps:

$$Y = \sum_{i=1}^{t} w_i \times \bar{p}_i \tag{15}$$

4 Experiments

4.1 Dataset and Experimental Setup

Dataset: The BraTS2020 [12] dataset comprises 369 pre-aligned MRI scans across four modalities: T1, T1ce, T2, and FLAIR. These scans are accompanied by expert-annotated segmentation masks, which identify GD-enhancing tumors, peritumoral edema, and tumor core regions. Each modality is represented as a 155×240×240 volume, with all images resampled and co-registered to ensure uniformity. The segmentation task focuses on delineating the whole tumor (WT), enhancing tumor (ET), and tumor core (TC) areas. For model training and

Table 1. Results on conventional networks with and without 3D-DDA

Validation set	Dice Score		
	WT	TC	ET
V-Net [13]	0.8010	0.7010	0.6840
Swin-UNETR [4]	0.8221	0.7482	0.7365
SegResNet [14]	0.8958	0.7911	0.7275
DynUnet [6]	0.8809	0.7857	0.7339
SegResNet + 3D-DDA [2]	0.8959	0.7911	0.7361
Diffusion+3D-DDA(Our)	**0.9155**	**0.8475**	**0.7364**

evaluation, the dataset is divided into training, validation, and test sets with a ratio of 0.7, 0.1, and 0.2, respectively.

Training Process: We randomly cropped the input scans and corresponding ground-truth labels to a size of 128×128×128, transformed them to a spacing of 1 mm, and applied random flips along the axial, sagittal, and coronal axes. The intensity of each voxel was adjusted by 10% and normalized across each channel. Labels were converted to one-hot encoding. The training was conducted over 150 epochs using the Adam optimizer with an initial learning rate of 0.0001, batch size 1. The learning rate schedule followed a Cosine Annealing approach, with a minimum learning rate of 0.00001. All experiments were performed on NVIDIA Tesla P100 16GB hardware, utilizing the PyTorch and MONAI frameworks. The visualization of the results is presented in Fig. 4.

4.2 Comparison with SOTA Methods

Table 1 presents the validation set Dice scores for the whole tumor (WT), tumor core (TC), and enhancing tumor (ET) regions across different network architectures, both with and without the 3D Dual Domain Attention (3D-DDA) block. Our proposed method, Diffusion+3D-DDA, demonstrates superior performance, achieving Dice scores of 0.9155, 0.8475, and 0.7364 for WT, TC, and ET, respectively. This outperforms all other conventional networks evaluated. For instance, the SegResNet with 3D-DDA achieves a Dice score of 0.8959 for WT, significantly lower than our model's score of 0.9155. The inclusion of the 3D-DDA block generally enhances the performance of the networks, as seen in the improvement of the SegResNet's ET Dice score from 0.7275 to 0.7361 when 3D-DDA is added. Compared to SegResNet with 3D-DDA, our proposed method shows an improvement of approximately 2.2% for WT, 7.1% for TC, and 0.04% for ET. These results underline the effectiveness of integrating the 3D-DDA block, which captures both local and global features, thus improving segmentation accuracy.

Table 2 compares our proposed method against state-of-the-art (SOTA) techniques using the validation set Dice scores for WT, TC, and ET. Our method

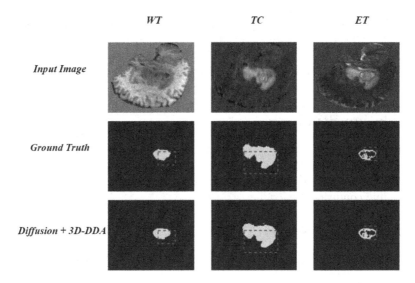

Fig. 4. Visualization of segmentation results for Whole Tumor (WT), Tumor Core (TC), and Enhancing Tumor (ET) on the BraTS2020 dataset.

achieves the highest Dice scores for WT (0.9155) and TC (0.8475), surpassing recent models such as GLF-Net and LGMSU-Net, which score 0.8930 and 0.8735 for WT, respectively. Additionally, our method achieves comparable performance for ET with a Dice score of 0.7364, closely matching the best ET score of 0.7443 achieved by Zou et al. (2020). Compared to the best-performing method in Table 2, our proposed method shows an improvement of approximately 2.5% for WT and 4.1% for TC, with a slight decrease of about 1.1% for ET. These results demonstrate the robustness and superior performance of our proposed Adaptive Dual Attention Diffusion (ADAD) framework, particularly in accurately segmenting the WT and TC regions. The inclusion of the 3D-DDA block significantly enhances the network's ability to capture comprehensive spatial and contextual information, leading to more precise segmentation outcomes.

Table 2. Comparison on the state-of-the-art methods

Validation set	Year	Dice Score		
		WT	TC	ET
Nuechterlein et al. [15]	2019	0.8834	0.8141	**0.7369**
Zou et al. [21]	2020	0.8658	0.7688	0.7443
GLF-Net [11]	2020	0.8930		
LGMSU-Net [16]	2022	0.8735		
SegResNet + 3D-DDA [2]	2023	0.8959	0.7911	0.7361
Our proposed method		**0.9155**	**0.8475**	0.7364

In summary, the results from Tables 1 and 2 highlight the significant improvements our ADAD framework brings to 3D medical image segmentation. The integration of the 3D-DDA block and the diffusion model enables our method to achieve higher segmentation accuracy and robustness compared to both conventional networks and state-of-the-art methods.

5 Ablation Studies

To validate the contributions of each component in the proposed ADAD framework, we conducted ablation studies on the BraTS2020 dataset. Specifically, we evaluated the impact of the 3D Dual Domain Attention (3D-DDA) block and the dynamic conditional encoding strategy.

5.1 Experimental Setup

We designed several experiments to isolate the effects of the 3D-DDA block and the dynamic conditional encoding:

- **Baseline Model**: The diffusion model without the 3D-DDA block and without dynamic conditional encoding.
- **Baseline + 3D-DDA**: The diffusion model with the 3D-DDA block but without dynamic conditional encoding.
- **Baseline + Dynamic Encoding**: The diffusion model with dynamic conditional encoding but without the 3D-DDA block.
- **Full Model (ADAD)**: The complete proposed method with both the 3D-DDA block and dynamic conditional encoding.

5.2 Results and Discussion

Table 3. Ablation study results on the BraTS2020 validation set

Model	WT Dice	TC Dice	ET Dice
Baseline Model	0.8820	0.7945	0.7220
Baseline + 3D-DDA	0.8985	0.8120	0.7295
Baseline + Dynamic Encoding	0.8910	0.8065	0.7260
Full Model (ADAD)	**0.9155**	**0.8475**	**0.7364**

As shown in Table 3, incorporating the 3D-DDA block into the baseline model improved Dice scores across all tumor regions, demonstrating the effectiveness of capturing local and global contextual information. Similarly, adding dynamic conditional encoding enhanced the model's performance, highlighting its role in refining segmentation accuracy at each diffusion step.

The full ADAD model, combining the 3D-DDA block and dynamic conditional encoding, achieved the highest Dice scores, confirming that each component contributes positively to the overall performance.

5.3 Analysis

The ablation studies indicate that:

- The **3D-DDA block** significantly improves the model's ability to handle complex medical images by effectively capturing spatial and contextual features.
- The **dynamic conditional encoding** strategy enhances the conditioning of the diffusion process, leading to more accurate segmentation results.
- Combining both components yields the best performance, suggesting that they complement each other in the ADAD framework.

6 Conclusion

This paper introduced ADAD framework for 3D medical image segmentation. Through extensive experiments and ablation studies, we demonstrated that the 3D-DDA block and the dynamic conditional encoding strategy individually and collectively enhance segmentation performance. The ablation studies confirmed that the 3D-DDA block effectively captures local and global contextual information, improving the network's capability to handle the complexity of medical images. The dynamic conditional encoding strategy further refines the segmentation by incorporating current-step information into the conditioning process. Our proposed ADAD framework achieved superior results on the BraTS2020 dataset, outperforming state-of-the-art methods. These findings highlight the importance of each component in our model and their contributions to advancing 3D medical image segmentation. Future work will explore integrating additional attention mechanisms and applying the ADAD framework to other medical imaging modalities.

References

1. Austin, J., Johnson, D.D., Jonathan, H., Daniel, T., Van Den Berg, R.: Structured denoising diffusion models in discrete state-spaces. Adv. Neural. Inf. Process. Syst. **34**, 17981–17993 (2021)
2. Do, N.T., Vo-Thanh, H.S., Nguyen-Quynh, T.T., Kim, S.H.: 3D-DDA: 3D dual-domain attention for brain tumor segmentation. In: 2023 IEEE International Conference on Image Processing (ICIP), pp. 3215–3219. IEEE (2023)
3. Gal, Y., Ghahramani, Z.: Dropout as a Bayesian approximation: representing model uncertainty in deep learning. In: International Conference on Machine Learning, pp. 1050–1059. PMLR (2016)
4. Hatamizadeh, A., Nath, V., Tang, Y., Yang, D., Roth, H.R., Xu, D.: Swin UNETR: Swin transformers for semantic segmentation of brain tumors in MRI images. In: Crimi, A., Bakas, S. (eds.) Brainlesion: Glioma, Multiple Sclerosis, Stroke and Traumatic Brain Injuries BrainLes 2021. LNCS, vol. 12962, pp. 272–284. Springer, Cham (2021). https://doi.org/10.1007/978-3-031-08999-2_22

5. He, Y., Nath, V., Yang, D., Tang, Y., Myronenko, A., Xu, D.: Swinunetr-v2: stronger Swin transformers with stagewise convolutions for 3d medical image segmentation. In: Greenspan, H., et al. (eds.) Medical Image Computing and Computer Assisted Intervention MICCAI 2023. LNCS, vol. 14223, pp. 416–426. Springer, Cham (2023). https://doi.org/10.1007/978-3-031-43901-8_40
6. Isensee, F., Jaeger, P.F., Kohl, S.A., Petersen, J., Maier-Hein, K.H.: nnU-net: a self-configuring method for deep learning-based biomedical image segmentation. Nat. Methods **18**(2), 203–211 (2021)
7. Jiang, P., Gu, F., Wang, Y., Tu, C., Chen, B.: DifNet: semantic segmentation by diffusion networks. In: Advances in Neural Information Processing Systems, vol. 31 (2018)
8. Kim, D.: Fine-grained human hair segmentation using a text-to-image diffusion model. IEEE Access **12**, 13912–13922 (2024). https://doi.org/10.1109/access.2024.3355542
9. Lee, H.H., et al.: Scaling up 3d kernels with Bayesian frequency re-parameterization for medical image segmentation. In: Greenspan, H., et al. (eds.) Medical Image Computing and Computer Assisted Intervention MICCAI 2023. LNCS, vol. 14223, pp. 632–641. Springer, Cham (2023). https://doi.org/10.1007/978-3-031-43901-8_60
10. Litjens, G., Kooi, T., Bejnordi, B.E., Setio, A.A.A., Ciompi, F., Ghafoorian, M., Van Der Laak, J.A., Van Ginneken, B., Sánchez, C.I.: A survey on deep learning in medical image analysis. Med. Image Anal. **42**, 60–88 (2017)
11. Liu, C., et al.: Brain tumor segmentation network using attention-based fusion and spatial relationship constraint. In: Crimi, A., Bakas, S. (eds.) Brainlesion: Glioma, Multiple Sclerosis, Stroke and Traumatic Brain Injuries BrainLes 2020. LNCS, vol. 12658, pp. 219–229. Springer, Cham (2020). https://doi.org/10.1007/978-3-030-72084-1_20
12. Menze, B.H., Jakab, A., Bauer, S., Kalpathy-Cramer, J., Farahani, K., Kirby, J., Burren, Y., Porz, N., Slotboom, J., Wiest, R., et al.: The multimodal brain tumor image segmentation benchmark (brats). IEEE Trans. Med. Imaging **34**(10), 1993–2024 (2014)
13. Milletari, F., Navab, N., Ahmadi, S.A.: V-net: fully convolutional neural networks for volumetric medical image segmentation. In: 2016 Fourth International Conference on 3D Vision (3DV), pp. 565–571. IEEE (2016)
14. Myronenko, A.: 3d MRI brain tumor segmentation using autoencoder regularization. In: Crimi, A., Bakas, S., Kuijf, H., Keyvan, F., Reyes, M., van Walsum, T. (eds.) Brainlesion: Glioma, Multiple Sclerosis, Stroke and Traumatic Brain Injuries BrainLes 2018. LNCS, vol. 11384, pp. 311–320. Springer, Cham (2019). https://doi.org/10.1007/978-3-030-11726-9_28
15. Nuechterlein, N., Mehta, S.: 3D-ESPNet with pyramidal refinement for volumetric brain tumor image segmentation. In: Crimi, A., Bakas, S., Kuijf, H., Keyvan, F., Reyes, M., van Walsum, T. (eds.) Brainlesion: Glioma, Multiple Sclerosis, Stroke and Traumatic Brain Injuries BrainLes 2018. LNCS, vol. 11384, pp. 245–253. Springer, Cham (2019). https://doi.org/10.1007/978-3-030-11726-9_22
16. Pang, X., Zhao, Z., Wang, Y., Li, F., Chang, F.: LGMSU-Net: local features, global features, and multi-scale features fused the u-shaped network for brain tumor segmentation. Electronics **11**(12), 1911 (2022)
17. Savioli, N., Montana, G., Lamata, P.: V-FCNN: volumetric fully convolution neural network for automatic atrial segmentation, pp. 273–281 (2019). https://doi.org/10.1007/978-3-030-12029-0_30

18. Wu, J., et al.: MedSegDiff: medical image segmentation with diffusion probabilistic model. In: Medical Imaging with Deep Learning, pp. 1623–1639. PMLR (2024)
19. Xue, F., Guo, L., Bialkowski, A., Abbosh, A.: Training universal deep-learning networks for electromagnetic medical imaging using a large database of randomized objects. Sensors **24**(1), 8 (2023)
20. Zhang, K., et al.: Practical blind image denoising via Swin-conv-UNet and data synthesis. Mach. Intell. Res. **20**(6), 822–836 (2023)
21. Zhou, Z., He, Z., Jia, Y.: AFPNet: a 3d fully convolutional neural network with Atrous-convolution feature pyramid for brain tumor segmentation via MRI images. Neurocomputing **402**, 235–244 (2020)

The Application of Graph Attention Mechanism in the Automation of Analog Circuit Design

Xinpeng Li[1], Minglei Tong[1(✉)], and Yongqing Sun[2]

[1] EE School, Shanghai University of Electric Power,
Shanghai, People's Republic of China
tongminglei@shiep.edu.cn
[2] College of Humanities and Sciences, Nihon University, Chiyoda, Japan
nakahara.eisei@nihon-u.ac.jp

Abstract. Automated annotation of circuit structures can generate hierarchical representations of analog circuit networks, thereby advancing the development of automated analog circuit design tasks. This paper introduces a graph attention network-based model that transforms circuit netlists into graph structures, proposes a feature extraction strategy to learn and predict the circuit structures composed of nodes in the netlists, and presents a method for quickly generating a large number of SPICE circuit netlists to provide ample data for training the graph model. Experiments compared the recognition effects of graph convolutional networks, graph isomorphism networks, and GraphSAGE on the same dataset. The results show that the GAT model outperforms the other models in accuracy, precision, and mean average precision, achieving 90.9%, 91.6%, and 91.9%, respectively. These results demonstrate the superiority of the GAT model in capturing circuit connections, especially in terms of its effectiveness in processing complex circuit diagrams.

Keywords: Automated Annotation · Graph attention networks · Circuit structure · Feature extraction

1 Introduction

Analog IC layout automation is one of the key research directions in the field of integrated circuit design, with the goal of fully automating the design of Analog circuits [14]. Existing automated design methods are seldom adopted on a large scale due to the high complexity of Analog circuits' typologies and their variants [7,18], and in previous attempts to solve the problem, matching circuits to pre-specified templates involves traversing all the possible topologies in a database [19,20] , and identification through fixed circuit rules requires rule declarations by experienced researchers [6,24] , so it is difficult to comprehensively cover all topological variants. Graph neural networks [22] have attracted much attention due to their excellent performance in identifying target variants

[4,25]. The application of graph neural networks has also been demonstrated in the field of electronic design automation (EDA). The pins of components in a circuit can be modelled as vertices in circuit design [26], but the isomorphism of the graph can lead to an excessively large set of vertices, which in turn causes loss of information [5], and defining components and network nodes as disjoint two-part sets [11] reduces the number of vertices while increasing the complexity of detecting adjacent nodes. Further, the development of graph convolution networks can classify circuit components by performing sub-graph isomorphism [2,8], and related studies have demonstrated that graphical representation of circuit netlist followed by graph convolution networks training can achieve circuit classification [15], Analog circuit structure identification [9,10,12], mixed digital-Analog circuit constraints extraction [1,16], layout parasitic parameters prediction [21], and Analog circuit layout constraint annotation [3].

In this study, we propose a structural annotation framework based on graph attention networks, which, after converting circuit netlist into graph structures, learns the structural features of these graphs by introducing an attention mechanism, which in turn enables us to dynamically focus on the relative importance between nodes in the graphs, and thus identify critical circuit elements and connections more accurately. The main contributions of this paper can be summarised as follows:

(1) An efficient method is proposed to convert Analog circuit netlist into graph representations, laying the foundation for subsequent graph learning tasks.

(2) A graph attention mechanism is introduced to identify and emphasise the role of important nodes in the circuit diagram, which further improves the accuracy and robustness of the model in recognising circuit structures.

2 Related Work

2.1 Circuit Netlist Representation and Structural Recognition

B. Xu et al. abstracted the circuit netlist into a graphical representation in order to achieve constraint extraction, preserving the pin information in the representation, where devices, pins and networks are represented as nodes, pins are connected to the circuit elements they belong to, and networks are connected to the pins according to the circuit netlist's connection to the pins [26]. All edges in this representation are not directly connected to circuit elements, but are first connected to edges. This representation makes the set of circuit vertices obtained too large and there is a risk of information loss.

W. Hamilton et al. used node feature information to generate node embedding for previously unseen data avoided the drawback of large vertex set by generating embedding by learning to sample and aggregate features from the local neighbourhood of the node [5]. K. Kunal et al. reduced the vertex set by defining component vertices and nodes in the circuit netlist as a bipartite set that does not intersect [11]. Using a trained graph convolution neural network to identify netlist elements for designing hierarchical circuit blocks, the study first created a graph representation of a flat network table based on a custom dataset, and then used graph isomorphism techniques to determine the structure of lower hierarchical structures, and then used the graph convolution network to identify sub-circuits in order to achieve approximate sub-graph isomorphism [9].

Graph Convolution Networks are highly dependent on the graphical representation and feature engineering of circuits [27] during model design, which may limit the ability of the model to generalise to different types or novel circuit designs, and different circuit structures may be isomorphic in graphical representation but functionally different [17]. This isomorphism may lead to ambiguity in the labelling process of GCNs, affecting the accuracy and reliability of the labelling.

2.2 Graph Attention Networks

Graph Attention Network (GAT) [23] is a neural network structure proposed by Veličković et al. It overcomes some of the limitations of graph neural networks through the mechanism of attention. GAT aggregates the neighbourhood information of a node by assigning different attention weights to the neighbouring nodes of each node in the graph, thus assigning different importance to each node's neighbours in the graph [13], which improves the predictive and generalisation capabilities of the network. The attention mechanism is defined as shown in Eq. 1.

$$\text{Attention}(\text{Query}, \text{Source}) = \sum_i \text{similarity}(\text{Query}, \text{Key}_i) \cdot \text{Value}_i \quad (1)$$

Source refers to the data source that needs to be processed, *Query* represents the prior information, and $similarity(Query, Key_i)$ indicates the correlation between the *Query* vector and the *Key* vector. *Attention* refers to the process of extracting information from the "Source" based on the condition of the *Query* using the attention mechanism. The various pieces of information in the "Source" are represented in a $Key - Value$ format. Finally, all the *Value* information is weighted and summed, with the weights determined by the correlation between the *Query* vector and the *Key* vector.

3 Works

3.1 Graphical Representation of Netlists

Data Preprocessing. In this paper, the SPICE format has been chosen as the standard format for circuit netlist data because it is widely used in the field of electronic circuit design and provides a wealth of information to describe the physical properties and connectivity relationships of circuits. Directly analysing the original SPICE netlist faces certain challenges, especially the complexity caused by the hierarchical design of circuits, in order to solve this problem, this paper adopts the pre-processing step of flattening processing to simplify the structure of the circuit netlist.

The core aim of the flattening process is to transform the hierarchical circuit netlist into a form that is simpler in structure and easier to analyse. By eliminating the hierarchical structure of the circuit and flattening all components and their connections to the same level, it provides an intuitive and unified view of the circuit representation. This transformation not only simplifies the data analysis and processing process, but also facilitates the implementation and application of circuit rules.

Following the flattening of the circuit netlist, the next stage involves a process of converting it into a graph structure, which involves extracting components from the netlist as nodes of the graph, and considering the electrical connections between the components as edges of the graph, thereby constructing a graph representation reflecting the topology of the circuit. Further, feature vectors and adjacency matrices are extracted from this graph structure to build the necessary data structure base for the inputs to the graph neural network model. The feature vector carries descriptive information about each node (i.e., circuit element), while the adjacency matrix reveals the connection patterns between nodes. The combined use of these two data structures provides a rich set of information resources for the graph attention network model, which enables the model to effectively learn the structural properties of the circuit and improves the accuracy and efficiency of the analysis.

Circuit Netlist Generation. The lack of datasets is an important factor affecting the development of Analog circuit design automation, this section focuses on the generation process of custom datasets, for the structural annotation of Analog circuit netlists, this paper proposes a hybrid netlist generation method based on circuit rules, and the flow is shown in Fig. 1. The method is mainly used to automate the generation of circuit netlists, especially for the comprehensive design of operational amplifiers and bias circuits. The method

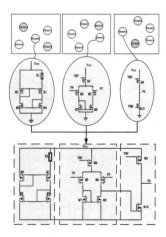

Fig. 1. Schematic diagram of circuit netlist generation process

adopts an innovative algorithmic flow to automatically read the SPICE format sub-circuit netlist files stored in a specified directory, which are designed based on certain circuit rules with the same naming constraints rules, e.g., the output pins of a bias circuit need to be connected to the bias input pins of the operational amplifier, and by applying these matching rules, the program dynamically generates connection By applying these matching rules, the program dynamically generates connection commands to automatically connect the corresponding pins together, eliminating the hierarchical structure of the circuit and flattening all components and their connections to the same level, simplifying the structure of the circuit netlist, thus making the subsequent analysis more efficient and direct. The naming constraints of the netlist are shown in Table 1.

Table 1. Naming Constraints of Circuit Netlists.

Naming Element	Description
M	MOSFET
R	Resistor
C	Capacitor
Vcc	Power
Gnd	Ground

Graphical Representation of Circuit Netlists. In this paper, we use an undirected bipartite graph G(V,E) to represent a network table of circuits, and all the nodes in the network table as a set of vertices, which are divided into two categories: components and electrical nodes. The set E consists of edges between nodes. Figure 2 shows an example of a graphical representation of a circuit netlist.

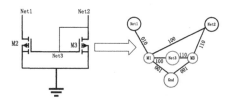

Fig. 2. Graphical representation of circuit netlist

The circuit diagram consists of two MOS tubes and multiple electrical nodes, the electrical nodes and components are set as nodes, in which the purple, yellow and red nodes represent different electrical nodes, respectively, and different features are assigned during feature extraction, and the blue vertices represent the components, and when the graph representation is carried out, in order to be able to show the connection relationship between transistors and the surrounding nodes more clearly, the connection to the transistors is provided. Each edge is assigned a three-bit label, which represents the connection relationship between the three stages of the transistor and its neighbouring nodes in the order of lgldls, as shown in Fig. 2, the leakage stage of M2 is connected to the Net1 node, so the label of the edge between M2 and Net1 is 010, and these labels will be used as a part of the node feature matrix in the subsequent node feature extraction.

Node Feature Extraction. The node features in this context are represented in the form of an adjacency matrix and an n×d feature matrix, where d represents the number of features. To better capture the overall structural characteristics of the circuit netlist, each node's features are associated with a 22-dimensional feature vector, as outlined below:

- 5 features are related to the type of component that the node represents.
- 5 features correspond to the type of electrical node the component is connected to.
- 3 features describe the circuit structure in which the node is located.
- 9 features are linked to the labels of the edges connected to the node.

3.2 Graph Attention Model Construction

The Graph Attention Model proposed in this paper is composed of three stages: graphical representation of the netlist, feature extractor, and predictor, as shown in Fig. 3. The feature extractor consists of two layers of Graph Attention Networks (GAT), with each layer comprising graph convolution and graph pooling operations. The predictor is a fully connected neural network, also known as a Multi-Layer Perceptron (MLP).

Node Feature Extraction. In order to better represent the circuit netlist through the graph and avoid circuits with different topologies having homogeneous graph representation, so for different nodes within the netlist to construct their special node features, based on the feature extraction rules in Subsect. 3.1, the global attributes of the node are added on top of it to indicate whether it belongs to more than one sub-circuit at the same time as well as whether it is a global signalling node or not, which will be comprehensively represented into the feature matrix.

GAT. Node feature aggregation is implemented using a two-layer GAT network, where for each node, GAT first calculates the attention coefficient between that node and each of its neighbours through a learnable weight matrix and an attention mechanism that calculates the attention coefficient based on the features

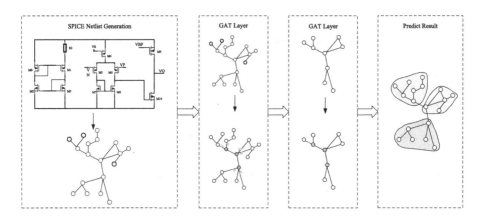

Fig. 3. Graph attention network for circuit structure recognition

of the current node and its neighbours. The input to the GAT network in this paper is a set of circuit netlist node feature matrices:

$$X = \left[\vec{h_1}, \vec{h_2}, \ldots, \vec{h_m}\right] \quad (2)$$

After input to the GAT model, the attention coefficients are calculated by node features $\vec{h_i}$ and neighbour node features $\vec{h_j}$:

$$e_{ij} = a^T LeakyReLU\left(\left[\vec{h_i} \middle\| \vec{h_j}\right] W^T\right) \quad (3)$$

where W is a linear transformation applied to the features of each node to obtain more expressive feature vectors, e_{ij} is the computed attention coefficient, representing the importance of neighboring node j to node i, and a^T is the attention weight matrix. The attention coefficients are normalized using the softmax function, yielding:

$$\alpha_{ij} = \frac{\exp(e_{ij})}{\sum_{k \in N_i} \exp(e_{ik})} \quad (4)$$

where k is the information of neighbour nodes other than neighbour node j. The normalized attention coefficient α_{ij} is obtained by Eq. (3). For the ith node feature in the feature matrix X, the output of the GAT layer is.

$$\vec{h_i'} = \sigma\left(\sum_{j \in N_i} \alpha_{ij} W \vec{h_j}\right) \quad (5)$$

where $\vec{h_i'}$ represents the output feature of node i, N_i represents all neighboring nodes of node i, and σ is the activation function. Through the GAT, the features of neighboring nodes are further learned, resulting in the output feature matrix X'.

$$X' = \left[\vec{h_1'} \ \vec{h_2'} \cdots \vec{h_m'} \right] \tag{6}$$

Through the above method, the attention mechanism is introduced to dynamically learn the interaction weights between nodes, which is more flexible in dealing with graph structure data, captures the complex dependencies between nodes, and is able to adapt to different graph structures, which improves the adaptability and prediction accuracy of the model to novel circuit designs to a certain extent.

Result Prediction. A fully connected layer is set up after the output of the GAT layer for predicting the output of the model, along with the nodes and edges contained in each circuit structure.

4 Experimental Results and Analysis

4.1 Experimental Environment

The experimental environment used in this paper is as follows:

- GPU: RTX 3060
- CPU: Core i5-12600KF
- Memory: 16GB
- Python: 3.8.17
- PYG (PyTorch Geometric): 2.3.1
- PyTorch: 2.0.0
- CUDA: 11.6

4.2 Evaluation Metrics

In the detection results, TP (True Positive) indicates the number of correctly predicted subcircuit nodes, i.e., the predicted nodes belong to the subcircuit and are predicted correctly, FP (False Positive) indicates that the predicted nodes belong to the subcircuit but are predicted incorrectly, TN (True negative), indicates that the predicted nodes do not belong to the subcircuit and are predicted correctly, and FN (False Negative), which indicates that the predicted node does not belong to the subcircuit, but the prediction is wrong.

Accuracy. Accuracy is one of the most common evaluation criteria, which indicates the ratio of the number of correctly classified nodes to the total number of nodes, and is usually applied to the case of balanced category distribution. The calculation formula is as follows:

$$Accuracy = \frac{TP + TN}{TP + TN + FP + FN} \tag{7}$$

Precision. The proportion of correct prediction results for all nodes belonging to a subcircuit is calculated as follows:

$$Precision = \frac{TP}{TP+FP} \tag{8}$$

Mean Average Precision. AP is the average precision of the prediction results of each subcircuit. mAP calculates the average of the average precision AP of the detection results of all categories of subcircuits with the following formula:

$$AP = \int_0^1 P(R)dR \tag{9}$$

$$mAP = \sum_{n-1}^{N} AP(n)/N \tag{10}$$

The formula states that $P(R)$ represents precision as a function of recall, and $AP(n)$ denotes the average precision for the nth query.

4.3 Dataset Preparation

The circuit netlist data converted to graphical representation is encapsulated through the Torch_Geometric library, and the connection matrix A is encapsulated in the DATA data along with the feature matrix X, which is saved in the local directory in the form of a pt-file, and the rules for obtaining the connection matrix A and the feature matrix X are described in detail in Subsect. 3.1, and the main parameters of the customised dataset in this paper are shown in Table 2.

Table 2. Custom Dataset Parameters.

Circuits	Nodes	Edges	Features
1480	44962	135460	22

4.4 Experimental Results

In this experiment, structural labelling experiments have been conducted on a custom dataset, and comparative experiments have been conducted for several existing graph models in order to test the effectiveness of the graph attention mechanism in the field of structural labelling of Analog circuits.

The experiment divides the samples into training set, validation set, and test set in the ratio of 7:2:1, which are independent of each other, and uses two layers of GAT for training. Through the application of two consecutive attention mechanisms, we are able to consider the neighbours of nodes as well as the neighbours of neighbours, so as to capture a wider range of information about the graph structure, to enhance the model's understanding of the global structure of graphs, and to improve the ability of feature representation. The size of the input dimension of the GAT layer is 22 dimensions, the number of hidden layer channels is 16 dimensions, using Adam optimiser and binary cross-entropy loss function, configuring weight_decay in the optimiser, using Dropout layer to randomly ignore a group of neurons to prevent overfitting, and after testing it is learnt that the performance of using ReLU activation function is better, a total of 500 epochs are trained, and the optimal results are obtained in the validation set when the Save model parameters.

In order to measure the performance of graph attention network in labelling circuit structures, it is compared with other graph neural network algorithms and the results obtained are shown in Table 3.

Table 3. Experimental Results.

Method	Accuracy	Precision	mAP
GCN	0.893	0.906	0.908
GIN	0.895	0.902	0.909
GraphSAGE	0.902	0.915	0.918
GAT	**0.909**	**0.916**	**0.919**

As shown in Fig. 4 is the circuit diagram of this design after importing into Cadence via the circuit netlist generated in Sect. 3.1 and after manual placement, the actual structure of the circuit is a fully differential amplifier, which is used as an input to output the three sub-structures of the circuit via the GAT network. From the above experimental results, it can be seen that the introduction of the graph attention network makes the accuracy of structure recognition improve

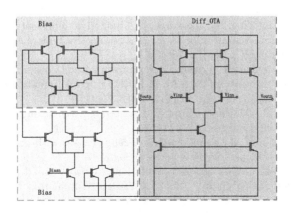

Fig. 4. Circuit structure recognition results

on each of the other graph models, suggesting that the complex relationships between nodes can be captured more effectively through the use of the graph attention mechanism.

5 Conclusion

To address the challenge posed by the limited availability of large-scale circuit netlists for training and the recognition of analog circuit structures, this paper presents a method for the rapid generation of integrated circuit netlists. Additionally, a Graph Attention Network (GAT) is introduced to facilitate circuit structure recognition. Experimental results demonstrate that the GAT exhibits superior performance and generalization capabilities when processing circuit topologies, effectively capturing critical structural information. This leads to improved accuracy and adaptability in graph classification tasks. Future work may focus on expanding the variety of circuits and incorporating actual circuit component parameters for constraint annotation, thereby enabling fully automated analog layout design.

References

1. Chen, H., Zhu, K., Liu, M., Tang, X., Sun, N., Pan, D.Z.: Universal symmetry constraint extraction for analog and mixed-signal circuits with graph neural networks. In: 2021 58th ACM/IEEE Design Automation Conference (DAC), pp. 1243–1248 (2021). https://doi.org/10.1109/DAC18074.2021.9586211
2. Defferrard, M., Bresson, X., Vandergheynst, P.: Convolutional neural networks on graphs with fast localized spectral filtering (2017). https://arxiv.org/abs/1606.09375
3. Gao, X., Deng, C., Liu, M., Zhang, Z., Pan, D.Z., Lin, Y.: Layout symmetry annotation for analog circuits with graph neural networks. In: 2021 26th Asia and South Pacific Design Automation Conference (ASP-DAC), pp. 152–157 (2021)
4. Hamilton, W.L., Ying, R., Leskovec, J.: Representation learning on graphs: Methods and applications (2018). https://arxiv.org/abs/1709.05584
5. Hamilton, W.L., Ying, Z., Leskovec, J.: Inductive representation learning on large graphs. In: Neural Information Processing Systems (2017). https://api.semanticscholar.org/CorpusID:4755450
6. Harjani, R., Rutenbar, R., Carley, L.: A prototype framework for knowledge-based analog circuit synthesis. In: 24th ACM/IEEE Design Automation Conference, pp. 42–49 (1987). https://doi.org/10.1145/37888.37894
7. Huang, G., et al.: Machine learning for electronic design automation: a survey (2021). https://arxiv.org/abs/2102.03357
8. Kipf, T.N., Welling, M.: Semi-supervised classification with graph convolutional networks (2017). https://arxiv.org/abs/1609.02907
9. Kunal, K., et al.: GANA: graph convolutional network based automated netlist annotation for analog circuits. In: 2020 Design, Automation & Test in Europe Conference & Exhibition (DATE), pp. 55–60 (2020https://doi.org/10.23919/DATE48585.2020.9116329

10. Kunal, K., et al.: Invited: align - open-source analog layout automation from the ground up. In: 2019 56th ACM/IEEE Design Automation Conference (DAC), pp. 1–4 (2019)
11. Kunal, K., Poojary, J., Dhar, T., Madhusudan, M., Harjani, R., Sapatnekar, S.S.: A general approach for identifying hierarchical symmetry constraints for analog circuit layout (2020). https://arxiv.org/abs/2010.00051
12. Li, H., Jiao, F., Doboli, A.: Analog circuit topological feature extraction with unsupervised learning of new sub-structures. In: 2016 Design, Automation & Test in Europe Conference & Exhibition (DATE), pp. 1509–1512 (2016)
13. Li, Q., Wang, D., Feng, S., Niu, C., Zhang, Y.: Global graph attention embedding network for relation prediction in knowledge graphs. IEEE Trans. Neural Netw. Learn. Syst. **33**, 6712–6725 (2021). https://api.semanticscholar.org/CorpusID:235412138
14. Lin, Y., Gao, X., Zhang, H., Wang, R., Huang, R.: Intelligent and interactive analog layout design automation. In: 2022 IEEE 16th International Conference on Solid-State & Integrated Circuit Technology (ICSICT), pp. 1–4 (2022).https://doi.org/10.1109/ICSICT55466.2022.9963217
15. Liou, G.H., Wang, S.H., Su, Y.Y., Lin, M.P.H.: Classifying analog and digital circuits with machine learning techniques toward mixed-signal design automation. In: 2018 15th International Conference on Synthesis, Modeling, Analysis and Simulation Methods and Applications to Circuit Design (SMACD), pp. 173–176 (2018). https://doi.org/10.1109/SMACD.2018.8434884
16. Liu, M., et al.: S3det: detecting system symmetry constraints for analog circuits with graph similarity. In: 2020 25th Asia and South Pacific Design Automation Conference (ASP-DAC), pp. 193–198 (2020).https://doi.org/10.1109/ASP-DAC47756.2020.9045109
17. Liu, M., et al.: Towards decrypting the art of analog layout: placement quality prediction via transfer learning. In: 2020 Design, Automation & Test in Europe Conference & Exhibition (DATE), pp. 496–501 (2020).https://doi.org/10.23919/DATE48585.2020.9116330
18. Lopera, D.S., Servadei, L., Kiprit, G.N., Hazra, S., Wille, R., Ecker, W.: A survey of graph neural networks for electronic design automation. In: 2021 ACM/IEEE 3rd Workshop on Machine Learning for CAD (MLCAD), pp. 1–6 (2021). https://api.semanticscholar.org/CorpusID:237427133
19. Massier, T., Graeb, H., Schlichtmann, U.: The sizing rules method for CMOS and bipolar analog integrated circuit synthesis. IEEE Trans. Comput. Aided Des. Integr. Circuits Syst. **27**(12), 2209–2222 (2008). https://doi.org/10.1109/TCAD.2008.2006143
20. Meissner, M., Hedrich, L.: Feats: framework for explorative analog topology synthesis. IEEE Trans. Comput. Aided Des. Integr. Circuits Syst. **34**(2), 213–226 (2015). https://doi.org/10.1109/TCAD.2014.2376987
21. Ren, H., Kokai, G.F., Turner, W.J., Ku, T.: Paragraph: layout parasitics and device parameter prediction using graph neural networks. In: 2020 57th ACM/IEEE Design Automation Conference (DAC) pp. 1–6 (2020). https://api.semanticscholar.org/CorpusID:221679424
22. Scarselli, F., Gori, M., Tsoi, A.C., Hagenbuchner, M., Monfardini, G.: The graph neural network model. IEEE Trans. Neural Netw. **20**(1), 61–80 (2009). https://doi.org/10.1109/TNN.2008.2005605
23. Veličković, P., Cucurull, G., Casanova, A., Romero, A., Liò, P., Bengio, Y.: Graph attention networks (2018). https://arxiv.org/abs/1710.10903

24. Wu, P.H., Lin, M.P.H., Chen, T.C., Yeh, C.F., Li, X., Ho, T.Y.: A novel analog physical synthesis methodology integrating existent design expertise. IEEE Trans. Comput. Aided Des. Integr. Circ. Syst. **34**, 199–212 (2015). https://api.semanticscholar.org/CorpusID:14927270
25. Wu, Z., Pan, S., Chen, F., Long, G., Zhang, C., Yu, P.S.: A comprehensive survey on graph neural networks. IEEE Trans. Neural Netw. Learn. Syst. **32**(1), 4–24 (2021). https://doi.org/10.1109/TNNLS.2020.2978386
26. Xu, B., et al.: Magical: toward fully automated analog ic layout leveraging human and machine intelligence: invited paper. In: 2019 IEEE/ACM International Conference on Computer-Aided Design (ICCAD), pp. 1–8 (2019). https://doi.org/10.1109/ICCAD45719.2019.8942060
27. Ying, R., He, R., Chen, K., Eksombatchai, P., Hamilton, W.L., Leskovec, J.: Graph convolutional neural networks for web-scale recommender systems. In: Proceedings of the 24th ACM SIGKDD International Conference on Knowledge Discovery & Data Mining (2018). https://api.semanticscholar.org/CorpusID:46949657

CROCODILE: Crop-Based Contrastive Discriminative Learning for Enhancing Explainability of End-to-End Driving Models

Chenkai Zhang[✉], Daisuke Deguchi, Jialei Chen, Zhenzhen Quan, and Hiroshi Murase

Nagoya University, Nagoya Aichi, Japan
zhang1354558057@gmail.com

Abstract. In autonomous driving, visual features play a crucial role. End-to-end driving models (E2EDMs) extract numerous visual features from the driving environment to solve driving tasks. However, these visual features are often difficult for humans to understand, leading to explainability issues. This study aims to improve the explainability of E2EDMs by enhancing their ability to extract semantically meaningful and driving-related visual features, like vehicles, pedestrians, and traffic signals. The training process of E2EDMs involves leveraging a backbone that is pre-trained on large datasets and subsequently fine-tuned for driving tasks. To address the explainability issue of E2EDMs, previous studies have designed complex E2EDMs during the fine-tuning stage. In this paper, we enhance the explainability by improving the backbone's ability to recognize driving-related features, *i.e.*, object features. We propose **CRO**p-based **CO**ntrastive **DI**scriminative **LE**arning (**CROCODILE**), an additional pre-training method for the backbone. CROCODILE improves the backbone's ability to preserve driving-related features while suppressing irrelevant features. Then, during fine-tuning, only driving-related features will be used for driving action prediction, thereby achieving high explainability. In addition, CROCODILE eliminates the need for complex structures in the fine-tuning stage.

1 Introduction

End-to-end driving models (E2EDMs) are the most popular autonomous driving models, they automatically extract and select visual features directly from the driving environment. These models are capable of learning optimal visual features tailored to specific driving tasks, resulting in higher prediction accuracy. However, these extracted visual features are difficult for humans to understand,

Supplementary Information The online version contains supplementary material available at https://doi.org/10.1007/978-981-96-2644-1_3.

Fig. 1. In the original image, the green arrow indicates the driving action is available, the red arrow indicates that it is not. The heatmap indicates the importance of each pixel in the prediction. Our approach could help E2EDMs better focus on semantically meaningful visual features than baseline. (Color figure online)

leading to explainability challenges. This study aims to improve the explainability of E2EDMs by enhancing their ability to extract more semantically meaningful and driving-related visual features, making the decision-making process more transparent and understandable.

Explanation methods are used to generate explanations for E2EDMs [2,6,15]. There are textual-based [1,11,22] and visual-based explanation methods [17,21,26], the former generates natural language to explain why the driving models perform a specific driving action, and the latter uses visual information, i.e., images to offer intuitive explanations. Visual-based explanations are particularly advantageous in telling what and where are the responsible features for the driving action, thus in this paper, we focus on visual-based explanations. Among various visual-based explanation methods, attribution-based methods [2,6,28] are most prevalent, they calculate the importance score of each input element in the model's prediction. As shown in Fig. 1, since the basis of the human recognition system lies in objects, the heatmap that highlights driving-related object features is more persuasive.

Like many downstream tasks, the training of E2EDMs is also based on fine-tuning a pre-trained backbone [3,16,20]. The purpose of pre-training the backbone is to prepare a feature extractor capable of processing images of the driving environment. Therefore, during the fine-tuning stage, the E2EDMs could use the extracted features to predict driving actions. In other words, the features extracted by the backbone have a significant impact on the prediction method of the E2EDMs, i.e., the explainability of E2EDMs.

As shown in Fig. 2a, previous studies focused on enhancing explainability during the fine-tuning stage of E2EDMs. Specifically, they added an object detection module after the backbone, which required the E2EDMs to use object information for driving action prediction, thereby enhancing explainability. However, such a side task requires significant modifications to the architecture, deviating from their inherent end-to-end nature. Additionally, this side task demands aux-

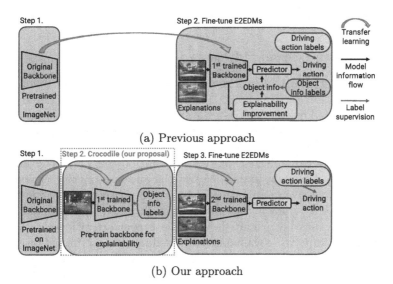

Fig. 2. The comparison of the previous approach and our approach, the previous approach enhances explainability during the fine-tuning stage. On the other hand, we enhance explainability in an additional pre-training stage.

iliary data, specifically, labels for object information. This requirement imposes stricter demands on the datasets in fine-tuning, as they must include not only labels for driving actions but also for object information [12,21,25].

To address this, we take a different path. As a visual feature extractor, the backbone plays a crucial role in the explainability of E2EDMs. Therefore, we improve explainability by enhancing the backbone's ability to process visual features through an additional pre-training stage. As shown in Fig. 2b, between the original pre-training stage and the fine-tuning stage, we introduce **CRO**p-based **CO**ntrastive **DI**scriminative **LE**arning (**CROCODILE**), CROCODILE pre-trains the backbone to preserve the driving-related features and suppress the irrelevant features. During fine-tuning, only driving-related features will be used for driving action prediction (as shown in Fig. 1), achieving high explainability.

The contributions of this paper are:

- The novelty of this paper lies in shifting the focus of enhancing the explainability of E2EDMs from the fine-tuning stage to the pre-training stage, which allows us to maintain the simplicity of the E2EDMs.
- CROCODILE decouples the simultaneous requirements for object and driving information. Specifically, we first enhance explainability on a dataset containing object information, then fine-tune driving tasks on another dataset.
- Our experiments demonstrate that CROCODILE is effective across different backbones and E2EDMs, then analyze this effectiveness in ablation study.

2 Related Work

2.1 Contrastive Learning Methods

Contrastive learning methods utilize positive and negative sample pairs to train the backbones [4,14,19]. He et al. [7] introduced a dynamic dictionary and a momentum-updated encoder to train the backbones to learn representations for image classification tasks. In object detection, contrastive learning also plays a crucial role. Zhang et al. [27] introduced Contrastive DeNoising Training, which aids the model in distinguishing between relevant objects and irrelevant background information, thereby stabilizing training and accelerating convergence.

Although contrastive learning methods have been successfully employed to enhance predictive accuracy in their respective tasks, their application in improving explainability has been limited. We believe contrastive learning methods can significantly improve explainability by guiding the discrimination between driving-related features and those that are not. Therefore, we deviate from traditional uses of contrastive learning and focus on enhancing explainability.

2.2 Approaches for Enhancing the Explainability of E2EDMs

Leveraging object information to enhance the explainability of E2EDMs has become a mainstream approach in the field. For instance, Wang et al. [17] utilized object features to predict driving actions. Xu et al. [21] developed a multitask model that incorporates object labels, while Zhang et al. [25] introduced an Objectification Branch (OB) into E2EDMs to improve explainability. However, these approaches complicate the fine-tuning process and compromise the end-to-end architecture since they require E2EDMs to solve the object detection and driving tasks simultaneously. Moreover, these approaches impose strict requirements on datasets, demanding that each driving scenario include both object and driving information, which forces researchers to propose additional datasets to meet the requirements, limiting the practical applications [12,21,25].

To address this problem, we propose a novel method to enhance the explainability of E2EDMs. We separate the enhancement of explainability from the fine-tuning stage, eliminating the need for object detection structures.

3 Proposed Approach

3.1 Basic Idea Behind Our Approach

If the backbone can only extract driving-related features and ignore irrelevant features, then during the fine-tuning stage of E2EDMs, only driving-related features will be used for driving action prediction, thereby achieving high explainability. Therefore, in this paper, we propose CROCODILE to further pre-train the backbone to endow it with this capability. Existing pre-trained backbones could process object features in images. Therefore, in this paper, we focus on

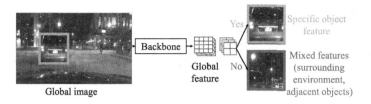

Fig. 3. Inside the global feature, we could locate the local feature corresponding to the driving-related object (green box). This important local feature should not be compromised by the unimportant background features (red box). (Color figure online)

further enhancing the backbone's ability to preserve driving-related object features and suppress irrelevant features. As shown in Fig. 2b, our approach further trains the pre-trained backbones [5], rather than training from scratch.

During the additional pre-training process for explainability, CROCODILE ensures that the driving-related object features are not compromised by the irrelevant features. For example, in the global image shown in Fig. 3, an object is located within a green box. After the backbone processes this global image, we obtain a global feature. Within this global feature, the local features corresponding to the object should still represent the specific object (green box), rather than being mixed with features from the surrounding environment (red box). This capability is fundamental for using the driving-related object features for driving tasks and is a prerequisite for the E2EDM's explainability.

3.2 Crop-Based Contrastive Discriminative Learning

To study how the driving-related object feature is affected by the driving environment during the backbone's transformation, we perform two crops: pre-backbone crop and post-backbone crop. As shown in Fig. 4, the pre-backbone crop happens before the backbone's transformation, resulting in a pure object feature that contains no environmental information. On the other hand, the post-backbone crop happens after the backbone's transformation, resulting in an impure object feature that inevitably includes environmental information. Overall, these two types of crops result in a pure object feature and an impure object feature, we use contrastive discriminative learning to eliminate the difference between these two features by denoting them as the positive pair, thereby enhancing the backbone's ability to preserve the features of the driving-related objects from being mixed with the environmental features.

For each driving scenario, we prepare an image with the bounding box information of the driving-related object. In this paper, we describe a bounding box using the coordinates of its center point, width, and height. As shown in Fig. 4, we crop the red car from the input image as

$$\hat{I}_i = I(x_i, y_i, w_i, h_i), \tag{1}$$

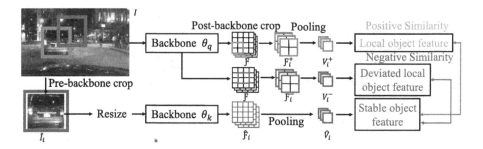

Fig. 4. The overview of **CROCODILE**. In the training process, there are two kinds of crops: pre-backbone crop and post-backbone crop. The pre-backbone crop cuts the driving-related object $\hat{I}_i$ out of the input image I, and then processes $\hat{I}_i$ with the backbone θ_k, obtaining pure driving-related object features $\hat{\mathcal{F}}_i$. The post-backbone crop cuts the global feature $\mathcal{F}$, which is obtained by processing the input image with the backbone θ_q. From $\mathcal{F}$, the post-backbone crop cuts out $\mathcal{F}_i^+$ and $\mathcal{F}_i^-$. We consider $\mathcal{F}_i^+$ and $\hat{\mathcal{F}}_i$ as positive pairs, $\mathcal{F}_i^-$ and $\hat{\mathcal{F}}_i$ are negative pairs. After training, the backbone θ_q is used for future fine-tuning.

where I is the input image, $\hat{I}_i$ is the i-th object image cropped (pre-backbone) from I. We use the coordinates of the object bounding box center point (x_i, y_i), along with width w_i and height h_i to crop $\hat{I}_i$ from I.

We then resize $\hat{I}_i$ to match the size of I. Then, I and $\hat{I}_i$ are processed by backbones θ_q and θ_k respectively, as

$$\mathcal{F} = \theta_q(I), \quad \hat{\mathcal{F}}_i = \theta_k(\hat{I}_i), \tag{2}$$

this process yields $\mathcal{F}$ representing the entire driving environment and $\hat{\mathcal{F}}_i$ representing the pure 2D object feature that is most related to driving tasks.

Within $\mathcal{F}$, we locate $\mathcal{F}_i^+$, the impure 2D object feature that is most related to driving tasks. We calculate the bounding box information of $\mathcal{F}_i^+$ as

$$\alpha_x = \frac{\widetilde{W}}{W}, \alpha_y = \frac{\widetilde{H}}{H} \tag{3}$$

$$\mathcal{F}_i^+ = \mathcal{F}(\alpha_x x_i, \alpha_y y_i, \alpha_x w_i, \alpha_y h_i), \tag{4}$$

where W and H are the width and height of the input image, $\widetilde{W}$ and $\widetilde{H}$ are the width and height of the feature map. $\mathcal{F}_i^+$ is cropped (post-backbone) from $\mathcal{F}$.

Similar to general contrastive discriminative learning [4,7,27], our approach involves positive and negative samples. We consider $\mathcal{F}_i^+$ and $\hat{\mathcal{F}}_i$ as a positive sample pair to guide the backbone to preserve the features of the driving-related object. On the other hand, to ensure that the driving-related object features are not compromised by environmental features, we design negative samples to teach the backbone to distinguish between $\hat{\mathcal{F}}_i$ and the surrounding mixed feature. We define $\mathcal{F}_i^-$, the 2D surrounding mixed feature, by keeping the size of bounding box unchanged while randomly deviating the center point of the $\mathcal{F}_i^+$ as

$$\mathcal{F}_i^- = \mathcal{F}(\alpha_x x_i \pm \epsilon * \alpha_x w_i, \alpha_y y_i \pm \epsilon * \alpha_y h_i, \alpha_x w_i, \alpha_y h_i). \ \epsilon \in [0.25, 0.5] \quad (5)$$

We apply global average pooling (GAP) to $\hat{\mathcal{F}}_i$, $\mathcal{F}_i^+$, and $\mathcal{F}_i^-$ to obtain corresponding vectors: $\hat{V}_i$, V_i^+, and V_i^-. We use these vectors to calculate the cosine similarity of the positive pair and negative pair, and then define the loss as

$$\mathcal{L} = 2 + S_i^- - S_i^+ = 2 + \frac{V_i^- \cdot \hat{V}_i}{\|V_i^-\|\|\hat{V}_i\|} - \frac{V_i^+ \cdot \hat{V}_i}{\|V_i^+\|\|\hat{V}_i\|}, \quad (6)$$

where S_i^+, S_i^- are the cosine similarities for the positive and negative pairs. The range of $\mathcal{L}$ is $0 \sim 2$. As a positive pair, V_i^+ should be close to $\hat{V}_i$; as a negative pair, the V_i^- should diverge from $\hat{V}_i$. Therefore, as the learning target of positive and negative pairs, $\hat{\mathcal{F}}_i$ has to be stable to facilitate training and convergence. To achieve this, we adopt a momentum update method [7] where backbone θ_q is normally trained by back-propagation, while backbone θ_k learns from θ_q as

$$\theta_k \leftarrow m\theta_k + (1-m)\theta_q, \quad (7)$$

where the momentum coefficient m is set to 0.999.

4 Experiments

4.1 Dataset

In this paper, we use two driving scene datasets. The first kind contains object information labels, which are used to train the backbones with CROCODILE. The second kind contains driving action labels, which are used to fine-tune E2EDMs.

Additionally Pre-train the Backbone: In the BDD-100K dataset [23], there is a collection gathered for object-tracking tasks. It comprises videos shot from a driver's perspective, each frame is annotated with the location of every object within the scene. The collection provides approximately 200K images.

Fine-Tune the E2EDMs: In the BDD-3AA (3 Available Actions) [26] dataset, which annotates each scenario with the availabilities of three driving actions: acceleration, left steering, and right steering. The BDD-3AA dataset considers the driving task as a multi-label classification problem. The BDD-3AA dataset comprises 500 video clips. When presented with successive images capturing the driving surroundings, the objective of the E2EDMs is to determine the availabilities for these three driving actions. For example, in a typical scene depicted in Fig. 5, the ground truth is represented as $A = [1, 1, 0]^T$, where 1 signifies an

Fig. 5. A typical scene in the dataset.

available action and 0 is an unavailable one. We utilized the macro F1 score to evaluate prediction accuracy as

$$Macro\ F_1 = \frac{F_1(\hat{A}_a, A_a) + F_1(\hat{A}_l, A_l) + F_1(\hat{A}_r, A_r)}{3}, \quad (8)$$

where A_a, A_l, A_r are the acceleration, steering left, and steering right actions, the A and $\hat{A}$ denote the ground truth label and the prediction result.

4.2 The Subjective Persuasibility Evaluation Method

As the persuasibility evaluation method for explanations proposed in [25], we gathered 5 participants who possess driver's licenses and assessed their satisfaction level with the explanations. We show the heatmaps to participants, each heatmap is evaluated by 5 participants. Participants rate each heatmap from 1 to 5, with 1 being low persuasibility and 5 being high persuasibility, and then we calculate the average value as the final score for this heatmap.

4.3 The Objective Persuasibility Evaluation Method

The objectification degree OD [24] represents the extent to which driving-related objects are utilized in the explanations. Given that the human recognition system relies on objects, the objectification degree determines the persuasibility of the explanation. Without using humans as participants, this method objectively evaluates the explanations generated by E2EDMs as

$$OD = \frac{\sum_{p \in O_{all}} Imp(p)}{\sum_p Imp(p)}, \quad (9)$$

where $Imp(p)$ represents the importance score of a pixel in the explanations, O_{all} is the mask of all objects' areas. $\sum_{p \in O_{all}} Imp(p)$ represents the summation of all pixels' importance scores inside the object area, $\sum_p Imp(p)$ represents the summation of all pixels' importance scores. The OD represents the proportion of the object's features among all the features important for driving actions.

4.4 Implementation Details

In the driving scenario, the biggest object is typically closest to the ego vehicle and thus has a significant impact on driving decisions. Therefore, we consider the biggest object in the images as the most driving-related object, and the training of CROCODILE is focused on this biggest object. The bounding box information for all objects, including the largest object, is annotated.

In this paper, the backbones in E2EDMs are pre-trained. As shown in Fig. 2b, there are two training approaches for these backbones. In the previous approaches, the backbones are pre-trained on ImageNet [5]. On the other hand, in our approach, the backbones pre-trained on ImageNet will be further pre-trained with CROCODILE. Then, the E2EDMs utilize the trained backbones and undergo overall fine-tuning on the BDD-3AA dataset [26]. We apply 5-fold cross-validation to train each E2EDM for 50 epochs and evaluate the average accuracy on corresponding test datasets. As shown in Fig. 2b, there are also two fine-tuning approaches for E2EDMs. In the previous studies [12,21,25], the E2EDMs are fine-tuned with the driving action labels and the object info labels. On the other hand, in our approach, the E2EDMs are fine-tuned with only the driving action labels. For the training of backbones and E2EDMs, the Adam optimizer is utilized with a weight decay of 1×10^{-4} and a learning rate of 0.001.

Based on previous research [26], explanations for the E2EDMs' should based on the high-level features used to predict the driving action. Therefore, all E2EDMs in this paper are integrated with an attention mechanism [10,13,18] applied to these high-level features, allowing us to generate faithful explanations by overlaying the attention mask on the input images.

5 Experimental Results and Discussion

To verify the effectiveness of CROCODILE across different backbones and E2E-DMs, we conducted two sets of experiments, comparing each with the corresponding baselines. In the first set of experiments, we train multiple backbones using the CROCODILE and then fine-tune the same E2EDM to present the effectiveness of CROCODILE on various backbones. Based on the first set of experiments, we identify the backbone on which CROCODILE performs best. In the second set of experiments, we fine-tune various E2EDMs on this backbone to present the effectiveness of CROCODILE on various E2EDMs. Furthermore, we present a detailed ablation study to analyze the source of the effectiveness.

Most results are averaged over five runs to minimize the randomness of our experimental results. For instance, each backbone trained with CROCODILE is trained five times, and an E2EDM is fine-tuned on each backbone. The average accuracy and explainability of the five E2EDMs are then presented as the final result. However, due to the high cost of some experiments, we fine-tuned the E2EDM on only one backbone, and we will explicitly mention such cases.

5.1 The Effectiveness of CROCODILE Across Different Backbones

For the baseline, ZHANG et al. [25] proposed the Objectification Branch (OB), which uses object information during fine-tuning to enhance the explainability of

Table 1. All methods and their configurations in Sect. 5.1. The names have a certain pattern: $a-b-c-d$, a is the backbone trained status, and it could be $Ours$ (additionally pre-trained by CROCODILE), $FRCNN$ (additionally pre-trained by Faster RCNN), or omitted (no additional pre-training); b is the backbone name; c is the E2EDM name; d is whether there is an objectification branch (OB) during fine-tuning, and it could be O (OB) or omitted (no OB).

Method	Backbone	E2EDM	OB [25]	CROCODILE
R18-CBAM	ResNet18 [8]	CBAM [18]		
R18-CBAM-O			✓	
Ours-R18-CBAM				✓
R101-CBAM	ResNet101 [8]			
R101-CBAM-O			✓	
FRCNN-R101-CBAM				
Ours-R101-CBAM				✓
D-CBAM	DenseNet201 [9]			
D-CBAM-O			✓	
Ours-D-CBAM				✓

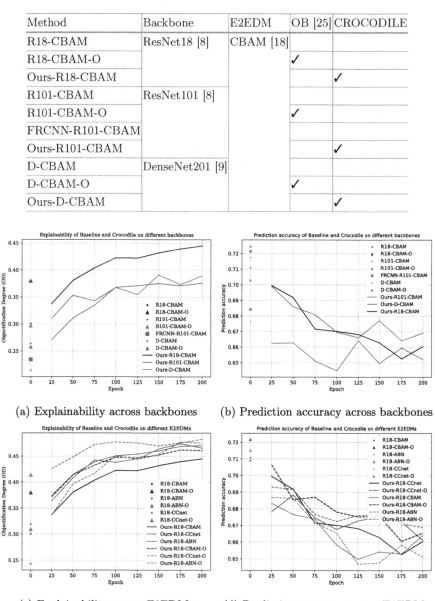

(a) Explainability across backbones (b) Prediction accuracy across backbones

(c) Explainability across E2EDMs (d) Prediction accuracy across E2EDMs

Fig. 6. CROCODILE and baseline across backbones and E2EDMs.

E2EDMs. Since their E2EDMs used the Convolutional Block Attention Mechanism [18], we refer to their two E2EDMs as CBAM and CBAM-O, the CBAM-O has OB and CBAM does not.

We train a backbone with CROCODILE for 200 epochs, saving the backbone every 25 epochs, resulting in eight backbones with different levels of training. Then, we fine-tune the CBAM based on this series, resulting in 8 E2EDMs. To demonstrate the effectiveness of CROCODILE across different backbones, as shown in Table 1, we train 3 series of E2EDMs and their corresponding baselines. In addition, we add another baseline for Ours-R101-CBAM. Since CROCODILE enhances the explainability of the E2EDM by enabling the backbone to preserve important object features, we further prove the effectiveness of CROCODILE by comparing it with a traditional object detection task. In Ours-R101-CBAM, we replace CROCODILE with an object detection task. Specifically, after obtaining the backbone pre-trained on ImageNet, the backbone is trained on the BDD-100K dataset for object detection using Faster RCNN. Then, we fine-tune the CBAM based on this backbone and refer to it as FRCNN-R101-CBAM.

As shown in Fig. 6a, we present the objective evaluation of the explainability of the aforementioned E2EDMs. The horizontal axis represents training duration, and the vertical axis represents the OD of each E2EDM, which indicates how much object feature information the E2EDM uses to make driving action predictions. Since the human cognitive system is object-based, a higher OD indicates higher explainability. Based on the performance of CROCODILE on 3 backbones, we could see that for any backbone, as the training duration increases, the explainability of the E2EDM gradually improves and outperforms all baselines. To further prove the effectiveness of CROCODILE, we evaluated the human subjective persuasibility of the explanations from Ours-R101-CBAM (200-th epoch), R101-CBAM, and R101-CBAM-O. Due to the high cost of this experiment, these E2EDMs are trained only once. As shown in Table 2, the explanations generated by Ours-R101-CBAM are easier to understand.

Table 2. The subjective persuasibility of explanations.

Method	R101-CBAM	R101-CBAM-O	Ours-R101-CBAM
Subjective persuasibility	2.48	2.70	**3.44**

In addition, we can see that Ours-R101-CBAM consistently outperforms those of FRCNN-R101-CBAM. This indicates that the object detection task cannot replace CROCODILE in enhancing the explainability of the E2EDMs. This outcome may be expected because what benefits the E2EDMs is not a backbone that excels at extracting all object information but the important objects.

Finally, we analyze the prediction accuracy of all E2EDMs. As shown in Fig. 6b, prediction accuracy decreases as the training duration increases. There is a typical trade-off between prediction accuracy and explainability.

Table 3. All methods and their configurations in Sect. 5.2. Compared to Table 1, the backbone is ResNet18, and we present more E2EDMs by replacing CBAM [18] with ABN [13] and CCnet [10]. In addition, to investigate the impact on explainability when combining CROCODILE with the baseline [25], we train 3 series of E2EDMs, *e.g.*, for Ours-R18-CBAM, there is Ours-R18-CBAM-O.

Method	Backbone	E2EDM	OB [25]	CROCODILE
R18-CBAM	ResNet18 [8]	CBAM [18]		
R18-CBAM-O			✓	
Ours-R18-CBAM				✓
Ours-R18-CBAM-O			✓	✓
R18-CCnet		CCnet [10]		
R18-CCnet-O			✓	
Ours-R18-CCnet				✓
Ours-R18-CCnet-O			✓	✓
R18-ABN		ABN [13]		
R18-ABN-O			✓	
Ours-R18-ABN				✓
Ours-R18-ABN-O			✓	✓

5.2 The Effectiveness of CROCODILE Across Different E2EDMs

In Sect. 5.1, we found that CROCODILE performed best on ResNet18, thus in Table 3, we fine-tune E2EDMs and their corresponding baselines on ResNet18.

As shown in Fig. 6c, for each E2EDM, the performance of CROCODILE consistently surpasses their corresponding baselines. Furthermore, when combining CROCODILE with the baseline methods, the Ours-R18-CBAM-O and Ours-R18-CCnet-O achieve superior performance than Ours-R18-CBAM and Ours-R18-CCnet-O, the Ours-R18-ABN-O achieve similar performance with Ours-R18-ABN. These results confirm the effectiveness of the CROCODILE. Next, we compare the heatmaps generated by different E2EDMs to intuitively understand the differences between CROCODILE and the baselines. As shown in Fig. 7, the E2EDMs from the CROCODILE have a stronger ability to utilize important object features compared to the baselines.

Finally, in Fig. 6d, as same as it does in Fig. 6b, prediction accuracy decreases as the training duration increases.

5.3 Ablation Study

We remove or replace certain components of the CROCODILE and observe whether the modified CROCODILE remains effective. Since we found that our approach performs best with ResNet18 and CCnet, we first train ResNet18 using different versions of CROCODILE and then fine-tune it on CCnet. As shown in Table 4, we analyze the effectiveness of various versions of the CROCODILE.

Table 4. Summary of all methods and their configurations in Sect. 5.3, there are three components in CROCODILE, DRF, CP, and Negative. For DRF, the BO represents the Biggest Object, RO represents the Random Object, and RC represents the Random Crop. For CP, the Crop represents using the original pre-backbone and post-backbone crops to make positive and negative pairs, the CDN represents using the CDN in [27] to make positive and negative pairs. For Negative, the check mark and the cross mark represent whether we consider the similarity of negative pairs in Eq. 6.

Methods	Backbone	E2EDM	CROCODILE components				
			DRF			CP		Negative
			BO	RO	RC	Crop	CDN	
Ours-R18-CCnet	ResNet18 [8]	CCnet [10]	✓			✓		✓
RO-R18-CCnet				✓		✓		✓
RC-R18-CCnet					✓	✓		✓
CDN-R18-CCnet			✓				✓	✓
w/o Negative-R18-CCnet			✓			✓		

- Driving-Related Feature (**DRF**): As introduced in Sect. 4.4, CROCODILE chooses the biggest object (**BO**) as the driving-related feature. We make two different modifications. **1.** We randomly select an object from all the objects. **2.** Instead of selecting objects, we randomly crop a region from the image.
- Contrastive Pairs (**CP**): CROCODILE determines the positive and negative samples in contrastive learning through two types of crops. To observe the impact of the crops, we replace it with another method to determine the positive and negative samples. In previous research, ZHANG et al. [27] propose Contrastive DeNoising (CDN). CDN randomly alters the coordinates of the top-left and bottom-right corners of an object's bounding box. A new bounding box with small changes is used as a positive sample, while a new bounding box with large changes is used as a negative sample, we replace the crops with CDN for selecting positive and negative samples.
- **Negative**: As introduced in Eq. 6, CROCODILE not only considers the similarity of the positive pairs but also considers negative pairs. We make a modified CROCODILE by only considering the similarity of positive pairs.

As shown in Fig. 8, for the first component, changing the BO to RO or RC significantly weakens the explainability of the E2EDM. More specifically, the RC results in worse explainability than the RO. This implies that object information, and particularly important object information, is crucial for explainability. For the second component, using CDN to select positive and negative samples significantly weakens the explainability. This indicates that the crops in CROCODILE is crucial. For the third component, without negative samples, the explainability is significantly weakened. This aligns well with the understanding of contrastive learning: having only positive samples without negative samples leads to the backbone taking shortcuts, *i.e.*, output the same feature

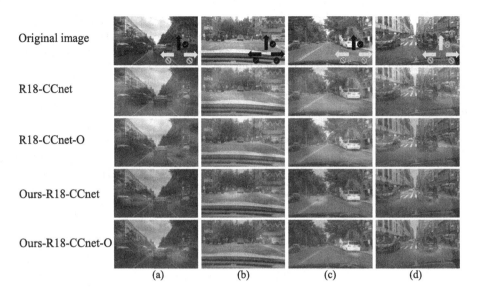

Fig. 7. Columns represent different driving scenes, rows represent E2EDMs.

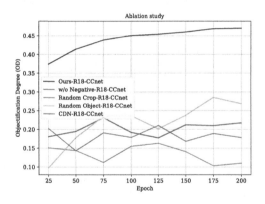

Fig. 8. The explainability of CROCODILE and its variations.

for any image. All explanations discussed in Sect. 5 are shown in supplementary materials.

6 Conclusion

In this paper, we proposed CROCODILE, a method that enhances the explainability of E2EDMs by adding an additional pre-training stage for the backbone. CROCODILE determines the positive and negative samples for contrastive learning through two types of crops, enabling the backbone to better process the driving environment image. Specifically, the important object features in the image are well-preserved, without being mixed with background information. This provides a foundation for fine-tuning the E2EDM, allowing it to only use important object features for driving action prediction, thereby improving explainability.

Although our method successfully enhances the explainability of E2EDMs, we observed a decline in prediction accuracy, showing the trade-off between these two aspects. Addressing the challenge of simultaneously improving both prediction accuracy and explainability is a key focus of our future research.

Acknowledgement. This work was supported by JST SPRING JPMJSP2125, JSPS KAKENHI Grant Number 23K28164, and JST CREST Grant Number JPMJCR22D1. The author Chenkai Zhang would like to take this opportunity to thank the "Interdisciplinary Frontier Next-Generation Researcher Program of the Tokai Higher Education and Research System."

References

1. Ben-Younes, H., Zablocki, É., Pérez, P., Cord, M.: Driving behavior explanation with multi-level fusion. Pattern Recogn. **123**, 108421 (2022)
2. Bojarski, M., et al.: Visualbackprop: visualizing CNNs for autonomous driving **2**, 1–2. arXiv preprint arXiv:1611.05418 (2016)
3. Bojarski, M., et al.: End to end learning for self-driving cars. arXiv preprint arXiv:1604.07316 (2016)
4. Chen, T., Kornblith, S., Norouzi, M., Hinton, G.: A simple framework for contrastive learning of visual representations. In: International Conference on Machine Learning, pp. 1597–1607. PMLR (2020)
5. Deng, J., Dong, W., Socher, R., Li, L.J., Li, K., Fei-Fei, L.: Imagenet: a large-scale hierarchical image database. In: 2009 IEEE Conference on Computer Vision and Pattern Recognition, pp. 248–255. IEEE (2009)
6. Guidotti, R., Monreale, A., Ruggieri, S., Turini, F., Giannotti, F., Pedreschi, D.: A survey of methods for explaining black box models. ACM Comput. Surv. (CSUR) **51**(5), 1–42 (2018)
7. He, K., Fan, H., Wu, Y., Xie, S., Girshick, R.: Momentum contrast for unsupervised visual representation learning. In: Proceedings of the IEEE/CVF Conference on Computer Vision and Pattern Recognition, pp. 9729–9738 (2020)
8. He, K., Zhang, X., Ren, S., Sun, J.: Deep residual learning for image recognition. In: Proceedings of the IEEE Conference on Computer Vision and Pattern Recognition, pp. 770–778 (2016)
9. Huang, G., Liu, Z., Van Der Maaten, L., Weinberger, K.Q.: Densely connected convolutional networks. In: Proceedings of the IEEE Conference on Computer Vision and Pattern Recognition, pp. 4700–4708 (2017)
10. Huang, Z., Wang, X., Huang, L., Huang, C., Wei, Y., Liu, W.: CCNet: criss-cross attention for semantic segmentation. In: Proceedings of the IEEE/CVF International Conference on Computer Vision, pp. 603–612 (2019)
11. Jin, B., et al.: Adapt: action-aware driving caption transformer. In: 2023 IEEE International Conference on Robotics and Automation (ICRA), pp. 7554–7561. IEEE (2023)
12. Kim, J., Rohrbach, A., Darrell, T., Canny, J., Akata, Z.: Textual explanations for self-driving vehicles. In: Ferrari, V., Hebert, M., Sminchisescu, C., Weiss, Y. (eds.) ECCV 2018. LNCS, vol. 11206, pp. 577–593. Springer, Cham (2018). https://doi.org/10.1007/978-3-030-01216-8_35

13. Mori, K., Fukui, H., Murase, T., Hirakawa, T., Yamashita, T., Fujiyoshi, H.: Visual explanation by attention branch network for end-to-end learning-based self-driving. In: 2019 IEEE Intelligent Vehicles Symposium (IV), pp. 1577–1582. IEEE (2019)
14. Oord, A.v.d., Li, Y., Vinyals, O.: Representation learning with contrastive predictive coding. arXiv preprint arXiv:1807.03748 (2018)
15. Ras, G., Xie, N., Van Gerven, M., Doran, D.: Explainable deep learning: a field guide for the uninitiated. J. Artif. Intell. Res. **73**, 329–396 (2022)
16. Tampuu, A., Matiisen, T., Semikin, M., Fishman, D., Muhammad, N.: A survey of end-to-end driving: architectures and training methods. IEEE Trans. Neural Netw. Learn. Syst. **33**(4), 1364–1384 (2020)
17. Wang, D., Devin, C., Cai, Q.Z., Yu, F., Darrell, T.: Deep object-centric policies for autonomous driving. In: 2019 International Conference on Robotics and Automation (ICRA), pp. 8853–8859. IEEE (2019)
18. Woo, S., Park, J., Lee, J.-Y., Kweon, I.S.: CBAM: convolutional block attention module. In: Ferrari, V., Hebert, M., Sminchisescu, C., Weiss, Y. (eds.) ECCV 2018. LNCS, vol. 11211, pp. 3–19. Springer, Cham (2018). https://doi.org/10.1007/978-3-030-01234-2_1
19. Wu, Z., Xiong, Y., Yu, S.X., Lin, D.: Unsupervised feature learning via nonparametric instance discrimination. In: Proceedings of the IEEE Conference on Computer Vision and Pattern Recognition, pp. 3733–3742 (2018)
20. Xu, H., Gao, Y., Yu, F., Darrell, T.: End-to-end learning of driving models from large-scale video datasets. In: Proceedings of the IEEE Conference on Computer Vision and Pattern Recognition, pp. 2174–2182 (2017)
21. Xu, Y., et al.: Explainable object-induced action decision for autonomous vehicles. In: Proceedings of the IEEE/CVF Conference on Computer Vision and Pattern Recognition, pp. 9523–9532 (2020)
22. Xu, Z., et al.: Drivegpt4: Interpretable end-to-end autonomous driving via large language model. arXiv preprint arXiv:2310.01412 (2023)
23. Yu, F., et al.: Bdd100k: a diverse driving dataset for heterogeneous multitask learning. In: IEEE/CVF Conference on Computer Vision and Pattern Recognition (CVPR), June 2020
24. Zhang, C., Deguchi, D., Chen, J., Murase, H.: Toward explainable end-to-end driving models via simplified objectification constraints. IEEE Trans. Intell. Transp. Syst. **25**, 14521–14534 (2024)
25. Zhang, C., Deguchi, D., Murase, H.: Refined objectification for improving end-to-end driving model explanation persuasibility. In: 2023 IEEE Intelligent Vehicles Symposium (IV), pp. 1–6. IEEE (2023)
26. Zhang, C., Deguchi, D., Okafuji, Y., Murase, H.: More persuasive explanation method for end-to-end driving models. IEEE Access **11**, 4270–4282 (2023)
27. Zhang, H., et al.: DINO: DETR with improved denoising anchor boxes for end-to-end object detection. arXiv preprint arXiv:2203.03605 (2022)
28. Zhang, Y., Tiňo, P., Leonardis, A., Tang, K.: A survey on neural network interpretability. IEEE Trans. Emerg. Topics Comput. Intell. **5**(5), 726–742 (2021)

Deterministic Guided Progressive Medical Image Cross-Modal Generation Based on Deep Learning

Chujie Zhang[1], Lanfen Lin[2], and Yen-Wei Chen[1]()

[1] Ritsumeikan University, Osaka, Japan
chen@is.ritsumei.ac.jp
[2] Zhejiang University, Hangzhou Zhejiang, China

Abstract. In this paper, we address critical challenges in medical image generation using deep learning techniques. While convolutional neural networks and generative adversarial networks (GANs) have achieved remarkable results in various image generation tasks, their application to medical imaging faces unique obstacles. These include the complexity and diversity of medical images, limitations in discriminator network structures, and the risk of model collapse and gradient vanishing in multi-scale discriminators. To overcome these issues, we propose a novel deterministic guided progressive GAN that specifically targets regions of interest (ROI) in medical images. Our approach progressively integrates adversarial generative networks, evolving from single-scale to multi-scale discriminators, to produce higher quality images. We demonstrate the efficacy of our model in generating high-precision cross-modal medical images through four comprehensive evaluation criteria, providing both quantitative and qualitative evidence of its performance compared to real images. This innovative method promises to significantly advance the field of medical image generation, potentially enhancing diagnostic accuracy and research capabilities in healthcare.

Keywords: Medical image synthesis · Deep learning · Progressive generative adversarial network · Multi-scale discriminators

1 Introduction

Medical imaging is a crucial diagnostic and research tool that provides visual representations of anatomical structures, playing a vital role in disease diagnosis and surgical planning [2]. Computed tomography (CT) and magnetic resonance imaging (MRI) are the most commonly used techniques in current clinical practice. These imaging modes offer complementary information, and their effective integration can significantly enhance medical decision-making [1]. However, obtaining paired multi-modal images is challenging, creating an increasing need for advanced multi-modal image generation techniques to support clinical diagnosis and treatment.

Medical image generation techniques have evolved from traditional machine learning methods to deep learning approaches. Earlier methods relied on explicit feature representation, such as random forests and k-nearest neighbor algorithms, optimizing feature representation through iterative methods. In recent years, convolutional neural networks, particularly generative adversarial networks (GANs), have achieved state-of-the-art performance in various image generation tasks [4,9–11].

Current methods often employ conditional GAN architectures with deterministic outputs, typically using L1/L2-based loss functions to learn deterministic mappings. However, these approaches do not explicitly model robustness to outliers or predictive uncertainty, leading to performance degradation when encountering unseen out-of-distribution patterns during testing [3]. While these methods can produce synthetic images of high visual quality, the content may still deviate significantly from the corresponding ground-truth values [8]. This discrepancy can lead to overconfidence or misinterpretation, potentially resulting in negative consequences, especially in medical applications. Additionally, these methods often focus on generating entire images, which can cause deformation of target regions without prior knowledge and result in poor quality generation of local target areas, manifesting as blurriness or unreasonable textures.

The discriminator in existing architectures typically uses a single-scale (Markovian) discriminator due to the significant differences in data distribution between different modalities in cross-modal medical image generation [7]. While multi-scale discriminators are common in natural image generation, their use in medical image generation often leads to mode collapse or gradient disappearance. A stronger discriminator generally produces higher quality results, and recent studies have shown that using multiple discriminator ensembles can enhance output quality [5]. However, these methods still lack guided step-by-step generation of high-quality images and fail to give special attention to areas with poorer generated results.

This paper aims to present a deep learning-based method for cross-modality medical image generation that addresses the limitations of existing techniques. Our approach utilizes a progressive adversarial generation network that gradually transitions from a single discriminator to multiple discriminators of varying complexities. This method aims to deterministically guide the generation of higher quality images, focusing on tumors or other regions of interest (ROI). Our main contributions are:

- We develop a multi-modality medical image generation technique that operates independently of paired data. This innovative method significantly improves the local generation quality of target areas, overcoming the limitations of traditional approaches that require aligned image pairs.
- We implement a progressive multi-scale discriminator approach within the adversarial generation network. This strategy guides the generator to focus on critical target regions, resulting in higher quality image production. Our method effectively mitigates common GAN issues such as gradient vanishing

and mode collapse, while also enhancing the utility of generated images for downstream diagnostic and analytical tasks.
- We create a flexible and modular method that can be easily integrated with existing medical image generation algorithms. This design allows for performance enhancement without altering the original network structure, thereby improving the quality of target area generation across various imaging modalities and frameworks.

2 Methods

2.1 Overview

Our proposed method for progressive medical image cross-modal generation utilizes a deep learning approach based on three adversarial generative modules, all built upon the CycleGAN architecture. This model is designed to generate high-quality cross-modal medical images, focusing particularly on liver and tumor regions. The process involves a staged approach, progressively incorporating more specialized discriminators to refine the generated images.

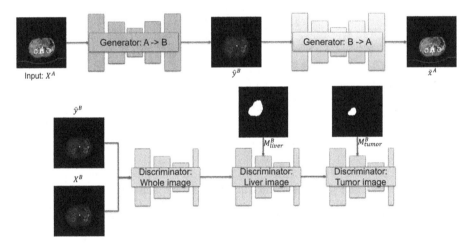

Fig. 1. Comprehensive Training Framework for Medical Image Cross-Modal Generation.

2.2 Progressive Multi-scale GAN Architecture

The Progressive Multi-Scale GAN Architecture, as illustrated in Fig. 1, consists of generators and multiple discriminators:

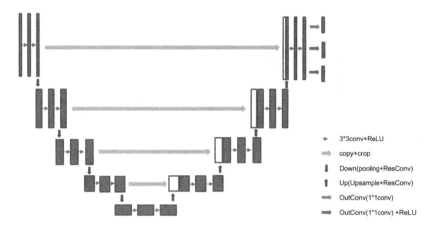

Fig. 2. Architecture of the Generator Network for Medical Image Cross-Modal Generation.

Generators. Two UNet-based generators are employed:

- For CT to MR: $G_{A \to B}$ (Generator).
- For MR to CT: $G_{B \to A}$ (Generator).

where A and B represent CT and MR modalities respectively,

Discriminators. The model incorporates three types of discriminators:

1. Whole image discriminators: D_{whole}.
2. Liver region multi-scale discriminators: D_{liver} (two scales).
3. Tumor region multi-scale discriminators: D_{tumor} (three scales).

The generator's network architecture, based on a modified UNet structure, is meticulously designed for cross-modal medical image generation, as illustrated in Fig. 2. This sophisticated design incorporates four essential sub-modules: ResConv, Down, Up, and OutConv. The ResConv module implements a residual structure with 3×3 convolutions and padding of 1, enhancing feature preservation. The Down module combines max pooling with ResConv for effective feature downsampling, while the Up module employs bilinear upsampling (doubling input size) followed by ResConv, ensuring precise alignment of corner pixels between input and output tensors. The OutConv module applies a 1×1 convolution for final refinement. A distinctive feature of this generator is its tri-headed output layer, which produces not only the generated image but also scale and shape maps. These additional outputs are crucial for implementing the zero-mean generalized Gaussian distribution loss function, significantly enhancing the model's capacity to capture intricate image characteristics and improve overall generation quality, particularly in the context of medical imaging where precision is paramount.

The discriminator network in our progressive multi-scale GAN architecture employs a sophisticated Markovian design, crucial for fine-grained image analysis. This structure utilizes 4 × 4 convolutions with padding of 1 and stride of 2, activated by LeakyReLU (negative slope 0.2). Departing from conventional CNN-based classifiers that typically end with fully connected layers, our Markovian discriminator consists entirely of convolutional layers, producing an n x n output matrix. The final classification is determined by averaging this matrix, allowing for more nuanced spatial assessment. As illustrated in Fig. 1, the model progressively integrates specialized discriminators across three GAN modules, each focusing on increasingly specific image regions. The first module introduces a whole-image discriminator, providing a global assessment. The second module adds a liver-focused multi-scale discriminator, operating at two scales: original and half-downsampled. The third module incorporates a tumor-centric multi-scale discriminator, functioning at three scales: original, half-downsampled, and quarter-downsampled. This progressive, multi-scale approach enables increasingly refined discrimination, particularly in diagnostically crucial regions such as the liver and tumors. By gradually narrowing the focus from whole images to specific anatomical structures, our model enhances its ability to generate highly accurate and detailed cross-modal medical images, addressing the unique challenges of medical imaging tasks.

2.3 Loss Functions

Our model builds upon the CycleGAN framework, incorporating a sophisticated loss function that combines multiple components to ensure high-quality image generation and domain transfer. The core of our loss function is the zero-mean generalized Gaussian distribution loss, defined as:

$$L_{\alpha\beta}^{G}(\hat{y}^B, \alpha, \beta, x^B) = \frac{1}{K}\sum(\frac{|\hat{y}^B - x^B|}{\alpha})^\beta - \log\frac{\beta}{\alpha} + \log\Gamma(\beta^{-1}) \quad (1)$$

In this equation, α and β represent the scale and shape maps generated by the generator, respectively. $\hat{y}^B$ denotes the generated MR image, while x^B is the real MR image. K signifies the total number of pixels in the data, and Γ is the gamma function.

To ensure cycle consistency, we employ:

$$L_{cyc}^G = E_{x\ P_{data}(x)}[||\hat{x}^A - x^A||_1] \quad (2)$$

Our model features three adversarial modules, each with its own generator and discriminator loss:

Whole-image discrimination:

$$L_{adv1}^G = L_2(D_{whole}(\hat{y}^B), 1) \quad (3)$$

$$L_{adv1}^D = L_2(D_{whole}(\hat{y}^B), 0) + L_2(D_{whole}(x^B), 1) \quad (4)$$

Liver-focused discrimination:

$$L_{adv2}^G = L_2(D_{liver}(\hat{y}^B, M_{liver}^B), 1) \tag{5}$$

$$L_{adv2}^D = L_2(D_{liver}(\hat{y}^B, M_{liver}^B), 0) + L_2(D_{liver}(x^B, M_{liver}^B), 1) \tag{6}$$

Tumor-centric discrimination:

$$L_{adv3}^G = L_2(D_{tumor}(\hat{y}^B, M_{tumor}^B), 1) \tag{7}$$

$$L_{adv3}^D = L_2(D_{tumor}(\hat{y}^B, M_{tumor}^B), 0) + L_2(D_{tumor}(x^B, M_{tumor}^B), 1) \tag{8}$$

Here, L_{adv1}^G, L_{adv1}^D, L_{adv2}^G, L_{adv2}^D, L_{adv3}^G, and L_{adv3}^D correspond to the generator and discriminator loss functions for each of the three adversarial modules.

The comprehensive loss function that guides our model's training is the sum of all these components:

$$L = L_{\alpha\beta}^G(\hat{y}^B, \alpha, \beta, x^B) + L_{cyc}^G + L_{adv1}^G + L_{adv1}^D + L_{adv2}^G + L_{adv2}^D + L_{adv3}^G + L_{adv3}^D \tag{9}$$

This multi-faceted loss function enables our model to generate high-fidelity cross-modal medical images while maintaining anatomical accuracy and preserving crucial diagnostic features.

3 Experimental Results

3.1 Dataset

In this paper, private data from a hospital was used, which includes magnetic resonance imaging (MRI) and computed tomography (CT) images of 305 patients, as well as corresponding tumor region masks. The dataset was divided into training, validation, and testing sets in a ratio of 6:2:2.

To train the progressive medical image cross-modality generation model, we proposed a preprocessing workflow for MR and CT images. The first step is to adjust the window width and window level. For CT images, we modified the image intensity by setting the window width and window level based on prior knowledge from doctors, in order to remove the histogram difference identified in the entire dataset. For MR images, we used the algorithm proposed by Manjón et al. [12]. The second step is pixel normalization. For CT images, we directly used linear normalization to scale the pixel values between −1 and 1. For MR images, we first used the z-score algorithm and then used linear normalization to scale the pixel values between −1 and 1. The final step is data selection. To select data with tumors, we calculated the index of the slice with the largest tumor in the tumor mask data and selected four slices above and below this slice, totaling nine slices as the dataset used for each patient. In total, 2745 pairs of CT and MRI images were created.

3.2 Evaluation Metrics

In the field of image generation, due to the limitation of human vision, the authenticity of generated images can only be subjectively evaluated. We used four different evaluation criteria to evaluate the model. The first one is based on the peak signal-to-noise ratio (PSNR) of the tumor area. This evaluation criterion is based on the characteristics of PSNR, mainly using the characteristics of PSNR to evaluate the tumor area of liver, as shown in formula 9.

$$PSNR = 10 \times \log_{10}(\frac{(2^n - 1)^2}{MSE}) \qquad (10)$$

The second evaluation metric is based on the structural similarity of the tumor region, as shown in Eq. 10. Structural similarity is a measure of the similarity between two images. The structural similarity algorithm is mainly used to detect the similarity between two images of the same size or to detect the degree of distortion in an image. In this paper, we only calculate the structural similarity of the tumor region to validate the model.

$$SSIM(x, y) = \frac{(2\mu_x \mu_y + c_1)(2\sigma_{xy} + c_2)}{(\mu_x^2 + \mu_y^2 + c_1)(\sigma_x^2 + \sigma_y^2 + c_2)} \qquad (11)$$

In which σ_x^2 is the variance of x, σ_y^2 is the variance of y, μ_x is the mean of x, μ_y is the mean of y, σ_{xy} is the covariance of x and y, and c_1 and c_2 are two constants.

The third one is based on learned image perceptual similarity, which was first proposed by Richard Zhang et al. in 2018 to measure the difference between two images [13]. This metric learns the inverse mapping from generated images to real ones, forcing the generator to learn the inverse mapping from fake images to real images and prioritizes the perceptual similarity between them.

The fourth one is Frechet Inception Distance (FID), which is one of the most popular metrics used to measure the feature distance between real and generated images [6]. Mathematically, Frechet Distance is used to calculate the distance between two "multivariate" normal distributions. In computer vision, especially in GAN evaluation, we use the Inception V3 model pre-trained on the Imagenet dataset. The specific algorithm details are shown in Formula 11:

$$FID = ||\mu_r - \mu_g||^2 + Tr(\sum_r + \sum_g - 2(\sum_r \sum_g)^{1/2}) \qquad (12)$$

where μ_g and μ_r are the means of the feature maps of the generated and real images, respectively. Tr represents the trace of the linear algebraic operation, and $\sum_g$ and $\sum_r$ represent the covariance matrices of the generated and real images, respectively.

3.3 Analysis

Our study focused on cross-modal generation between MR and CT images across three distinct data phases. We evaluated the performance of our trained models using four comprehensive metrics. The following analysis presents both quantitative and qualitative assessments of our results, providing a thorough examination of the model's effectiveness in generating high-quality cross-modal medical images. Our method performed MR-to-CT and CT-to-MR image translation

Table 1. Quantitative evaluation of cross-modal medical image generation performance across three different data types (ART, PV, and NC) for MR to CT and CT to MR conversions. The performance is measured using four metrics: TPSNR (Total Peak Signal-to-Noise Ratio), TSSIM (Total Structural Similarity Index), LPIPS (Learned Perceptual Image Patch Similarity), and FID (Fréchet Inception Distance).

Data type		TPSNR	TSSIM	LPIPS	FID
ART	MR →CT	33.76	0.85	0.395	21.75
	CT→ MR	34.64	0.88	0.410	20.98
PV	MR →CT	35.51	0.89	0.417	20.39
	CT→ MR	37.19	0.91	0.431	19.30
NC	MR →CT	30.91	0.84	0.366	24.59
	CT→ MR	37.20	0.85	0.387	23.67

tasks across three datasets: ART (Arterial), PV (Portal Venous), and NC (Non-Contrast). The results, evaluated using four metrics (TPSNR, TSSIM, LPIPS, and FID), are presented in Table 1. The PV dataset consistently demonstrated superior performance, while the NC dataset showed the lowest scores across most metrics. This variation can be attributed to the distinct image features characteristic of each phase. The model appears to perform better when there is a greater contrast in grayscale values between the source and target images. Notably, our approach achieved high-quality results for both MR-to-CT and CT-to-MR translations. The CT-to-MR translation in the PV dataset yielded the best overall performance, with the highest TPSNR (37.19), TSSIM (0.91), LPIPS (0.431), and the lowest FID (19.30). These scores represent state-of-the-art performance in cross-modal medical image translation using unpaired datasets. Interestingly, CT-to-MR translations generally outperformed MR-to-CT translations across all datasets, as evidenced by higher TPSNR and TSSIM values. This suggests that our model may be more adept at generating MR images from CT scans than vice versa. The ART dataset showed balanced performance between MR-to-CT and CT-to-MR translations, while the NC dataset exhibited the largest performance gap between the two translation directions. These results demonstrate the effectiveness of our approach in handling various types of medical imaging data and its potential for improving cross-modal image generation in clinical applications.

Figures 3 and 4 present the qualitative results of our cross-modal image generation for MR-to-CT and CT-to-MR conversions, respectively. The images are organized by the three phases: ART (Arterial), PV (Portal Venous), and NC (Non-Contrast). To highlight the model's performance in generating tumor

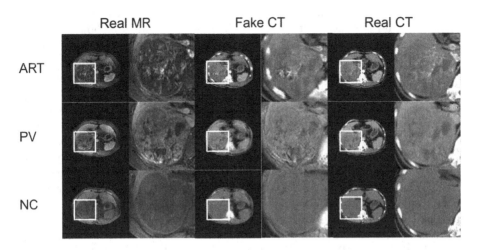

Fig. 3. Qualitative comparison of CT images generated from MR images across three different scan types: ART (Arterial), PV (Portal Venous), and NC (Non-Contrast). The figure shows real MR images (left column), generated "fake" CT images (middle column), and corresponding real CT images (right column) for each scan type. Red boxes highlight regions of interest to compare the quality and accuracy of the generated CT images against the real CT scans. (Color figure online)

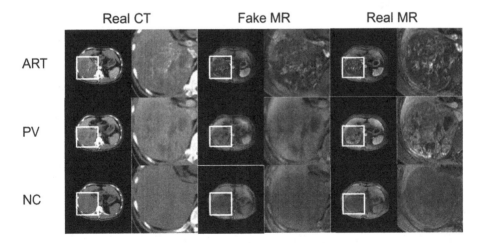

Fig. 4. Qualitative results of MR images generated from CT scans across three different phases: ART (Arterial), PV (Portal Venous), and NC (Non-Contrast). The figure displays real CT images (left column), synthetically generated "fake" MR images (middle column), and corresponding real MR images (right column) for each phase. Blue boxes highlight regions of interest to facilitate comparison between the generated MR images and the actual MR scans, demonstrating the effectiveness of the cross-modal generation technique. (Color figure online)

regions within the liver, we have marked areas of interest with red boxes (for MR-to-CT) and blue boxes (for CT-to-MR). These regions are shown enlarged alongside the original images for detailed comparison. Upon close examination, the generated images demonstrate remarkable fidelity to the ground truth in terms of both structural accuracy and spatial integrity. This is particularly evident in the enlarged tumor regions, where our model has successfully preserved fine details and subtle lesions without introducing blurriness or losing critical features. In the MR-to-CT results (Fig. 3), the fake CT images closely mimic the appearance and contrast of real CT scans across all three phases. The model accurately captures the higher contrast of CT imaging, especially in bone and soft tissue differentiation. Similarly, in the CT-to-MR results (Fig. 4), the generated MR images effectively replicate the characteristic soft tissue contrast and detail of real MR scans. The fake MR images maintain the complex textural patterns typical of MRI, particularly noticeable in the liver parenchyma and surrounding tissues. Notably, our model performs consistently well across all three phases (ART, PV, NC), adapting to the specific imaging characteristics of each. This demonstrates the robustness of our approach in handling various contrast phases commonly encountered in clinical imaging. These qualitative results corroborate our quantitative findings, showcasing the model's capability to generate high-quality, clinically relevant cross-modal medical images while preserving critical diagnostic information.

4 Conclusion

In this paper, we introduce a novel approach to medical image generation using a progressive generative adversarial network with multiple discriminators at different scales. Our method aims to provide deterministic guidance, focusing particularly on liver tumor regions, to generate higher quality images while overcoming common GAN training challenges. Through comprehensive qualitative and quantitative analyses, we have demonstrated the model's success in improving generation quality for target regions. The results show significant enhancements in structural accuracy, detail preservation, and overall image fidelity across various imaging modalities and contrast phases. Looking ahead, we plan to apply our method to state-of-the-art approaches to showcase its broader applicability, and we will refine the network structure of our multi-scale discriminators to better suit the specific needs of medical image generation. These future directions aim to further advance the field, ultimately contributing to improved diagnostic tools and patient care in clinical practice.

References

1. Chen, J., Wei, J., Li, R.: TarGAN: target-aware generative adversarial networks for multi-modality medical image translation. In: de Bruijne, M., et al. (eds.) MICCAI 2021. LNCS, vol. 12906, pp. 24–33. Springer, Cham (2021). https://doi.org/10.1007/978-3-030-87231-1_3

2. Ernst, P., Hille, G., Hansen, C., Tönnies, K., Rak, M.: A CNN-based framework for statistical assessment of spinal shape and curvature in whole-body MRI images of large populations. In: Shen, D., et al. (eds.) MICCAI 2019. LNCS, vol. 11767, pp. 3–11. Springer, Cham (2019). https://doi.org/10.1007/978-3-030-32251-9_1
3. Goodfellow, I., et al.: Generative adversarial networks. Commun. ACM **63**(11), 139–144 (2020)
4. Han, X.: MR-based synthetic CT generation using a deep convolutional neural network method. Med. Phys. **44**(4), 1408–1419 (2017)
5. Hardy, C., Le Merrer, E., Sericola, B.: MD-GAN: multi-discriminator generative adversarial networks for distributed datasets. In: 2019 IEEE International Parallel and Distributed Processing Symposium (IPDPS), pp. 866–877. IEEE (2019)
6. Heusel, M., Ramsauer, H., Unterthiner, T., Nessler, B., Hochreiter, S.: GANs trained by a two time-scale update rule converge to a local Nash equilibrium. In: Advances in Neural Information Processing Systems **30** (2017)
7. Isola, P., Zhu, J.Y., Zhou, T., Efros, A.A.: Image-to-image translation with conditional adversarial networks. In: Proceedings of the IEEE Conference on Computer Vision and Pattern Recognition, pp. 1125–1134 (2017)
8. Jin, C.B., et al.: Deep CT to MR synthesis using paired and unpaired data. Sensors **19**(10), 2361 (2019)
9. Kang, J., Kim, S., Lee, K.M.: Multi-modal/multi-scale convolutional neural network based in-loop filter design for next generation video codec. In: 2017 IEEE International Conference on Image Processing (ICIP), pp. 26–30. IEEE (2017)
10. Liu, B., Tang, R., Chen, Y., Yu, J., Guo, H., Zhang, Y.: Feature generation by convolutional neural network for click-through rate prediction. In: The World Wide Web Conference, pp. 1119–1129 (2019)
11. Liu, Z., Dou, Y., Jiang, J., Xu, J.: Automatic code generation of convolutional neural networks in FPGA implementation. In: 2016 International Conference on Field-programmable Technology (FPT), pp. 61–68. IEEE (2016)
12. Manjón, J.V.: MRI preprocessing. Imaging Biomarkers: Development and Clinical Integration, pp. 53–63 (2017)
13. Zhang, R., Isola, P., Efros, A.A., Shechtman, E., Wang, O.: The unreasonable effectiveness of deep features as a perceptual metric. In: Proceedings of the IEEE Conference on Computer Vision and Pattern Recognition, pp. 586–595 (2018)

Separate Guided Denoising Training for Human-Object Interaction Detection

Yuki Isoda[✉] and Daisuke Kobayashi

Corporate Research and Development Center, Toshiba Corporation, Minato City, Japan
{yuki2.isoda,daisuke32.kobayashi}@toshiba.co.jp

Abstract. Understanding scenes requires not only the detection objects but also the recognition of the interactions between them. Human-Object Interaction (HOI) detection plays a crucial role in enhancing contextual comprehension by identifying the interactions between humans and objects, which is essential for building more robust and intelligent vision systems. While DETR-based models have shown significant success in HOI detection, they are hindered by slow training convergence. The SOV-STG method has attempted to address this challenge in previous research. To further improve the learning efficiency and accuracy of SOV-STG, we introduce a novel Separate Guided Denoising training strategy specifically designed for HOI detection. Our approach separates the denoising of noised ground truth data for both the human-object decoder and the verb decoder, enabling more efficient and targeted training. Furthermore, we enhance training performance by merging redundant human-object pair annotations, and filtering and regenerating noised bounding boxes. The proposed method was validated on the HICO-DET dataset, achieving state-of-the-art results. Our contributions include a novel training strategy that improves accuracy and ablation studies demonstrating its effectiveness.

Keywords: Human-Object Interaction Detection · Denoising training · Visual Relationship Detection

1 Introduction

Recent advancements in computer vision and machine learning have significantly enhanced our ability to understand the visual semantics of complex interactions between humans and objects in images. Human-Object Interaction (HOI) detection, which aims to identify triplets of <human, verb, object>, is a crucial task with applications in various domains such as human activity recognition [43], image retrieval [17], and visual question answering [38].

Models based on DETR [3] (Detection Transformer) have recently achieved impressive results in HOI detection. However, DETR-based models are known for their slow training convergence and require many epochs to achieve good performance. Originally designed for object detection, DETR framework has prompted extensive research to improve its efficiency, resulting in approaches such as DAB-DETR [29] and DN-DETR [23]. DAB-DETR enhances performance by leveraging 4D coordinates (x, y, w, h) as anchor boxes and refining the mode through spatial positional training. Meanwhile, DN-DETR focuses on the instability of bipartite graph matching. Denoising (DN) training of queries is introduced to learn one-to-one matching between the outputs of DN queries and ground truth data, which stabilizes the bipartite graph matching during the training process.

HOI detection faces similar challenges with the slow convergence of DETR training. SOV-STG [5] has been proposed to extend these studies to HOI detection. SOV assigns a single decoder to detect humans, objects, and recognize verbs, while STG utilizes learnable object and verb label embeddings to guide training. SOV stabilizes training by focusing the decoders on specific targets, and STG speeds up convergence by connecting ground truth labeling information with predefined dataset labels. However, the STG DN training strategy presents challenges when applied to HOI detectios Fig. 1 show the differences in noise handling between the previous method and the proposed method. Specifically, it is unclear what is removed as noise and what remains for DN training. To address this, we propose a separate guided DN training strategy, where the detection of human-object pairs and the DN training for verb recognition are performed separately. This approach clarifies what should be denoised for each decoder, making the DN training is more effective. Furthermore, we focus on the existing dataset and add noise to make the DN training for HOI detection more effective. There are cases in existing datasets where verbs are annotated in redundant bounding boxes for the same human-object pair. As shown in Fig. 2, one output may recognize only "straddle" while another recognizes "ride" and "sit on" even though they refer to the same human-object pair. It is redundant to recognize split verb annotations for the same human-object pair, so we merge them into a single pair. Furthermore, we filter and regenerate noised bounding boxes to ensure they are closest to their respective reconstruction targets.

In summary, our contributions are threefold:

- We proposed a separate guided DN training strategy that allows each decoder to concentrate on its specific denoising task.
- We introduced methods for merging redundant human-object pair annotations and filtering and regenerating noised bounding boxes to enhance the effectiveness of DN training.
- We achieved state-of-the-art results in the HICO-DET benchmark.

2 Related Work

Two-Stage Methods. The two-stage approach [4,10,12,14,15,21,22,34,41, 42,44–46,52,53,55,57], using off-the-shelf detectors, first detects humans and objects and recognize interactions for detected human-object pairs.Since the

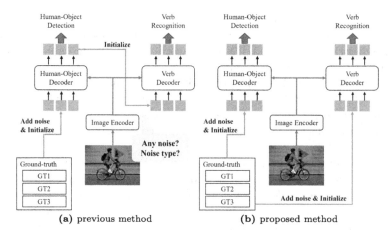

Fig. 1. Comparison between the previous method and the proposed method. In the previous method, DN queries initialized using noised ground truth and fed into the human-object decoder. The output of the human-object decoder is then passed to the verb decoder, making unclear which part of the noise is removed during the human-object decoding and what remains for the verb decoder. In contrast, our proposed method separately initializes DN queries for both the human-object decoder and the verb decoder from the noised ground truth, ensuring that each decoder handles its specific denoising task independently.

introduction of the multi-stream architecture in HO-RCNN [4], many methods have been proposed based on this framework. In HO-RCNN, human appearance features, object appearance features, and spatial features are extracted in each stream and then recognize interactions.Subsequent methods have incorporated human pose features [12,25,41,55], linguistic features [9,31], and graph structures [9,35,40,52,57] to improve recognize interactions.

One-stage Methods. One-stage methods [1,2,5,6,19,24,26,27,33,36,36,37, 39,47–51,54,59,60] typically perform object detection and recognize interactions simultaneously. Early methods used interaction keypoints [26,47] and join regions [18] as predefined anchors.Recently, DETR-based HOI detectors have gained attention, leading to significant performance improvements. However, these methods often suffer from slow learning convergence. Some approaches [7,27,51,56] have introduced multiple decoders for each subtask to address this issue, but they still face challenges in achieving fast convergence.

Effective Learning Methods with Ground Truth. In the DETR family of object detection methods [3,23,29,58] DN-DETR introduces query denoising(DN) to accelerate training by addressing the instability of bipartite graph matching. The DN queries are initialized by adding noise to both the ground truth bounding boxes and their associated labels. These noised queries are then fed into the Transformer decoder. The model is trained to reconstruct the original bounding boxes and labels, stabilizing the training process and improving

(a) Original annotation (b) Merged annotation

Fig. 2. Illustration of the problem of redundant human-Object pairs annotations in the HICO-DET [4] dataet for HOI detection. The left side displays the original annotations of the dataset, where human-object pairs are represented by red and green bounding boxes. In this example, the same human-object pair is annotated twice with different verbs, causing redundant annotations. The right side displays the result after merging these redundant annotations, combining the verb labels into a single instance for each human-object pair. This merging reduces redundancy and improves the accuracy of the model during denoising training by avoiding the learning of conflicting annotations.

convergence speed. In the HOI detection task, HQM [54] encodes shifted ground truth boxes as hard positive queries to guide training. However, HQM does not consider ground truth label information. DOQ [36] introduces an oracle query that implicitly encodes the ground truth boxes and object labels of human-object pairs, guiding the decoder to reconstruct the ground truth HOI instances during training. SOV-STG [5] encodes DN training queries from noised ground truth data to guide the reconstruction of the original ground truth. However, SOV-STG uses two decoders and inputs the DN training query to the human-object decoder. Since the verb decoder receives the output of the human-object decoder, the noise introduced to the verb decoder is not clearly defined. The unstable output of the human-object decoder during the early stages of training destabilizes the verb decoder's training. In addition, the unstable output of the human-object decoder during the early stages of training may destabilize the verb decoder's training. Furthermore, in the later stages of DN training process applied to the human-object decoder for verbs can interfere with the training of the verb recognition decoder. We propose a DN training method for the verb decoder that is not dependent on the results of the human-object decoder. Furthermore, we use annotation cleaning and added noise filtering to make the DN training more effective for HOI detection.

3 Method

Figure 3 shows the training pipeline of our framework. In the normal training and inference phases, learnable anchor boxes and label queries are used as inputs to the human-object decoder, which is responsible for detecting human-object pairs. The embeddings and anchor pairs generated by the human-object decoder are

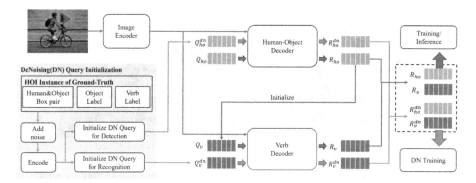

Fig. 3. The training pipeline of our proposal framework. During the inference and training phase, The human-object decoder and the verb decoder are connected in series to predict HOI. Queries Q_{ho} are initialized learnable parameters and queries Q_v are initialized Human-Object Decoder output. However, in the denoising training phase, DN queries Q_{ho}^{dn} and Q_v^{dn} are initialized and input in parallel to the human-object decoder and the verb decoder. This separation enables each decoder to focus on its specific task, improving training efficiency by reducing inference between decoders. By isolating the tasks of detection human-object pairs and recognizing verbs, the model effectively reconstruct ground truth HOI instances.

subsequently fed into the verb decoder to predict verb classes. The human-object embeddings and anchor pairs output from the human-object decoder are then input to the verb decoder to predict verb classes. For denoising(DN) training, we utilize DN queries that are initialized from ground truth HOI instances with added noise. These DN queries are separately input into both the human-object decoder and the verb decoder, allowing for the reconstruction of the ground truth HOI instances. This separation of DN queries enhances learning efficiency by clearly defining the reconstruction targets for each decoder. Section 3.1 provides a detailed description of our framework, while Sect. 3.2 explains the methods for redundant human-object pair annotations and filter and regenerate noised bounding boxes.

3.1 Separate Guided Denoising Training

In our training pipeline, we populate the human-object decoder and the verb decoder with their respective DN queries, thereby enhancing training efficiency by clearly defining the training targets for each decoder.

Human-Object Decoder and Verb Decoder. Our framework uses a image encoder extract global features, which are then input to a human-object decoder and a verb decoder. The image encoder leverages a hierarchical backbone and a deformable transformer encoder [58] to capture multi-scale global features $fg \in \mathbb{R}^{N_g \times D}$, where N_g represents the total number of pixels in the multi-scale feature map and D denotes the hidden dimension of the embedding within the entire transducer architecture. For the decoding process, we utilize the deformable transformer decoder as proposed in [29], capable of handling label queries and anchor boxes. The human-object decoder uses a label query $Q_o \in \mathbb{R}^{N_q \times D}$ as its input query, which is initialized from learnable parameters. The output from the human-object decoder is then used to predict the object class, object bounding box, and human bounding box. The output from the human-object decoder is then utilized to predict the object class, object bounding box, and human bounding box. Subsequently, the output is fed into the verb decoder to predict verb classes. During DN training, the human-object decoder and verb decoder receives a DN queries, which is generated by adding noise to the ground truth bounding boxes and verb labels.

Label-Specific Priors. We initialize the object and verb label embeddings based on the SOV-STG [5] framework. Specifically, the object label embeddings $t_o \in \mathbb{R}^{C_o \times D}$ serve as the object label priors, and the verb label embeddings $t_v \in \mathbb{R}^{C_v \times D}$ serve as the verb label priors. The query embeddings for object labels $q_o \in \mathbb{R}^{N_q \times D}$ are initialized through a linear combination of the object label embeddings t_o and the object coefficient matrix $A_o \in \mathbb{R}^{N_q} \times C_o$. Similarly, the query embeddings for verb labels $q_v \in \mathbb{R}^{N_q \times D}$ are initialized through a linear combination of t_v and $A_v \in \mathbb{R}^{N_q} \times C_v$. These label embeddings t_o and t_v are utilized in both the DN training and inference stages, and are trained jointly to enhance training efficiency.

Separate Guided Denoising Training Strategy. As shown in Fig. 3, DN training is conducted separately for the human-object decoder and the verb decoder. This separation clarifies the training targets for each decoder and enhances training efficiency. DN queries are generated for each decoder's input. The process of adding noise to the ground truth and creating DN label embeddings follows the SOV-STG framework. First, we explain the method of adding noise to the ground truth. Given a set of ground truth object labels $O_{gt} = o_i{}_{i=1}^{K}$ and a set of verb labels $V_{gt} = v_i{}^K i = 1$ for an image, where o_i and v_i are the object class and verb class labels, respectively, and K is the number of ground truth HOI instances. For the ith ground truth HOI instance, the noised object label is obtained by randomly changing the ground truth index of the object class o_i to another object class index. Since the verb labels v_i consist of co-occurring ground truth classes, the indices other than the ground truth verb labels are randomly changed to preserve the co-occurring ground truth indices that appear in the noised verb labels. Two flipping rate hyperparameters $\eta_o \in (0, 1)$ and $\eta_v \in (0, 1)$ control the percentage of noised HOI instances for object and verb labels, respectively. In addition, the verb class flipping rate hyperparameter $\lambda_v \in (0, 1)$ controls the class-specific flipping rate of verb labels.

Next, we describe how to initialize DN label embeddings using noised object and verb labels. The DN query embeddings are initialized using the indices of the noised label and label embeddings t_o and t_v. The DN query embedding of the object label $q\tilde{o}^i$ is initialized based on the object label embeddings t_o, which correspond to the index of the noised object label. Similarly, the DN query embedding of the verb label q_v^i is initialized based on the sum of the verb label embeddings t_v, which correspond to the indices of the noised verb labels.

Finally, we explain the method of initializing the DN queries for DN training. The detection DN queries Q_{ho}^{dn} used for training the human-object decoder are initialized from the DN query embeddings of the object labels q_o^i and the DN query embeddings of the verb labels q_v^i. Here, the DN process is trained to reconstruct the ground truth of the object while handling noised verb labels as anything. The detection DN queries Q_{ho}^{dn} used for training the human-object decoder is initialized from the DN query embeddings of object label q_o^i and the DN query embeddings of verb labels q_v^i. Here, the DN is trained to reconstruct the ground truth of the object. The recognition DN queries Q_v^{dn} used for training the verb decoder are initialized from the DN query embeddings of the verb labels q_v^i and the query embeddings of the ground truth object labels q_{ogt}^i. The query embeddings of the ground truth object labels enable DN training of verb decoder given the ground truth object information. Thus, by inputting detection DN queries into the human-object decoder and recognition DN queries into the verb decoder, the DN training of each decoder can be clarified to reconstruct targets.

3.2 For Effective Denoising Training of HOI Detection

Merging Redundant Human-Object Pair Annotations. In object detection, a single object in an image is typically assigned a single instance with a bounding box and class label. However, in HOI detection, a single human-object pair may have multiple bounding box pairs, as shown in Fig. 4. During DN training, the data is trained using one-to-one matching to reconstruct ground truth from noised data. If multiple bounding box pairs exist for the same human-object pair and if the verbs are split, only the verbs associated with each bounding box pair are learned as positive, while the rest are learned as negative. To ensure all verbs associated with a human-object pair are recognized and prevent them from being incorrectly learned as negative, we propose a method to unify ground truth instances for the same human-object pair. If the object labels are the same and the minimum value of Intersection over Union (IoU) for a human-object bounding box pair is above the threshold and does not contain the same verb label, it can be considered a split verb annotation for the same human-object pair. The formula for IoU is as follows.

$$IoU_{min}^{(i,j)} = min(IoU_{hum}^{(i,j)}, IoU_{obj}^{(i,j)}) \tag{1}$$

$$IoU_{hum}^{(i,j)}(\boldsymbol{B}_{hum}^i, \boldsymbol{B}_{hum}^j) = \frac{|\boldsymbol{B}_{hum}^i \cap \boldsymbol{B}_{hum}^j|}{|\boldsymbol{B}_{hum}^i \cup \boldsymbol{B}_{hum}^j|} \tag{2}$$

$$IoU_{obj}^{(i,j)}(\boldsymbol{B}_{obj}^i, \boldsymbol{B}_{obj}^j) = \frac{|\boldsymbol{B}_{obj}^i \cap \boldsymbol{B}_{obj}^j|}{|\boldsymbol{B}_{obj}^i \cup \boldsymbol{B}_{obj}^j|} \tag{3}$$

where i, j denote the ith or jth HOI instance in the same image, and $\boldsymbol{B}_{hum}$ and $\boldsymbol{B}_{obj}$ represent the human and object bounding box. If there are split verb annotations for the same human-object pair, the human and object bounding boxes are averaged, and the set of verb labels is unified into a single instance. Eventually, the process is repeated for HOI instances in the image, ensuring there are no more split annotations for the same human-object pair.

Filtering and Regenerating Noised Bounding Boxes. In DN training for HOI detection, ground truth bounding boxes of human-object pair are reconstructed from noised bounding boxes. When object detection is based on anchors, it is common to detect objects that are close to the anchor. However, adding noise to the bounding box may bring it closer to objects other than the target object for reconstruction. For example, if the noised bounding box as shown in Fig. 4, is used as an anchor for DN training, it will learn to detect objects that are not the closest to the anchor. To address this issue, we introduce to filter out the noised bounding box and regenerate it.

We compute the *IoU* between the ground truth bounding boxes and the noised bounding boxes. If the *IoU* of a noised bounding box with any bounding box other than the target for reconstruction is the largest, regenerate it. We distinguish between the bounding boxes of humans and objects, and repeat this calculation only for the bounding boxes of humans and only for the objects. Continue regenerating the noised bounding boxes until all are closest to their respective reconstruction targets. In this way, DN training more effective.

Fig. 4. The effect of noised bounding boxes in the HICO-DET [4] dataset. The left side shows the ground truth instances, while the right side shows the effect of adding noise. The red and green boxes represent humans, and the blue box represents the object (a bench). The dashed circle represents the original center, and the arrow represents the displacement required adjust the noised bounding box to match correct one. When a bounding box is used as an anchor, the model tends to detect the object closest to it. During denoising training, the ground truth and noised bounding boxes are learned through one-to-one matching. If the noised box is closer to non-target object than the actual target, the model learns incorrect predictions by focusing on the wrong object. To resolve this, we regenerate the noised bounding boxes to ensure that the target object is closest to the anchor, improving training efficiency. (Color figure online)

3.3 Training and Inference

As shown in Fig. 3, our proposed method performs DN training simultaneously with normal training. For the inference queries Q_{ho} and Q_v, the Hungarian algorithm is used to match ground truth HOI instances with predicted HOI instances, and the matching cost and the learning loss of predicted HOI instances are similar to previous transformer-based methods [6]. For DN queries Q_{ho}^{dn} and Q_v^{dn}, the ground truth index used in query initialization is used for matching with predictive HOI instances, and the loss function is the same as for inference queries.

4 Experiment

Based on the HICO-DET [4] and V-COCO [11] datasets, the proposed method was evaluated and compared with the previous method, STG. In addition, experiments were carried out on state-of-the-art method RLIPv2. Furthermore, an ablation study was conducted to analyse the contribution of each element and to demonstrate the effectiveness of the proposed method. Through these experiments, we were able to validate the improvements brought by our proposed approach.

4.1 Experimental Settings

Dataset and Metrics. The HICO-DET dataset contains 38,118 images for training and 9,658 images for testing. The 117 verb classes and 80 object classes in HICO-DET form a total of 600 HOI classes. Based on the number of HOI instances appearing in the dataset, the HOI classes in HICO-DET are classified into two categories, '*rare*' and '*non-rare*'. The V-COCO dataset contains 5,400 training images and 4,946 test images. In V-COCO, 80 object classes and 29 verb classes are annotated and two scenarios are considered: scenario 1 with 29 verb classes and scenario 2 with 25 verb classes. The mean Average Precision (mAP) scores are reported according to standard evaluations [4].

Details of Implementation. We have applied and investigated the proposed method in the SOV-STG [5] and RLIPv2 [50] frameworks in order to develop an optimal approach for denoising(DN) training for HOI detection. All experiments were performed on 8 NVIDIA A40 GPUs.

SOV-STG Setups. The SOV-STG framework comprises a human decoder, an object decoder, and a verb decoder, along with the STG DN training strategy. The weights of the image encoder, human decoder, and object decoder were initialized using the DAB-DeformableDETR model trained on the COCO dataset [28].The human and object decoders were fed the same detection DN query, and the corresponding indices of the decoder outputs represented human-object pairs. The verb decoder, which combines the outputs of the human and object decoders using the SO-Attention module, was then used to predict verb classes.

The feature image encoder consists of a ResNet-50 [13] backbone and a 6-layer deformable transformer encoder.The total number of backbones and decoders is based on the SOV-STG paper set-up, while ResNet-50 and 3-layer decoders were validated in SOV-STG-S.The hidden dimension of the transformer is $D = 256$ and the number of queries is $N_q = 64$.In the DN part, a $2N_p = 6$ group of noised labels is generated for each ground truth HOI instance. The dynamic DN scale is set to $\gamma = \frac{2}{3}$, the box noisification rate is set to $\delta_b = 0.4$, the object label flipping rate to $\eta_o = 0.3$ and the verb noisification rate to $\eta_v = 0.6$, The maximum noisification level is defined by setting the flipping rate of verb labels to $\lambda_v = 0.6$.The model is trained by the AdamW optimiser with a learning rate of 2e-4 and weight decay of 1e-4.The backbone was fixed in the SOV validation to reduce training time.The batch size is set to 32, the training epochs are 30, and learning rate drops at the 20th epoch.

RLIPv2 Setups. Since RLIPv2 does not use the label-specific priors of SOV-STG, we replaced the label embeddings used for initializing the DN queries with language features obtained from RLIPv2's Asymmetric Language-Image Fusion (ALIF). Without altering the content of the inference queries, we added new DN queries and fine-tuned the pre-trained model on the dataset. The basic setup was similar to that of SOV-STG.We verified this by finetuning the pre-trained models of RLIPv2-ParseDA ResNet-50 and Swin-Large [32] on each dataset. The batch size is set to 16, the training epochs are 20, and learning rate drops at the 15th epoch.The other setups was similar to that of SOV-STG setups.

4.2 Comparison to State-of-the-Arts

Table 1 presents a comparison of our proposed method with recent state-of-the-art (SOTA) methods on the HICO-DET dataset. Our method, when integrated with SOV, shows an improvement of 0.4% points in mean Average Precision (mAP) over the experimental results of SOV-STG in the full category under default settings. Furthermore, when our method is applied to the pre-trained model of the SOTA method RLIPv2-ParSeDA and fine-tuned on the HICO-DET dataset, we achieve an improvement of 0.80% points for the ResNet-50 (R50) model and 1.00% points for the Swin-Large (Swin-L) model. Table 2 compares the results on the V-COCO dataset, demonstrating that our proposed method improves accuracy in both scenario 1 and scenario 2. Specifically, our method enhances the performance of RLIPv2-ParSeDA, leading to higher accuracy scores in both scenarios.

4.3 Ablation Study

Contributions of Proposed Component

Table 3 shows the contributions of each proposed component using the HICO-DET dataset. The columns "Separate Guided" "Merge Annotations" and "Noise Filtering" indicate whether DN training is separated, merging redundant human-object pair annotations, filtering and regenerating noised bounding boxes,

Table 1. Comparisons with previous methods on HICO-DET. R50 denote ResNet-50 [13]. Swin-L denote Swin-Large [32]. * denotes evaluation results using publicly available models or models we have trained, and unmarked denotes results from paper.

Method	Backbone	Default Setting		
		Full	Rare	Non-Rare
CATN [8]	R50	31.86	25.15	33.84
Liu et al. [30]	R50	33.51	30.30	34.46
QAHOI [6]	R50	26.18	18.06	28.61
QPIC [37]	R50	29.07	21.85	31.23
CDN-S [51]	R50	31.44	27.39	33.53
DOQ(CDN-S) [36]	R50	33.28	29.19	34.50
GEN-VLKT-S [27]	R50	33.75	29.25	35.10
DiffHOI-S [48]	R50	34.41	31.07	35.40
CLIP4HOI [33]	R50	35.33	33.95	35.74
LOGICHOI [24]	R50	35.47	32.03	36.22
PViC w/DETR [53]	R50	34.69	32.14	35.45
DiffHOI-L [48]	Swin-L	41.50	39.96	41.96
PViC w/$\mathcal{H}$-DETR [53]	Swin-L	44.32	44.61	44.24
SOV-STG* [5]	R50	33.19	29.39	34.32
SOV+Ours*	R50	33.59	29.20	34.90
RLIPv2-ParSeDA* [50]	R50	34.60	30.07	36.82
RLIPv2-ParSeDA+Ours*	R50	35.40	31.43	37.36
RLIPv2-ParSeDA* [50]	Swin-L	45.12	45.33	44.70
RLIPv2-ParSeDA+Ours*	Swin-L	**46.12**	**45.58**	**47.22**

respectively. Row (1) represents the baseline result without the proposed method, using the SOV-STG R50 model. Both "Separate Guided" and "Merge Annotations" were effective on their own and improved accuracy. "Noise Filtering" needed to be combined with "Merge Annotations" to be effective. The combination of all elements resulted in the highest accuracy improvement.

Contributions of Merging Redundant Human-Object Annotations on HICO-DET. In Table 4, the effect of merging redundant human-object annotations is investigated. During DN training, the data is trained using one-to-one matching to reconstruct ground truth from noised data.If multiple bounding box pairs exist for the same human-object pair and the verbs are split, only the verbs associated with each bounding box pair are learned as positive, while the rest are learned as negative.Despite the same person-object pair annotations, it learns to recognise only different verbs for each bounding box pair.In merging annotations, if the minimum value of IoU for a rectangular human-object pair with the same object labels is above a threshold and does not contain the same verb labels,

Table 2. Comparisons with previous methods on V-COCO.

Method	Backbone	AP_{role}^{S1}	AP_{role}^{S2}
RLIP-ParSe [49]	R50	61.9	64.2
MSTR [20]	R50	62.0	65.2
ParSe [49]	R50	62.5	64.8
GEN-VLKT-M [27]	R101	63.3	65.6
GEN-VLKT-L [27]	R101	63.6	65.9
CDN-L [51]	R101	63.9	65.9
SSRT [16]	R101	65.0	67.1
SOV-STG* [5]	R50	63.1	64.6
SOV+Ours*	R50	63.4	65.2
RLIPv2-ParSeDA* [50]	R50	65.9	68.1
RLIPv2-ParSeDA+Ours*	R50	66.4	68.5
RLIPv2-ParSeDA* [50]	Swin-L	72.0	74.1
RLIPv2-ParSeDA+Ours*	Swin-L	**72.4**	**74.8**

Table 3. Ablation studies for proposal component on HICO-DET.

#	Separate Guided	Merging Annotations	Noise Filtering	Default Setting Full	Rare	Non-Rare
(1)				33.19	29.39	34.32
(2)	✓			33.48	29.23	34.75
(3)		✓		33.48	**29.82**	34.44
(4)			✓	33.34	29.61	34.46
(7)	✓	✓	✓	**33.59**	29.20	**34.90**

it is considered a segmented annotation for the same human-object pair and is combined into a single annotation.merging threshold represents the threshold of IoU and (1) represents the result without merging duplicate human-object annotations. The accuracy improved the most when $threshold = 0.8$. Otherwise, for example, when $threshold = 0.4$, the object class and verb class are looked at for pairs with IoU greater than 0.4 and a decision is made whether to combine them into one.If the threshold is low, different human-object pairs are combined, which reduces accuracy.In HICO-DET, there is noise in the form of redundant human-object pairs in the dataset annotations. The proposed method reduces this noise and improves accuracy.

Contributions of Filter and Regenerate Noised Bounding Boxes on HICO-DET. Table 5 examines the effects of noise filtering. As noised bounding boxes are used as anchors, undesired training may occur if they are close to a human or object other than the reconstruction target. Therefore, the noise-added bounding box is monitored, and the noise is regenerated when a human

Table 4. Ablation studies for Merging Annotations on R50 model.

Method	Merging threshold	Default Full mAP
(1)	1.0	33.19
(2)	0.8	**33.48**
(3)	0.6	33.37
(4)	0.4	33.16

or object other than the reconstruction target is closest. Rows (1) and (3) show the results when noise filtering is not applied. Applying noise filtering improved accuracy. Additionally, in row (4), where the parameter that generates noise in the bounding box is increased, the improvement in accuracy is greater, verifying the effectiveness of noise filtering.

Table 5. Ablation studies for Noise Filtering on HICO-DET.

Method	Noise Filtering	Noise Parameter δ	Default Full mAP
(1)		0.4	33.19
(2)	✓	0.4	**33.34**
(3)		0.8	32.66
(4)	✓	0.8	33.21

5 Conclusion

This paper introduces a novel separate guided denoising (DN) training strategy for Human-Object Interaction (HOI) detection, where the human-object decoder and the verb decoder are trained independently. This approach allows for the application of explicit noise to each decoder, enhancing the effectiveness of DN training and demonstrating superior performance compared to previous methods. Additionally, our method includes merging redundant human-object annotations and filtering and regenerating noised bounding boxes, which further improve the efficiency of DN training for HOI detection. This strategy can be seamlessly integrated into DETR-based one-stage methods, incorporating both a human-object decoder and a verb decoder, thereby enhancing the performance of state-of-the-art models on relevant benchmarks.

References

1. Cao, Y., et al.: Detecting any human-object interaction relationship: Universal HOI detector with spatial prompt learning on foundation models. In: NeurIPS (2023)
2. Cao, Y., et al.: Re-mine, learn and reason: Exploring the cross-modal semantic correlations for language-guided HOI detection. In: ICCV (2023)
3. Carion, N., Massa, F., Synnaeve, G., Usunier, N., Kirillov, A., Zagoruyko, S.: End-to-end object detection with transformers. In: Vedaldi, A., Bischof, H., Brox, T., Frahm, J.-M. (eds.) ECCV 2020. LNCS, vol. 12346, pp. 213–229. Springer, Cham (2020). https://doi.org/10.1007/978-3-030-58452-8_13
4. Chao, Y., Liu, Y., Liu, X., Zeng, H., Deng, J.: Learning to detect human-object interactions. In: WACV (2018)
5. Chen, J., Wang, Y., Yanai, K.: Focusing on what to decode and what to train: efficient training with HOI split decoders and specific target guided denoising. CoRR **abs/2307.02291** (2023)
6. Chen, J., Yanai, K.: QAHOI: query-based anchors for human-object interaction detection. In: MVA (2023)
7. Chen, M., Liao, Y., Liu, S., Chen, Z., Wang, F., Qian, C.: Reformulating HOI detection as adaptive set prediction. In: CVPR (2021)
8. Dong, L., et al.: Category-aware transformer network for better human-object interaction detection. In: CVPR (2022)
9. Gao, C., Xu, J., Zou, Y., Huang, J.-B.: DRG: dual relation graph for human-object interaction detection. In: Vedaldi, A., Bischof, H., Brox, T., Frahm, J.-M. (eds.) ECCV 2020. LNCS, vol. 12357, pp. 696–712. Springer, Cham (2020). https://doi.org/10.1007/978-3-030-58610-2_41
10. Gao, C., Zou, Y., Huang, J.: ICAN: instance-centric attention network for human-object interaction detection. In: BMVC (2018)
11. Gupta, S., Malik, J.: Visual semantic role labeling. CoRR **abs/1505.04474** (2015)
12. Gupta, T., Schwing, A.G., Hoiem, D.: No-frills human-object interaction detection: factorization, layout encodings, and training techniques. In: ICCV (2019)
13. He, K., Zhang, X., Ren, S., Sun, J.: Deep residual learning for image recognition. In: CVPR (2016)
14. Hou, Z., Peng, X., Qiao, Yu., Tao, D.: Visual compositional learning for human-object interaction detection. In: Vedaldi, A., Bischof, H., Brox, T., Frahm, J.-M. (eds.) ECCV 2020. LNCS, vol. 12360, pp. 584–600. Springer, Cham (2020). https://doi.org/10.1007/978-3-030-58555-6_35
15. Hou, Z., Yu, B., Qiao, Y., Peng, X., Tao, D.: Affordance transfer learning for human-object interaction detection. In: CVPR (2021)
16. Iftekhar, A.S.M., Chen, H., Kundu, K., Li, X., Tighe, J., Modolo, D.: What to look at and where: semantic and spatial refined transformer for detecting human-object interactions. In: CVPR (2022)
17. Johnson, J., et al.: Image retrieval using scene graphs. In: CVPR (2015)
18. Kim, B., Choi, T., Kang, J., Kim, H.J.: UnionDet: union-level detector towards real-time human-object interaction detection. In: Vedaldi, A., Bischof, H., Brox, T., Frahm, J.-M. (eds.) ECCV 2020. LNCS, vol. 12360, pp. 498–514. Springer, Cham (2020). https://doi.org/10.1007/978-3-030-58555-6_30
19. Kim, B., Lee, J., Kang, J., Kim, E., Kim, H.J.: HOTR: end-to-end human-object interaction detection with transformers. In: CVPR (2021)
20. Kim, B., Mun, J., On, K., Shin, M., Lee, J., Kim, E.: MSTR: multi-scale transformer for end-to-end human-object interaction detection. In: CVPR (2022)

21. Kim, D.-J., Sun, X., Choi, J., Lin, S., Kweon, I.S.: Detecting human-object interactions with action co-occurrence priors. In: Vedaldi, A., Bischof, H., Brox, T., Frahm, J.-M. (eds.) ECCV 2020. LNCS, vol. 12366, pp. 718–736. Springer, Cham (2020). https://doi.org/10.1007/978-3-030-58589-1_43
22. Lei, T., Caba, F., Chen, Q., Jin, H., Peng, Y., Liu, Y.: Efficient adaptive human-object interaction detection with concept-guided memory. In: ICCV (2023)
23. Li, F., Zhang, H., Liu, S., Guo, J., Ni, L.M., Zhang, L.: DN-DETR: accelerate DETR training by introducing query denoising. In: CVPR, pp. 13609–13617 (2022)
24. Li, L., Wei, J., Wang, W., Yang, Y.: Neural-logic human-object interaction detection. In: NeurIPS (2023)
25. Li, Y., Liu, X., Wu, X., Huang, X., Xu, L., Lu, C.: Transferable interactiveness knowledge for human-object interaction detection. IEEE Trans. Pattern Anal. Mach. Intell. **44**(7) (2022)
26. Liao, Y., Liu, S., Wang, F., Chen, Y., Qian, C., Feng, J.: PPDM: parallel point detection and matching for real-time human-object interaction detection. In: CVPR (2020)
27. Liao, Y., Zhang, A., Lu, M., Wang, Y., Li, X., Liu, S.: GEN-VLKT: simplify association and enhance interaction understanding for HOI detection. In: CVPR (2022)
28. Lin, T.-y, et al.: Microsoft COCO: common objects in context. In: Fleet, D., Pajdla, T., Schiele, B., Tuytelaars, T. (eds.) ECCV 2014. LNCS, vol. 8693, pp. 740–755. Springer, Cham (2014). https://doi.org/10.1007/978-3-319-10602-1_48
29. Liu, S., et al.: DAB-DETR: dynamic anchor boxes are better queries for DETR. In: ICLR (2022)
30. Liu, X., Li, Y., Wu, X., Tai, Y., Lu, C., Tang, C.: Interactiveness field in human-object interactions. In: CVPR (2022)
31. Liu, Y., Yuan, J., Chen, C.W.: Consnet: learning consistency graph for zero-shot human-object interaction detection. In: ACMMM (2020)
32. Liu, Z., Lin, Y., Cao, Y., Hu, H., Wei, Y., Zhang, Z., Lin, S., Guo, B.: Swin transformer: Hierarchical vision transformer using shifted windows. In: ICCV (2021)
33. Mao, Y., Deng, J., Zhou, W., Li, L., Fang, Y., Li, H.: CLIP4HOI: towards adapting CLIP for practical zero-shot HOI detection. In: NeurIPS (2023)
34. Park, J., Park, J., Lee, J.: VIPLO: vision transformer based pose-conditioned self-loop graph for human-object interaction detection. In: CVPR (2023)
35. Qi, S., Wang, W., Jia, B., Shen, J., Zhu, S.: Learning human-object interactions by graph parsing neural networks. In: ECCV (2018)
36. Qu, X., Ding, C., Li, X., Zhong, X., Tao, D.: Distillation using oracle queries for transformer-based human-object interaction detection. In: CVPR (2022)
37. Tamura, M., Ohashi, H., Yoshinaga, T.: QPIC: query-based pairwise human-object interaction detection with image-wide contextual information. In: CVPR, pp. 10410–10419 (2021)
38. Tsai, T.J., Stolcke, A., Slaney, M.: A study of multimodal addressee detection in human-human-computer interaction. IEEE Trans. Multim. **17**(9), 1550–1561 (2015)
39. Tu, D., Sun, W., Zhai, G., Shen, W.: Agglomerative transformer for human-object interaction detection. In: ICCV (2023)
40. Ulutan, O., Iftekhar, A.S.M., Manjunath, B.S.: VSGNet: spatial attention network for detecting human object interactions using graph convolutions. In: CVPR (2020)
41. Wan, B., Zhou, D., Liu, Y., Li, R., He, X.: Pose-aware multi-level feature network for human object interaction detection. In: ICCV (2019)

42. Wang, H., Zheng, W., Yingbiao, L.: Contextual heterogeneous graph network for human-object interaction detection. In: Vedaldi, A., Bischof, H., Brox, T., Frahm, J.-M. (eds.) ECCV 2020. LNCS, vol. 12362, pp. 248–264. Springer, Cham (2020). https://doi.org/10.1007/978-3-030-58520-4_15
43. Wang, L., Zhao, X., Si, Y., Cao, L., Liu, Y.: Context-associative hierarchical memory model for human activity recognition and prediction. IEEE Trans. Multim. **19**(3), 646–659 (2017)
44. Wang, S., Yap, K., Ding, H., Wu, J., Yuan, J., Tan, Y.: Discovering human interactions with large-vocabulary objects via query and multi-scale detection. In: ICCV (2021)
45. Wang, S., Yap, K., Yuan, J., Tan, Y.: Discovering human interactions with novel objects via zero-shot learning. In: CVPR (2020)
46. Wang, T., et al.: Deep contextual attention for human-object interaction detection. In: ICCV (2019)
47. Wang, T., Yang, T., Danelljan, M., Khan, F.S., Zhang, X., Sun, J.: Learning human-object interaction detection using interaction points. In: CVPR (2020)
48. Yang, J., Li, B., Yang, F., Zeng, A., Zhang, L., Zhang, R.: Boosting human-object interaction detection with text-to-image diffusion model. CoRR **abs/2305.12252** (2023)
49. Yuan, H., et al.: RLIP: relational language-image pre-training for human-object interaction detection. In: NeurIPS (2022)
50. Yuan, H., et al.: Rlipv2: Fast scaling of relational language-image pre-training. In: ICCV, pp. 21592–21604 (2023)
51. Zhang, A., et al.: Mining the benefits of two-stage and one-stage HOI detection. In: NeurIPS (2021)
52. Zhang, F.Z., Campbell, D., Gould, S.: Spatially conditioned graphs for detecting human-object interactions. In: ICCV (2021)
53. Zhang, F.Z., Yuan, Y., Campbell, D., Zhong, Z., Gould, S.: Exploring predicate visual context in detecting of human-object interactions. In: ICCV (2023)
54. Zhong, X., Ding, C., Li, Z., Huang, S.: Towards hard-positive query mining for detr-based human-object interaction detection. In: Avidan, S., Brostow, G., Cissé, M., Farinella, G.M., Hassner, T. (eds.) Computer Vision ECCV 2022. LNCS, vol. 13687, pp. 444–460. Springer, Cham (2022). https://doi.org/10.1007/978-3-031-19812-0_26
55. Zhong, X., Ding, C., Qu, X., Tao, D.: Polysemy deciphering network for robust human-object interaction detection. Int. J. Comput. Vis. **129**(6), 1–10 (2021)
56. Zhou, D., et al.: Human-object interaction detection via disentangled transformer. In: CVPR (2022)
57. Zhou, P., Chi, M.: Relation parsing neural network for human-object interaction detection. In: ICCV (2019)
58. Zhu, X., Su, W., Lu, L., Li, B., Wang, X., Dai, J.: Deformable DETR: deformable transformers for end-to-end object detection. In: ICLR (2021)
59. Zhuang, Z., Qian, R., Xie, C., Liang, S.: Compositional learning in transformer-based human-object interaction detection. In: ICME (2023)
60. Zou, C., et al.: End-to-end human object interaction detection with HOI transformer. In: CVPR, pp. 11825–11834 (2021)

GTA: Global Tracklet Association for Multi-object Tracking in Sports

Jiacheng Sun, Hsiang-Wei Huang, Cheng-Yen Yang(✉),
Zhongyu Jiang, and Jenq-Neng Hwang

University of Washington, Seattle, WA, USA
{sjc042,hwhuang,cycyang,zyjiang,hwang}@uw.edu

Abstract. Multi-object tracking in sports scenarios has become one of the focal points in computer vision, experiencing significant advancements through the integration of deep learning techniques. Despite these breakthroughs, challenges remain, such as accurately re-identifying players upon re-entry into the scene and minimizing ID switches. In this paper, we propose an appearance-based global tracklet association algorithm designed to enhance tracking performance by splitting tracklets containing multiple identities and connecting tracklets seemingly from the same identity. This method can serve as a plug-and-play refinement tool for any multi-object tracker to further boost their performance. The proposed method achieved a new state-of-the-art performance on the SportsMOT dataset with HOTA score of 81.04%. Similarly, on the SoccerNet dataset, our method enhanced multiple trackers' performance, consistently increasing the HOTA score from 79.41% to 83.11%. These significant and consistent improvements across different trackers and datasets underscore our proposed method's potential impact on the application of sports player tracking. We open-source our project codebase at (https://github.com/sjc042/gta-link.git).

Keywords: Multi-Object Tracking in Sports · Tracklet Refinement

1 Introduction

In recent years, advancements in computer vision and deep learning have revolutionized sports analytics, offering unprecedented insights into player performance and strategy. For example, sports video understanding [6,10], sports field registration [13,20], and 2D/3D human pose estimation for sports [21]. Among these innovations, sports player tracking systems have emerged as a cornerstone, providing coaches and analysts with valuable data on player movements, positioning, and interactions during game play [12]. These systems have become integral to modern sports, enabling data-driven decision-making and performance optimization across various disciplines.

However, despite significant progress in multi-object tracking technologies, challenges persist in accurately tracking players, including the irregular movements and similar appearances of sports players, and the lack of re-identification

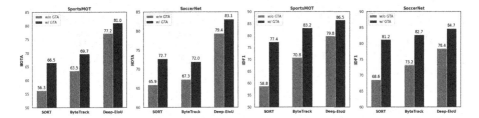

Fig. 1. Our proposed Global Tracklet Association (GTA) method significantly boosts the HOTA and IDF1 score of existing trackers, such as SORT, ByteTrack, and Deep-EIoU, on sports tracking datasets, including SportsMOT and SoccerNet

algorithm in handling re-entry situation after leaving camera field of view after certain amount of time.

Current state-of-the-art on-line tracking algorithms often fail in long-term object re-identification with the aforementioned challenges. While these trackers perform well in controlled environments or short-term scenarios, they struggle to maintain consistent player identities throughout entire matches or when players re-enter the field after substantial absences. This limitation significantly impacts the accuracy and reliability of player performance analysis, tactical evaluations, and automated game statistics.

To address these persistent issues, our paper proposes an effective plug-and-play post-processing algorithm named Global Tracklet Association (GTA), designed specifically for sports player tracking applications. GTA aims to refine the tracking results of on-line or off-line trackers by improving long-term re-identification capabilities and handling the unique challenges posed by sports environments. By leveraging global temporal information and advanced association techniques, GTA enhances the robustness and accuracy of player tracking, potentially bridging the gap between current tracking technologies and the demanding requirements of professional sports analytics.

2 Related Work

2.1 Sports Player Tracking

With progress made in object detection and tracking, recent studies have focused on challenging multi-object tracking (MOT) scenarios like sports [7,8] and dancing [24]. Tracking sports players is more difficult than tracking pedestrians due to the complex nature of sports scenarios, including frequent occlusions, rapid direction changes, varying player densities, and similar appearance as in Fig. 2, and re-entries to

Fig. 2. Examples of **different players** on the same teams, highlighting the challenge of distinguishing between players with similar appearances in sports tracking.

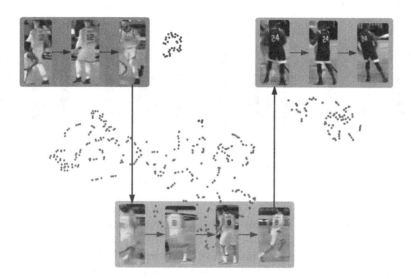

Fig. 3. An example of a mix-up error in a single tracklet. The tracklet output by the online tracking system contains **three** different identities, represented by purple, green, and red points. The figure illustrates the tracklet's features extracted by a ReID model and clustered using the DBSCAN clustering algorithm. (Color figure online)

the camera view. Several works [4,18,19,29] have proposed methods to handle irregular object motion, improving tracking performance compared to traditional Kalman filter-based methods [3].

There are two predominant types of errors in sports player tracking: the mix-up error (Fig. 3) and the cut-off error (Fig. 4). The first type, mix-up errors, occur due to irregular movements and occlusions during tracking, leading to a single tracklet mistakenly including multiple players (Fig. 3). The second type, cut-off errors, arise because, unlike targets in traditional pedestrian tracking datasets [23], sports players often re-enter the camera's view after exiting. Assigning a new tracking ID to a previously tracked player results in a cut-off error, which fragments a single player's tracklet into multiple parts (Fig. 4).

2.2 Person Re-identification

When a player re-appears in the scene, a re-identification method is needed to assign the correct tracking ID. Although some trackers can re-identify players who re-enter shortly after exiting by lengthening the tracking buffer, cut-off errors from extended absences and re-entries are most prevalent in sports tracking. Person re-identification (ReID) is crucial in multi-object tracking for identifying individuals across video instances or different cameras.

Several methods have been proposed to address ReID challenges. OSNet [32] introduced a lightweight CNN-based method with state-of-the-art performance on various ReID datasets. Many tracking methods incorporate ReID models in the data association stage, such as DeepSORT [26] and BoTSORT [1]. In sports multi-player tracking, Deep-EIoU [19] has shown that using a ReID model can significantly improve tracking performance in sports player tracking scenarios.

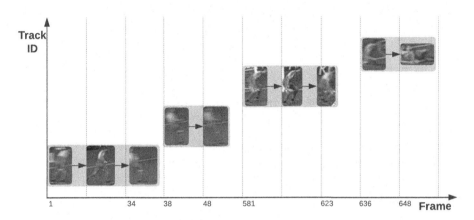

Fig. 4. An example of a cut-off error, where a player's tracklet is fragmented into four separate segments due to the player exiting and re-entering the camera view multiple times throughout the video sequence

2.3 Global Link Models

To address cut-off errors, where a tracker incorrectly assigns different tracking IDs to the same target, several previous works have proposed global linking models that utilize various types of information, such as appearance, spatial, and temporal cues, to associate fragmented tracklets and reassign the correct tracking IDs after the online tracking process. These models either utilize motion or appearance features of tracklets for tracklet-level association. For example, Translink [30] incorporates a CNN and temporal attention network to extract and encode a tracklet's appearance features, treating the merging process of tracklet pairs as a binary classification task. AFLink [9] uses only spatial-temporal information. Some methods [5,17,28] utilize feature clustering methods to merge tracklets and boost the performance on multi-camera tracking scenarios. MambaTrack [15] proposed a motion model that serves as a motion predictor and extracts tracklet motion features for further global tracklet association.

Additionally, some methods exploit object moving direction [14,16] or metadata [27] as clues to conduct global tracklet association and enhance tracking performance. [25] proposed a universal tracklet booster based on CNN and temporal attention to address both mix-up and cut-off errors.

In this work, we propose a novel plug-and-play box-grained global tracklet association model, including a tracklet splitter and a connector. The proposed

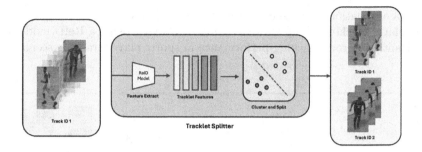

Fig. 5. Illustration of tracklet splitter

method is specifically designed to conduct player re-identification and boost the tracking performance in various sports scenarios.

3 Methods

Drawing inspiration from global link models presented in recent works [18] [29], we propose the Global Tracklet Association (GTA) method, a novel plug-and-play approach designed to address both mix-up and cut-off errors in multi-object tracking for sports scenarios. Our method consists of two key modules: a **Tracklet Splitter** and a **Tracklet Connector**, which leverage deep feature representations and spatial constraints to enhance tracking accuracy and robustness.

Our proposed post-processing method follows a two-stage process to enhance tracking accuracy. Prior to post-processing, box-grained embedding features from online tracking results are generated by a CNN-based ReID model [32] for each tracklet. In the first stage of our tracklet association model, these tracklets are processed through a **Tracklet Splitter** to address mix-up errors, ensuring that instances of different identities are correctly separated, as illustrated in Fig. 5. In the second stage, the split tracklets belonging to the same identities are further merged by the proposed **Tracklet Connector** to correct cut-off errors, as depicted in Fig. 6.

3.1 Tracklet Splitter

The proposed tracklet splitter addresses mix-up errors within a single tracklet, $T = \{t_0, \ldots, t_n\}$, by splitting the tracklet into multiple fragments, t_i, ensuring that each fragment contains only bounding boxes with similar appearance features, measured by close cosine distances in the feature embedding space. We employ DBSCAN clustering [11] to split the tracklet into multiple clusters (tracklet fragments) based on their box-grained appearance embedding feature, which are generated by an OSNet ReID model [32]. The pipeline for our tracklet splitter is illustrated in Fig. 5.

DBSCAN (Density-Based Spatial Clustering of Applications with Noise) is a powerful clustering algorithm that operates by grouping points that are closely packed together while marking points that lie alone in low-density regions as outliers. It is important to note that, unlike traditional DBSCAN implementations, our adapted version assigns outliers to the nearest clusters at the end of the clustering process. This modification is based on the assumption that each bounding box contains a valid detection instance, ensuring that no potentially valuable data points are discarded. This approach allows for a more comprehensive analysis of the tracklet data, preserving information that might otherwise be lost. When applying DBSCAN to the process of tracklet splitting, we incorporate three crucial hyperparameters:

Minimum Samples (s): The Minimum Samples parameter, denoted as s, specifies the minimum number of points required to establish a densed region cluster. This parameter serves two critical purposes in the clustering process. First, it ensures that clusters are only formed when there is a sufficient concentration of points, effectively preventing the creation of clusters with too few instances. Second, it aids in noise reduction by initially labeling points that do not meet this threshold as noise or outliers, especially in the case of sport tracking.

Maximum Neighbor Distance (ϵ): The Maximum Neighbor Distance, represented by ϵ (epsilon), determines the radius within which points are considered neighbors. This parameter is fundamental in controlling the density requirement for cluster formation. In our implementation, we utilize cosine similarity as the distance metric between fragment features, offering several advantages in the context of tracklet splitting. It controls the density requirement for cluster formation, balancing the need to capture variations within a single identity while distinguishing between different identities.

Maximum Clusters (k): Unlike the original algorithm, we introduce a maximum clusters parameter, k, to limit the number of clusters and prevent excessive fragmentation of the tracklet. If the total number of final clusters exceeds k, clusters are progressively merged until only k clusters remain. This modification ensures that a given tracklet is not split into an excessive number of fragments, maintaining a balance between accuracy and fragmentation.

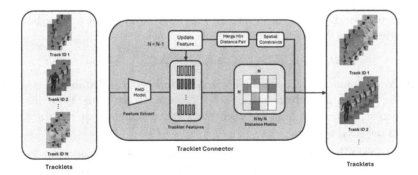

Fig. 6. Illustration of tracklet connector

3.2 Tracklet Connector

Our tracklet connector is designed to merge fragmented tracklets from the same identity using a hierarchical clustering approach with a distance threshold α, addressing the cut-off errors that occur when players exit and re-enter the field of view, or the ID switch during the tracking process. Our approach, as depicted in Fig. 6, consists of tracklet clustering based on tracklets distance while applying temporal and spatial constraints to ensure accurate and reliable tracklet merging, and the three main components for the tracklet connector are listed below:

Constructing Tracklet Feature Distance Matrix. The first step of our tracklet connector is the initial construction of a symmetric cosine distance matrix that measures the similarity between all tracklet pairs within a video sequence. The distance matrix is constructed as follows:

$$D_{i,j} = \begin{cases} 1, & \text{if } i \neq j \ \& \ \Pi_i \cap \Pi_j \neq \emptyset \\ \frac{1}{N_i N_j} \sum_{i \in \Pi_i} \sum_{j \in \Pi_j} (1 - \frac{F_m^i \cdot F_n^j}{\|F_m^i\| \, \|F_n^j\|}), & \text{otherwise} \end{cases} \quad (1)$$

where $D_{i,j}$ denotes the tracklet distance between tracklet pair T_i and T_j; Π_i and Π_j represent temporal spans of tracklet T_i and T_j; F_m^i and F_n^j are tracklet T_i and T_j's embedding feature at frame m and n respectively; N_i and N_j are the length of the tracklet. This distance enables a nuanced representation of tracklet similarities based on the rich nature of deep features and temporal clues.

Enforcing Spatial Constraints. In a common sports game video where the camera position remains fixed, player movement is constrained by the field boundaries, and players do not exit and re-enter from opposite sides of the field. To reflect this, our method enforces spatial constraints for merging tracklets using:

$$\theta_{\text{hor}} = \beta \Delta_{\text{max,hor}}, \quad (2)$$

$$\theta_{\text{ver}} = \beta \Delta_{\text{max,ver}}, \quad (3)$$

where $\Delta_{\text{max,hor}}$ and $\Delta_{\text{max,ver}}$ denote the maximum horizontal and vertical distances from all bounding boxes in the current video. The spatial factor, $\beta \in (0,1]$, sets thresholds of θ_{hor} and θ_{ver}, limiting the association distances threshold between temporally adjacent tracklet's exit and entry points using bounding box center. Then, $D_{i,j}$ will be updated.

$$D_{i,j} = 1, \text{ if } \Delta_{i,j,hor} > \theta_{hor} \text{ or } \Delta_{i,j,ver} > \theta_{ver}, \quad (4)$$

where $\Delta_{i,j,hor}$ and $\Delta_{i,j,ver}$ are the horizontal and vertical distance between the beginning and ending of tracklets i and j, respectively. This approach filters out unreasonable associations between tracklets, enhancing the accuracy of the tracklets merging process.

Hierarchical Clustering. After the distance matrix is obtained using Eq. 1, we further conduct hierarchical clustering following [14] to merge fragment tracklets. We continuously merge the tracklets until no tracklet pair's distance is larger than the merging threshold α.

4 Experiments

4.1 Datasets

We evaluate our method on two large-scale sports player tracking datasets: SportsMOT [8] and SoccerNet [7]. These datasets are representative of athlete tracking in team sports scenarios, presenting unique challenges such as players with similar appearances, frequent re-entries into the camera's field of view, and abrupt changes in motion.

SportsMOT is a multi-object tracking dataset that contains over 240 video sequences spanning three team sports: basketball, football, and volleyball. Each sport presents its unique challenges, such as the fast-paced, close-quarters action of basketball, the wide-field dynamics of football, and the rapid vertical movements in volleyball. The dataset provides a robust foundation for developing and testing tracking algorithms in complex, real-world sports environments and has been widely used in benchmarking MOT for sport tracking.

SoccerNet focuses exclusively on videos captured from soccer matches, providing a collection of over 100 high-quality video clips extracted from professional games. For our experiments, we utilize the test set from the SoccerNet tracking dataset published in 2023.

4.2 Implementation Details

Detector. For the SportsMOT test set, we use YOLOX as the detection model following [19], and for the SoccerNet 2023 test set, we directly apply oracle detection following others' implementations [19] for fair comparison.

Tracker. In our work, we test our tracklet refinement method on three trackers: SORT [3], ByteTrack [31], and Deep-EIoU [19]. SORT and ByteTrack are both implemented to track with spatial and motion cues. Deep-EIoU incorporates both spatial and appearance information to achieve state-of-the-art performance with a HOTA score of 77.2% on the SportsMOT test set and 85.4% on the SoccerNet test set published in 2022.

ReID Model. For our experiments, we use the OSNet [32] model trained on SportsMOT dataset. OSNet is chosen for its capability to capture discriminative features suitable for re-identification of athletes with similar appearances, which is critical for our tracklet refinement process.

Hyperparameters. We set minimum cluster samples s to 5, maximum neighbor distance threshold ϵ to 0.6, maximum clusters k to 3, merging threshold α to 0.4, and β to 1 for SportsMOT dataset and 0.7 for SoccerNet dataset, respectively.

4.3 Performance

Evaluation Metrics. We utilize commonly used tracking metrics, including HOTA [22] for its comprehensive evaluation of both detection and association

Table 1. Tracking performance on SportsMOT before and after applying our Global Tracklet Association (GTA) method

Method	HOTA↑	AssA↑	IDF1↑	DetA↑	MOTA↑	IDs↓
SORT [26]	56.28	42.67	58.83	74.30	85.11	5180
SORT + GTA	**66.52** (+10.24)	**59.59** (+16.92)	**77.37** (+18.54)	74.29	**85.27**	**3547** (-1633)
ByteTrack [31]	63.46	51.81	70.76	77.81	94.91	3147
ByteTrack + GTA	**69.74** (+6.28)	**62.61** (+10.80)	**83.16** (+12.40)	77.72	**95.01**	**2107** (-1040)
Deep-EIoU [19]	77.21	67.63	79.81	88.22	96.30	2909
Deep-EIoU + GTA	**81.04** (+3.83)	**74.51** (+6.88)	**86.51** (+6.70)	88.21	**96.32**	**2737** (-172)

Table 2. Tracking performance on SoccerNet before and after applying our Global Tracklet Association (GTA) method

Method	HOTA↑	AssA↑	IDF1↑	DetA↑	MOTA↑	IDs↓
SORT [26]	65.89	57.15	68.56	76.11	82.59	3281
SORT + GTA	**72.73** (+6.84)	**69.62** (+12.47)	**81.24** (+12.68)	76.04	**82.93**	**1374** (-1907)
ByteTrack [31]	67.30	60.38	73.22	75.14	84.66	4558
ByteTrack + GTA	**71.97** (+4.67)	**69.03** (+8.65)	**82.67** (+9.45)	75.10	**84.91**	**3149** (-1409)
Deep-EIoU [19]	79.41	71.55	78.40	88.14	87.92	2803
Deep-EIoU + GTA	**83.11** (+3.70)	**78.38** (+6.83)	**84.66** (+6.26)	88.13	**88.03**	**2188** (-615)

accuracy (DetA and AssA); and CLEAR metrics [2], where IDF1 and MOTA serve as the standard benchmark for tracking performance across various scenarios, and IDs to verify the effectiveness of our method in reducing and associating the correct identities.

Performance on SportsMOT. In Table 1, our proposed method demonstrates significant performance improvements over existing trackers across various tracking metrics like HOTA, AssA, IDF1, IDs, and Frag. For the SportsMOT dataset, GTA achieved the highest HOTA improvement of 10.24% for SORT and 3.83% for Deep-EIoU, reaching a state-of-the-art HOTA of 81.04%. Demonstrating the GTA method is applicable for diverse kinds of sports tracking.

Performance on SoccerNet. In Table 2, GTA improved HOTA by 6.84% for SORT and 3.7% for Deep-EIoU. These results demonstrate the effectiveness of our tracklet refinement method in enhancing tracker performance on challenging sports player tracking datasets. Figure 8 illustrates the qualitative results, showing GTA effectively connecting tracklet fragments from online tracking.

4.4 Ablation Study

To evaluate the effectiveness of the splitter and connector modules of our proposed method, we conduct ablation studies on the performance gain of SORT

Table 3. Effectiveness of *Splitter* and *Connector* on different MOT algorithms and datasets

Dataset	Method	Connector	Splitter	HOTA↑	AssA↑	IDF1↑
SportsMOT	SORT[3]			56.28	42.67	58.83
		✓		65.43 (+9.15)	57.77 (+15.10)	76.13 (+17.30)
		✓	✓	66.52 (+10.24)	59.59 (+16.92)	77.37 (+18.54)
	ByteTrack[31]			63.46	51.81	70.76
		✓		69.51 (+6.05)	62.21 (+10.40)	82.88 (+12.12)
		✓	✓	69.74 (+6.28)	62.61 (+10.80)	83.16 (+12.40)
	DeepEIoU[19]			77.21	67.63	79.81
		✓		80.48 (+3.27)	73.50 (+5.87)	85.76 (+5.95)
		✓	✓	81.04 (+3.83)	74.51 (+6.88)	86.51 (+6.70)
SoccerNet	SORT			65.89	57.15	68.56
		✓		71.74 (+5.85)	67.74 (+10.59)	79.73 (+11.17)
		✓	✓	72.73 (+6.84)	69.62 (+12.47)	81.24 (+12.68)
	ByteTrack			67.30	60.38	73.22
		✓		71.05 (+3.75)	67.30 (+6.92)	78.22 (+4.61)
		✓	✓	71.97 (+4.67)	69.03 (+8.65)	82.67 (+9.45)
	DeepEIoU			79.41	71.55	78.40
		✓		82.01 (+2.60)	76.32 (+4.77)	83.13 (+4.73)
		✓	✓	83.11 (+3.70)	78.38 (+6.83)	84.66 (+6.26)

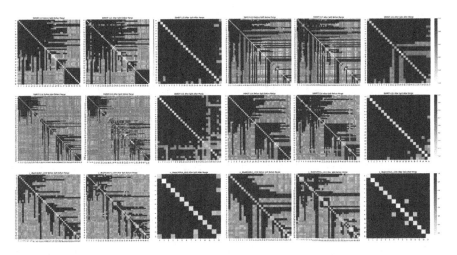

Fig. 7. The cosine distance matrix of the embeddings of three stages: (1) Before Split, (2) After Split, and (3) After Connect. Both the x-axis and y-axis represent the IDs, while the darker color represents the farther distance

[3], ByteTrack [31], and DeepEIoU [19] after applying each module. The ablation study summarized in Table 3 highlights the effectiveness of different modules of the Global Tracklet Association (GTA) method on the performance of the three

trackers across the SportsMOT and SoccerNet datasets. The proposed Connector alone results in notable improvements for SORT in HOTA, IDF1, and AssA scores, with an increase of 9.15% in HOTA on SportsMOT and 5.85% on SoccerNet. When the Splitter and Connector are both applied for SORT, we obtained HOTA improvements of 10.24% on SportsMOT and 6.84% on SoccerNet, along with substantial increases in IDF1 and AssA, demonstrating the effectiveness of both modules.

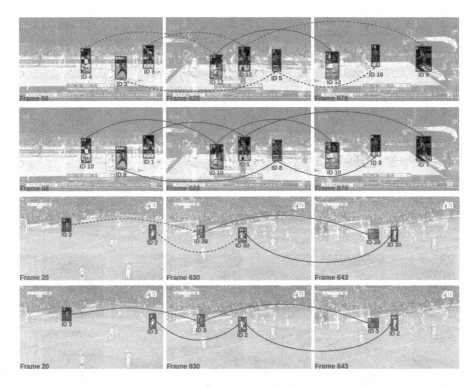

Fig. 8. Tracking visualization of athletes before and after applying Global Tracklet Association (GTA). In rows one and three, dashed lines indicate association errors, showing inconsistent athlete IDs across frames. In contrast, solid lines represent correct associations with consistent IDs after applying GTA (rows two and four). The comparison highlights how the algorithm improves ID continuity across frames in both basketball (first two rows) and soccer (second two rows) sequences

5 Conclusion

In this paper, we proposed the Global Tracklet Association (GTA), a novel tracklet refinement method to enhance the performance of existing trackers in challenging sports player tracking scenarios. Our approach effectively addresses common issues such as mix-up errors and cut-off errors by leveraging a combination of ReID model and unsupervised clustering techniques for tracklet splitting and merging. The integration of our GTA method with trackers like SORT, ByteTrack, and Deep-EIoU has demonstrated significant improvements in various tracking performance metrics, particularly in metrics that related to associations like HOTA, AssA, and IDF1, while also reducing the number of ID switches (IDs) and tracklet fragments (Frag). Our proposed module achieves state-of-the-art performance on SportsMOT and SoccerNet datasets. Future work will focus on exploring additional enhancements and adaptations of the GTA method for other multi-object tracking scenarios beyond sports.

References

1. Aharon, N., Orfaig, R., Bobrovsky, B.Z.: Bot-sort: robust associations multi-pedestrian tracking. arXiv preprint arXiv:2206.14651 (2022)
2. Bernardin, K., Stiefelhagen, R.: Evaluating multiple object tracking performance: the clear mot metrics. J. Image Video Process. **2008** (2008)
3. Bewley, A., Ge, Z., Ott, L., Ramos, F., Upcroft, B.: Simple online and realtime tracking. In: 2016 IEEE International Conference on Image Processing (ICIP), pp. 3464–3468 (2016). https://doi.org/10.1109/ICIP.2016.7533003
4. Cao, J., Pang, J., Weng, X., Khirodkar, R., Kitani, K.: Observation-centric sort: rethinking sort for robust multi-object tracking. In: Proceedings of the IEEE/CVF Conference on Computer Vision and Pattern Recognition, pp. 9686–9696 (2023)
5. Cherdchusakulchai, R., et al.: Online multi-camera people tracking with spatial-temporal mechanism and anchor-feature hierarchical clustering. In: Proceedings of the IEEE/CVF Conference on Computer Vision and Pattern Recognition, pp. 7198–7207 (2024)
6. Cioppa, A., et al.: A context-aware loss function for action spotting in soccer videos. In: The IEEE Conference on Computer Vision and Pattern Recognition (CVPR) (June 2020)
7. Cioppa, A., et al.: Soccernet-tracking: multiple object tracking dataset and benchmark in soccer videos. In: Proceedings of the IEEE/CVF Conference on Computer Vision and Pattern Recognition, pp. 3491–3502 (2022)
8. Cui, Y., Zeng, C., Zhao, X., Yang, Y., Wu, G., Wang, L.: Sportsmot: a large multi-object tracking dataset in multiple sports scenes. In: Proceedings of the IEEE/CVF International Conference on Computer Vision, pp. 9921–9931 (2023)
9. Du, Y., et al.: Strongsort: make deepsort great again. IEEE Trans. Multimedia (2023)
10. Duan, H., Zhao, Y., Chen, K., Lin, D., Dai, B.: Revisiting skeleton-based action recognition. In: 2022 IEEE/CVF Conference on Computer Vision and Pattern Recognition (CVPR). IEEE (Jun 2022).https://doi.org/10.1109/cvpr52688.2022.00298
11. Ester, M., Kriegel, H.P., Sander, J., Xu, X.: A density-based algorithm for discovering clusters in large spatial databases with noise. In: Proceedings of the Second International Conference on Knowledge Discovery and Data Mining (KDD 1996), pp. 226–231. AAAI Press (1996)
12. Gade, R., Moeslund, T.B.: Constrained multi-target tracking for team sports activities. In: Proceedings of the IEEE/CVF Conference on Computer Vision and Pattern Recognition (CVPR) Workshops (June 2020)

13. Gutiérrez-Pérez, M., Agudo, A.: No bells just whistles: Sports field registration by leveraging geometric properties. In: Proceedings of the IEEE/CVF Conference on Computer Vision and Pattern Recognition (CVPR) Workshops, pp. 3325–3334 (June 2024)
14. Hsu, H.M., Huang, T.W., Wang, G., Cai, J., Lei, Z., Hwang, J.N.: Multi-camera tracking of vehicles based on deep features re-id and trajectory-based camera link models. In: CVPR workshops, pp. 416–424 (2019)
15. Huang, H.W., Yang, C.Y., Chai, W., Jiang, Z., Hwang, J.N.: Exploring learning-based motion models in multi-object tracking. arXiv preprint arXiv:2403.10826 (2024)
16. Huang, H.W., Yang, C.Y., Hwang, J.N.: Multi-target multi-camera vehicle tracking using transformer-based camera link model and spatial-temporal information. arXiv preprint arXiv:2301.07805 (2023)
17. Huang, H.W., et al.: Enhancing multi-camera people tracking with anchor-guided clustering and spatio-temporal consistency id re-assignment. In: Proceedings of the IEEE/CVF Conference on Computer Vision and Pattern Recognition (CVPR) Workshops, pp. 5239–5249 (June 2023)
18. Huang, H.W., et al.: Observation centric and central distance recovery for athlete tracking. In: Proceedings of the IEEE/CVF Winter Conference on Applications of Computer Vision, pp. 454–460 (2023)
19. Huang, H.W., et al.: Iterative scale-up expansioniou and deep features association for multi-object tracking in sports. In: Proceedings of the IEEE/CVF Winter Conference on Applications of Computer Vision, pp. 163–172 (2024)
20. Jiang, W., Higuera, J.C.G., Angles, B., Sun, W., Javan, M., Yi, K.M.: Optimizing through learned errors for accurate sports field registration. In: 2020 IEEE Winter Conference on Applications of Computer Vision (WACV). IEEE (2020)
21. Jiang, Z., Ji, H., Menaker, S., Hwang, J.N.: Golfpose: golf swing analyses with a monocular camera based human pose estimation. In: 2022 IEEE International Conference on Multimedia and Expo Workshops (ICMEW), pp. 1–6. IEEE (2022)
22. Luiten, J., et al.: Hota: a higher order metric for evaluating multi-object tracking. Int. J. Comput. Vis. **129**, 548–578 (2021)
23. Milan, A., Leal-Taixe, L., Reid, I., Roth, S., Schindler, K.: Mot16: A benchmark for multi-object tracking (2016), arXiv preprint arXiv:1603.00831
24. Sun, P., Cao, J., Jiang, Y., Yuan, Z., Bai, S., Kitani, K., Luo, P.: Dancetrack: multi-object tracking in uniform appearance and diverse motion. In: Proceedings of the IEEE/CVF Conference on Computer Vision and Pattern Recognition, pp. 20993–21002 (2022)
25. Wang, G., Wang, Y., Gu, R., Hu, W., Hwang, J.N.: Split and connect: a universal tracklet booster for multi-object tracking. IEEE Trans. Multimedia **25**, 1256–1268 (2022)
26. Wojke, N., Bewley, A., Paulus, D.: Simple online and realtime tracking with a deep association metric. In: 2017 IEEE International Conference on Image Processing (ICIP), pp. 3645-3649. IEEE (2017)
27. Yang, C.Y., et al.: Sea you later: metadata-guided long-term re-identification for uav-based multi-object tracking. In: Proceedings of the IEEE/CVF Winter Conference on Applications of Computer Vision (WACV) Workshops, pp. 805–812 (January 2024)
28. Yang, C.Y., et al.: An online approach and evaluation method for tracking people across cameras in extremely long video sequence. In: Proceedings of the IEEE/CVF Conference on Computer Vision and Pattern Recognition, pp. 7037–7045 (2024)

29. Yang, F., Odashima, S., Masui, S., Jiang, S.: Hard to track objects with irregular motions and similar appearances? make it easier by buffering the matching space. In: Proceedings of the IEEE/CVF Winter Conference on Applications of Computer Vision, pp. 4799–4808 (2023)
30. Zhang, Y., Wang, S., Fan, Y., Wang, G., Yan, C.: Translink: transformer-based embedding for tracklets' global link. In: ICASSP 2023-2023 IEEE International Conference on Acoustics, Speech and Signal Processing (ICASSP), pp. 1–5. IEEE (2023)
31. Zhang, Y., et al.: Bytetrack: multi-object tracking by associating every detection box. In: European conference on computer vision. pp. 1–21. Springer (2022). https://doi.org/10.1007/978-3-031-20047-2_1
32. Zhou, K., Yang, Y., Cavallaro, A., Xiang, T.: Omni-scale feature learning for person re-identification. In: Proceedings of the IEEE/CVF International Conference on Computer Vision (2019)

IsoTGAN: Spatial and Geometrical Constraints at GAN and Transformer for 3D Contour Generation

Thao Nguyen Phuong[1,2(✉)], Vinh Nguyen Duy[1], and Hidetomo Sakaino[1]

[1] AI-Image Group, Data Solution Department, (FCJ.ABC) FPT Consulting Japan, FPT Software, Minato, Japan
{ThaoNP47,VinhND1,HidetomoS}@fpt.com
[2] Department of Computer Science and Communications Engineering, Waseda University, Shinjuku, Japan
nguyenphuongthao@akane.waseda.jp

Abstract. Image generation in 2D and 3D has become an active research topic in Deep Learning. Single or multiple input images with non-orthogonal views are used for another shape and texture with different viewing angles. On the other hand, Computer-Aided Design (CAD) relies on handling-based 3D generation, i.e., isometric view images, from three orthographic view line drawings in 2D. However, since unique viewing pairs of such 2D and 3D images are required to train, SOTA models are insufficient to generate desirable images. More spatial and geometrical constraints remain undone due to less corresponding image features between images. This paper proposes IsoTGAN with a GAN-based generator with the Transformer in its generator, where three images of an object's front, side, and top view are input. The encoder is trained by the spatial and geometrical relation among three view images. A novel Gaussian Enhanced Euclidean attention mechanism and a geometry-constrained loss function are also proposed for further local image feature enhancement. Extensive experiments on the SPARE3D dataset demonstrate that the proposed IsoTGAN outperforms State-of-the-art (SOTA) models, i.e., DINO, regarding local and global image feature accuracy. This helps generate 3D isometric view images in auto-CAD system.

Keywords: Isometric view · attention · 3D generation · Transformer · GAN · CAD

1 Introduction

3D understanding is important in visual semantic analysis. In today's digital era, CAD software is essential for creating 3D shapes in various industrial sectors, including automotive, aerospace, manufacturing, and architectural design. The most commonly used technique for depicting 3D objects on a 2D surface is orthographic projection, also known as orthogonal projection or analemma.

This method projects lines from the object perpendicular to the projection plane, resulting in a parallel representation. Designers often use 2D orthographic drawings to showcase their concepts in the early stages of design. Nevertheless, to further explore and understand the object, a 3D model is necessary. Isometric view images are crucial for the 3D reconstruction process as they retain a significant amount of information about the 3D object. Therefore, there is a strong demand for a method that can convert three-view contour drawings into isometric view images, which would greatly facilitate the design process and increase overall efficiency.

In recent years, numerous research papers have introduced deep learning models for 2D/3D CAD modeling [11,15,19,22,33,37,44,46,49]. However, these studies typically treat the task as a sequence-to-sequence issue, outputting sequences of CAD commands for creating 2D drawings or reconstructing 3D objects. Only one study has explored the generation of isometric images from three-view orthographic line drawings [19], and this approach similarly generates CAD commands.

Isometric view images are crucial for reconstructing 3D objects. Generating these images from drawings of three orthographic views can be approached as a multi-images-to-image translation problem. To generate a comprehensive isometric view image, it is necessary to understand the relation between each input view and isometric view. Previous image-to-image translation methods exploit GAN [14,35], Autoregressive models [32], Variational AutoEncoders (VAEs) [26], Normalizing Flows [7], and recently, Diffusion models [38]. Nevertheless, these methods only take one image or text as input and map it to the target domain without the need to consider the relation between multiple input images. Although recent Diffusion methods [12,18] can take many input images to generate 3D scenes, they deal with natural 2D images, which are very different from images containing only flatted contour lines. Since many different objects have the same single-view image, these Diffusion frameworks [28,29,50] that produce 3D scenes from only one image may not generate correct isometric view images.

Originally developed for natural language processing (NLP), the Transformer architecture [42] uses multi-head self-attention and stacked feed-forward MLP layers to capture the long-term relationships between words. Dosovitskiy et al. [8] have adapted this architecture for highly competitive ImageNet classification by viewing an image as a sequence of 16 × 16 visual words. However, 3D isometric view image generation can be found in only a few papers by Transformer. If applied, transformer-based translation accuracy will be higher than that of CNN-based translation.

Many variants of GAN models with attention mechanism [3,5,9,41,54] have been proposed. However, they are weak at generating locally detailed texture images without sufficient constraints. Because these attention mechanisms deal with one image at a time. Moreover, they do not have any mechanism to tackle the important relation between each view of the object.

To address this problem, this paper proposes IsoTGAN with a GAN and Transformer-based model. IsoTGAN can translate 2D three orthographic view contour images to 3D isometric view contour images. Since orthogonal images only with contour become harder to match than most non-orthogonal images with color, texture, and surface, limited image features, i.e., spatial and geometrical relation between each orthographic view of the object, must be modeled and utilized than images with color, texture, and surface, unlike common image-to-image translation tasks. Spatial and geometrical image features are pixel-based image address and curvature/line, respectively.

This paper also proposes the spatial and geometrical relation between each view into a representation vector that is embedded in an encoder of GAN. For a further enhancement to complicated contour images, the Transformer blocks with cross-attention are equipped in the generator, which takes both the vector output by the above encoder and the image feature as input and models the attention between them. The cross-attention mechanism performs weighted product between relation vector and image features. This ensures long-range interactions across three orthographic views and constraint guidance between each input image patch. Moreover, although there is much research about spatial and channel attention, they cannot deal with geometrical matching enhancement. Therefore, the novel Gaussian Enhanced Euclidean norm (GEE) attention is proposed to focus more on important view relation features, which takes responsibility for efficiently mapping flattened projection to isometric projection. Here, we use a Gaussian function which takes the Euclidean norm of channel or spatial dimension as input to refine features output by the channel and spatial attention mechanism. Finally, the loss function is optimized by a geometry constraint for contour-to-contour transformation.

In summary, the contributions of this paper are as follows:

- IsoTGAN integrated the advantage of GAN and Transformer is proposed for enhancing 3D isometric view images with local and global image features. IsoTGAN inputs three orthogonal view images unlike SOTAs' GAN-Transformer models with one-to-one image generation framework.
- In proposed IsoTGAN, Gaussian Enhanced Euclidean norm (GEE) attention is newly presented for effectively recognizing each object view's relation. A modification of the GAN loss function by a geometry constraint is also proposed to generate the isometric view image for facilitating the 3D reconstruction process.
- To the best of our knowledge, the spatial and geometrical constraints between each orthographic view contour image of an object using an encoder network.
- Extensive experiments on SPARE3D [16] dataset show superior results on isometric view generation task compared to SOTAs. Owing to proposed IsoTGAN, industrial workers at CAD will carry out 2Dto3D CAD tasks without taking a number of tedious handling processes. Moreover, a new framework of proposed IsoTGAN can be used just in about 0.5 sec per image for the isometric view image generation, which allows a batch file translation.

2 Related Work

2.1 Transformer for Image Generation

Recently, many papers have incorporated Transformer into GAN architecture for image generation. [20] proposes GANformer, in which the Bipartite Transformer with simplex and duplex attention is applied between the image features and latent variables in the generator. In [23], a Transformer-based generator and discriminator with grid self-attention for efficient computation are presented. Another work [53] employs double attention in Swin Transformer [30] which additionally attends to the style tokens. Nevertheless, there is no other work that incorporates Transformer to reconstruct one image from multiple images.

2.2 Image-to-Image Translation

Numerous methods have utilized GANs for image-to-image translation tasks. This process involves converting an input image from one domain to another, using either paired or unpaired datasets for training. The initial pix2pix framework [21] employs image-conditional GAN [31] for various paired image-to-image translation tasks, such as turning semantic labels into synthetic images, given groundtruth image. Additionally, various techniques have been developed for unpaired translation [55], unsupervised cross-domain generation [40], multi-domain translation [4], and few-shot translation [27].

More recently, Diffusion models [38] have shown outstanding performance in areas like image generation [6,17], image super-resolution [13,51], image restoration [10,47], unpaired image-to-image translation [39,48], and image editing [2,24], surpassing GANs in these applications. However, Diffusion models need a lot of computation resources for training and inferencing, which hinders their ability in the industrial field. Moreover, typical image-to-image translation methods primarily aim to convert an image to one or many target domains without involving multiple input images and the relation between them.

While many of these models focus on generating realistic photos or animations, the generation of contour line drawings has been less explored. Contrary to these approaches, our framework addresses the challenge of converting three orthographic view contour drawings into their respective isometric view image, necessitating an understanding of the relation among multiple input images and the ability to produce contour images.

2.3 Deep Generative Models For CAD

Recently, several researches have focused on generating CAD commands either as 2D contour drawings [11,33,37,44] or 3D objects [15,19,22,46,49]. All of these methods employ Transformer-based encoder-decoder architecture, using 2D sketches to create CAD modeling programs. Among these, the work by Hu et al. (2023) [19] is most closely related to our study. They use a generative model to develop a method for reconstructing 3D CAD models from three orthographic views. However, this method deals with sequences rather than images and generates a sequence of CAD programs.

2.4 Isometric View Image Generation

Han et al. [16] apply the pix2pix framework [21] for generating isometric view images from three orthographic views, proposing a baseline that does not account for any relation between different views of object during training and generating. In this paper, we introduce a new approach that incorporates that relation into a vector, which is then used to generate the isometric view image.

3 Proposed Method

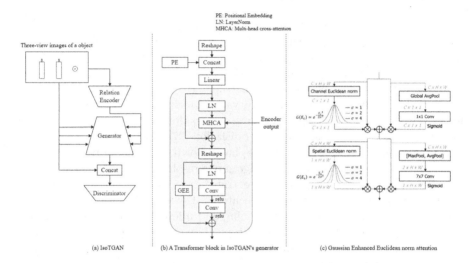

Fig. 1. The overall architecture of the proposed IsoTGAN. ⊗ represents broadcast element-wise multiplication and ⊕ denotes element-wise addition. $\mathbf{E_c}$ and $\mathbf{E_s}$ represent Channel Euclidean norm and Spatial Euclidean norm, respectively.

This section introduces the IsoTGAN framework, which is proposed to automatically generate isometric view image from three orthographic view contour drawings. Initially, an encoder is utilized to capture the spatial and geometric relation between each orthographic view of the object and embed these relations into a vector. This vector is then input into IsoTGAN along with orthographic view images at three different scales. The Transformer blocks are employed, which are able to model long-range interactions and constraints across three orthographic views of the object, to produce an isometric image. Additionally, the novel GEE attention mechanism and a modification to the GAN loss function are detailed. The overall architecture of the proposed method is illustrated in Fig. 1.

3.1 Three-View Contour Drawings Encoder

Our framework enhances the generation of isometric view image by using a three-view relation vector as input for the generator. This setup involves using an encoder to convert a triplet of front, side, and top view images into a vector, which is then provided to the generator. The image encoder comprises 6 stride-2 convolutional layers followed by two linear layers that calculate the mean and variance of the output distribution. The encoder and generator together form a VAE. In this framework, the encoder aims to capture the spatial and geometrical relation of the three views, while the generator combines the encoded vector with information from three-view images to reconstruct the isometric view image. During testing, the encoder acts as a relation guidance network to direct the generator in accurately constructing the isometric view image. We implement a KL-Divergence loss term [26] to facilitate training as follows:

$$\mathcal{L}_{\mathrm{KLD}} = \mathcal{D}_{\mathrm{KL}}(q(\mathbf{z} \mid \mathbf{x}) \| p(\mathbf{z})), \tag{1}$$

where the prior distribution $p(\mathbf{z})$ is a standard Gaussian distribution and the variational distribution q is fully determined by a mean vector and a variance vector [26].

3.2 IsoTGAN's Generator

Transformer Block. The conventional Transformer block consists of a multi-head self-attention and a feed-forward layer. In the self-attention layer, all pairwise relationships between input elements are considered, with each element being updated through attention to all others. In the task of generating isometric view image from three orthographic view contour drawings, only attending to image features is not adequate because the strong spatial and geometrical relation between each view are neglected. Therefore, we perform cross-attention between input image features and the vector output by the encoder to enhance reconstruction result. Formally, let $X^{n \times d}$ denote an input set of n image patch vectors of dimension d, and $Y^{m \times d}$ is a set of m spatial and geometrical vectors output by the relation encoder, the cross-attention in one head is then computed as follow:

$$\begin{aligned} Q &= XW_q \\ K &= YW_k \\ V &= YW_v \\ Attention(Q, K, V) &= softmax\left(\frac{QK^\top}{\sqrt{d}}\right)V, \end{aligned} \tag{2}$$

where Q, K, V represent query, key, value matrices, all keeping the dimension d, and W denote weight matrices. The positional embedding is also added to provide the distinct spatial position of each image patch.

After the multi-head cross-attention (MHCA) layer, two 3 × 3 convolution layers are utilized, then ReLU [1] activation function is applied for nonlinearity.

We use a sinusoidal positional encoding across horizontal and vertical dimensions for the image features X to ensure features location inside the image. The output of MHCA layer goes through the GEE attention (detailed in Sect. 3.2), then is added to the output of the convolution layer so that the model can focus more on important image features.

IsoTGAN is our adaptation of SPADE [34]. The network architecture follows the design of SPADE, except that SPADE blocks are replaced by Transformer blocks. With cross-attention, instead of intensively modeling interactions between all pixel pairs in an image, it enables adaptive long-range interactions between distant areas in an adequate manner, through a global spatial and geometrical vector. This vector selectively collects information from the entire input and distributes it to related regions, controlling the way the model generates isometric image, in which the output image must follow projection physics rules. Intuitively, information can move in both directions, from the local pixel level to the global high-level representation and back again. Given that soft-attention is inclined to group elements by proximity and content similarity, the transformer architecture allows the model to precisely modulate local semantic regions, avoiding discontinuity in generating the contour of object shape.

Gaussian Enhanced Euclidean Norm Attention. The relation representation vector in three-view drawings can carry extraneous information from axis contours, which traverse the object's center, but not the object outlines. To enhance understanding of each object view's relation as well as minimizing the influence of axes contour in isometric view image generation, we introduce the Gaussian Enhanced Euclidean norm (GEE) attention. This approach hypothesizes that smaller attention activations correlate with global contexts of higher absolute values [36].

The GEE framework includes both channel and spatial attention modules, each split into two branches, as depicted in Fig. 1. Unlike the Gaussian Context Transformer approach [36], the left branch of GEE directly uses the Euclidean norm of feature maps as input to a Gaussian function without the normalization operation of global average pooling (GAP). The intuition behind is that the Euclidean norm quantifies the magnitude of a vector or matrix, with larger Euclidean norm indicating greater deviation from the vector or matrix to its origin, so constraining the Euclidean norm with Gaussian function will improve model generalizability. Additionally, we argue that this principle applies to spatial dimensions as well, thus extending it to capture global spatial information.

The right branch is quite similar to the CBAM architecture [45], except that we utilize only GAP followed by a 1×1 convolution in its channel attention module to reduce model parameters.

Concretely, given a feature map $\mathbf{F} \in \mathbb{R}^{C \times H \times W}$ as input, C denotes the number of channels and H, W are spatial dimension, GEE computes attention as follows.

$$\begin{aligned} \mathbf{F}' &= \mathbf{M_c}(\mathbf{F}) \otimes \mathbf{F} \oplus \mathbf{G}(\mathbf{E_c}) \otimes \mathbf{F} \\ \mathbf{F}'' &= \mathbf{M_s}(\mathbf{F}') \otimes \mathbf{F}' \oplus \mathbf{G}(\mathbf{E_s}) \otimes \mathbf{F}', \end{aligned} \quad (3)$$

where $\otimes$ represents broadcast element-wise multiplication and $\oplus$ denotes element-wise addition. $\mathbf{F}''$ is the final output. $\mathbf{E_c} \in \mathbb{R}^{C \times 1 \times 1}$ and $\mathbf{E_s} \in \mathbb{R}^{1 \times H \times W}$ represents Channel Euclidean norm and Spatial Euclidean norm, respectively, and is formulated as follows.

$$\mathbf{E_c} = \left\{ e_{ck} = \sqrt{\sum_{i=1}^{W} \sum_{j=1}^{H} \mathbf{F}_k(i,j)^2} : k \in \{1, \ldots, C\} \right\}, \tag{4}$$

$$\mathbf{E_s} = \left\{ e_{sij} = \sqrt{\sum_{k=1}^{C} \mathbf{F}'_{ij}(k)^2} : i \in \{1, \ldots, W\}, j \in \{1, \ldots, H\} \right\}. \tag{5}$$

A Gaussian function $\mathbf{G}(x) = \exp(-\frac{x^2}{2\sigma^2})$ processes the input x with its maximum value at 1, a mean of 0, and a standard deviation σ, aligning with the hypothesis about the relationship between global contexts and attention activations. A larger σ leads to a more uniform distribution among attention activations. The effect of σ on the model's effectiveness will be investigated in the experiment section.

$\mathbf{M_c} \in \mathbb{R}^{C \times 1 \times 1}$ and $\mathbf{M_s} \in \mathbb{R}^{1 \times H \times W}$ is channel attention map and spatial attention map, respectively, and computed as

$$\mathbf{M_c}(\mathbf{F}) = \text{Sigmoid}(f^{1 \times 1}(\text{GAP}(\mathbf{F}))), \tag{6}$$

$$\mathbf{M_s}(\mathbf{F}) = \text{Sigmoid}(f^{7 \times 7}([\text{MaxPool}(\mathbf{F}); \text{AvgPool}(\mathbf{F})])), \tag{7}$$

where $f^{k \times k}$ represents a convolution operation with kernel size of $k \times k$.

The left branch enhances the output from the right branch by focusing more on significant spatial and geometrical relationships between each view of the object. The output from the GEE block is then added element-wise to the output of the convolution layer.

Loss Function. The conventional conditional GAN framework for image-to-image translation consists of a generator G and a discriminator D. In this paper, the generator G is responsible for converting three orthographic views drawings into isometric view image, whereas the role of discriminator D is to differentiate between real isometric images and those produced by the generator. This framework operates under a supervised learning manner, where the training dataset consists of three views-isometric view image pairs $\{(\mathbf{x_i}, \mathbf{y_i})\}$, with $\mathbf{x_i}$ representing a triplet of object's front, side, and top view image and $\mathbf{y_i}$ being the corresponding isometric view image. Conditional GANs work by modeling the conditional distribution of real images based on the input images through the following minimax game:

$$\min_{G} \max_{D} \mathcal{L}_{\text{GAN}}(G, D). \tag{8}$$

For our task, the generator G takes the relation vector output by the encoder E as an extra input, hence, the objective function $\mathcal{L}_{\text{GAN}}(G, D)$ is computed as

$$\mathbb{E}_{(\mathbf{x},\mathbf{y})}[\log D(\mathbf{x}, \mathbf{y})] + \mathbb{E}_{\mathbf{x}}[\log(1 - D(\mathbf{x}, G(\mathbf{x}, E(\mathbf{x}))))]. \tag{9}$$

The above objective function does not include the geometry constraint of isometric view image, which is crucial for reasonable generation, i.e. the generated image must follow projection physics rules. To incorporate the geometric constraints of the isometric view image, we use a geometric transformation function $f(\cdot)$. We input both the combined three-view images $\mathbf{x}$ and their transformed versions $\tilde{\mathbf{x}} = f(\mathbf{x})$ into the generator G. In this study, we apply vertical and horizontal flipping. The aim is to minimize the discrepancy between the generated isometric image $\mathbf{g} = G(\mathbf{x}, \mathbf{z})$ and its flipped version $\tilde{\mathbf{g}} = f(G(f(\mathbf{x}), \mathbf{z}))$, where $\mathbf{z} = E(\mathbf{x})$ is the output vector from the encoder. The geometry loss is calculated as follows:

$$\mathcal{L}_{geo}(G) = \mathbb{E}_{\mathbf{x},\mathbf{z}}\left[\|G(\mathbf{x},\mathbf{z}) - f(G(f(\mathbf{x}),\mathbf{z}))\|_1\right]. \tag{10}$$

In contrast to other image-to-image translation tasks, an isometric view image consists solely of the contour lines of the object. Consequently, a pixel can only assume one of two values, 0 or 255, where 0 indicates that the pixel is part of the object's contour, and 255 denotes that it belongs to the background. The total loss is added by both Cross-Entropy loss and geometry loss. Additionally, MoNCE [52] is incorporated to enhance versatility, leading to the following modification of the total loss function:

$$\mathcal{L} = \mathcal{L}_{\text{GAN}}(G, D) + MoNCE + \mathcal{L}_{geo}(G) + \ell\left(\hat{p}_i, p_i\right), \tag{11}$$

where $\ell(\cdot, \cdot)$ denotes the standard Cross-Entropy. $\hat{p}_i$ is the model's output probability whereas there is a contour point at pixel i, p_i is the ground truth contour point of the isometric image.

4 Experiments

4.1 Datasets

We conducted our experiments using the SPARE3D [16] dataset, which comprises 5000 pairs of three-view and isometric images. Typically, represented hidden lines in objects are not displayed in actual CAD drawings; thus, we preprocessed the images to remove all hidden lines. We concatenated three-view images along channel dimension so the input $C = 9$. The dataset was divided into 4000 training pairs and 1000 testing pairs.

4.2 Experiment Setup

We carried out all experiments using a single NVIDIA RTX 4090 GPU with 24 GB of VRAM. Our model was implemented using the PyTorch framework. The learning rates were set to 0.001 for the generator and 0.003 for the discriminator. We utilized the Adam optimizer [25] with parameters $\beta_1 := 0.001$ and $\beta_2 := 0.9$. The evaluation metrics employed in our experiments include:

- Frechet Inception Distance (FID) is employed to measure the distance between the distribution of generated results and the distribution of real images.
- Structural Similarity Index (SSIM) is used to measure similarity of generated results and real images.
- L1 and L2 calculates the absolute difference and square of the difference between generated and real images, respectively.

4.3 Experimental Results

Quantitative Results. Table 1 shows the quantitative comparison results of the proposed IsoTGAN with the baseline model pix2pix [16], SPADE [34], MoNCE [52], and DINO [43]. Overall, our model outperforms other methods with a large margin in all metrics scores. Compared to the runner-up model, a reduction of about 45% in FID and an increase of approximately 42% in SSIM are obtained. Standard reconstruction metrics L1 and L2 also significantly decrease by about 37% and 41%, respectively, proving the efficiency of our proposed IsoTGAN when modeling the spatial and geometrical relation between each view of the object and guiding the generator to reconstruct the corresponding isometric view image.

While other GAN-based methods, without Transformer blocks, are only strong at decomposing global attributes of the entire image, the cross-attention layer and GEE in the Transformer block help IsoTGAN enhance the ability to reconstruct local details of objects in isometric view. The encoder of IsoTGAN provides comprehensive spatial and geometrical relation vector for the generator to understand these relations; therefore, it is capable of generating reasonable isometric view images.

Proposed IsoTGAN consists of a number of new modules, i.e., GEE, with GAN and Transformer. However, computational complexity has been realized just in about 0.5 s per 3D isometric image generation.

Table 1. Quantitative generation performance comparisons with baseline model

Method	FID ↓	SSIM ↑	L1 ↓	L2 ↓
pix2pix [16] (baseline)	43.82	0.436	29.34	28.73
SPADE [34]	40.39	0.458	29.17	28.55
MoNCE [52]	41.29	0.477	27.11	26.20
DINO [43]	38.91	0.506	24.60	24.63
IsoTGAN (ours)	**21.18**	**0.723**	**15.44**	**14.37**

Qualitative Results. Figure 2 shows the qualitative results of IsoTGAN and other paired image-to-image translation models. It can be clearly seen that our proposed method generates more similar isometric view images compared to ground truth images. In other methods' generated results, the contour lines of isometric images are discontinued or incompleted, showing that typical GAN loss function is not adequate in reconstructing images containing only contour lines from images of the same type, even though their structure looks more simple compared to natural images and other generative models are good at converting drawing edges into color images.

The pix2pix [16] baseline generates isometric images that lack a lot of details and are incomprehensive. SAPDE [34] and MoNCE [52] perform better as they could capture the overall structure of objects, but small details are not reconstructed successfully. Especially, MoNCE's output of example seventh looks very similar to the groundtruth. Generated results of DINO [43] are discontinued at some parts in objects contour lines, and in example fifth, one small circle on top of the object is missing. Our proposed IsoTGAN successfully generates small details of objects, and the discontinuity in objects contour lines is minimized.

Some objects have the same front view and side view projection, other objects have top view projection similar to the rotation operation of front view or side view projection. Therefore, IsoTGAN has to distinguish between three views and discern their relations and constraints to produce comprehensive output. The modified loss function helps IsoTGAN predict contour lines better. By incorporating geometry loss, the proposed model has learned to reconstruct an isometric view image given its vertical flipped three-view image input; therefore, IsoTGAN is able to understand how projection translation affects an object's image on a 2D plane. The combination of shapes in the isometric view is also understood well by the model.

Ablation Studies. Ablation experiments were conducted on SPARE3D dataset to investigate the contribution of different components in the proposed method. First, we utilized only the encoder in our framework, while Transformer and GEE were replaced with conventional convolution, and the loss function contained only $\mathcal{L}_{GAN}$ and MoNCE. Then, we gradually added Transformer, GEE, $\mathcal{L}_{geo}$, and ℓ. In the next step, the encoder was discarded while other components were kept. When the encoder was not used, MCHA in Transformer was substituted by self-attention that performs attention on image patches. We also performed experiments where only Transformer and GEE were applied, and then only $\mathcal{L}_{geo}$ and ℓ were added.

Overall, it can be seen clearly that the encoder plays a crucial role. As Table 2 illustrates, without the encoder, despite keeping all other components, the generation metrics are significantly worse compared to when the encoder is equipped. Only utilizing the encoder (method a) yields better FID and SSIM than discarding it while keeping all other four components. Without the encoder, IsoTGAN becomes a conventional GAN model, so the model cannot understand the spatial and geometrical relation between each orthogonal view of the object. The Transformer with MHCA layers also contributes immensely to the effectiveness of the proposed method by incorporating a spatial and geometrical vector to

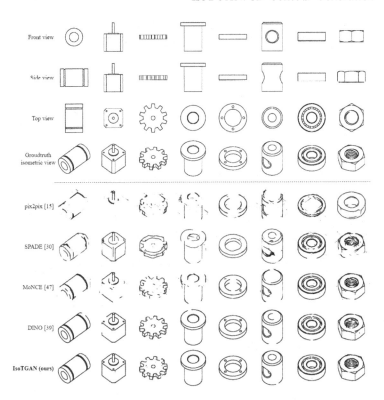

Fig. 2. Qualitative comparisons of isometric view generation performance with baseline model.

selectively distribute information from the entire input to appropriate regions, enhancing the generation of local parts of objects. The GEE attention mechanism in conjunction with Transformer provides extra information about noise contexts for a more accurate generation of isometric view image. Additionally, the modification of loss function gives more geometry constraint for the model to ensure that the generated output follows objects' physical rules.

Impact of Standard Deviation in GEE. In this section, we explore how the standard deviation σ in the Gaussian function $\mathbf{G}(x) = \exp(-\frac{x^2}{2\sigma^2})$ influences the performance result of IsoTGAN in generating isometric view image from three orthographic view contour drawings. The findings are presented in Table 3. It is observed that as σ increases, the network performance first improves and then decreases. The optimal performance is achieved when σ is set at 4. This behavior makes sense because a very high variance can reduce the differences in attention activations across channel and spatial dimension, which interferes with the effective reduction of global noise contexts. Conversely, a too low variance might limit the significance of other important features and incorrectly emphasize noise contexts.

Table 2. Ablation studies on SPARE3D dataset of the proposed method

Method	Encoder	Transformer	GEE	$\mathcal{L}_{geo}$	ℓ	FID ↓	SSIM ↑	L1 ↓	L2 ↓
(a)	✓					28.18	0.527	21.33	21.05
(b)		✓				31.15	0.508	23.86	22.56
(c)		✓	✓			30.12	0.512	23.77	22.66
(d)		✓	✓	✓	✓	29.47	0.520	21.02	20.53
(e)				✓	✓	31.21	0.480	24.50	23.21
(f)		✓		✓	✓	29.30	0.498	21.10	20.77
(g)	✓			✓	✓	24.20	0.605	17.89	16.37
(h)	✓	✓		✓	✓	21.82	0.695	15.66	14.49
(i)	✓	✓				22.89	0.587	16.70	15.52
(j)	✓	✓	✓			21.98	0.686	15.85	14.43
(k)	✓	✓	✓	✓		21.37	0.711	15.54	14.62
(l)	✓	✓	✓	✓	✓	**21.18**	**0.723**	**15.44**	**14.37**

Table 3. Generation result of IsoTGAN on SPARE3D dataset with different standard deviation σ in GEE

σ	FID ↓	SSIM ↑	L1 ↓	L2 ↓
1	22.56	0.648	16.13	15.31
2	22.51	0.674	15.98	15.22
4	**21.18**	**0.723**	**15.44**	**14.37**
6	21.89	0.705	16.07	15.16

Attention Visualization. To better understand the interpretability of the IsoTGAN's generator, we analyze the attention matrices generated by the Transformer blocks. We examine the attention weights output by the MHCA layer. Each element (i, j) in these matrices shows the degree of attention that token i gives to token j. Since the model employs multi-head attention, multiple attention matrices are produced, one per head.

For further reconfirmation, we calculate the average of these weights across all heads and target tokens for each block. Considering an input image size of 256×256, this results in an average attention vector with dimensions $w \times h$ (16×16). Each entry j-th in this vector indicates the average attention token j receives. By overlaying the attention heatmap on the input images, shown in Fig. 3, it can be seen clearly that each block pays attention to a specific part inside three-view images. It focuses on contour lines of the shape and by that, strengthens the generation of local details of the object.

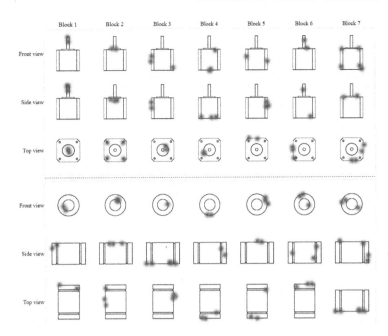

Fig. 3. Attention heatmap generated by attention weights from Transformer blocks in IsoTGAN's generator

5 Conclusion

This paper has proposed a novel IsoTGAN framework for effectively automatic isometric view image generation from three orthographic views contour drawings. Extensive experimental results demonstrate that our proposed method shows a strong capability in isometric view image generation, achieving state-of-the-art performance. Physical constraints for better refining the reconstruction of difficult features of objects will be considered.

References

1. Agarap, A.F.: Deep learning using rectified linear units (relu). arXiv preprint arXiv:1803.08375 (2018)
2. Brooks, T., Holynski, A., Efros, A.A.: Instructpix2pix: learning to follow image editing instructions. In: Proceedings of the IEEE/CVF Conference on Computer Vision and Pattern Recognition, pp. 18392–18402 (2023)
3. Cheng, Y., Gan, Z., Li, Y., Liu, J., Gao, J.: Sequential attention gan for interactive image editing. In: Proceedings of the 28th ACM International Conference on Multimedia, pp. 4383–4391 (2020)
4. Choi, Y., Choi, M., Kim, M., Ha, J.W., Kim, S., Choo, J.: Stargan: unified generative adversarial networks for multi-domain image-to-image translation. In: Proceedings of the IEEE Conference on Computer Vision and Pattern Recognition, pp. 8789–8797 (2018)

5. Daras, G., Odena, A., Zhang, H., Dimakis, A.G.: Your local gan: Designing two dimensional local attention mechanisms for generative models. In: Proceedings of the IEEE/CVF Conference on Computer Vision and Pattern Recognition, pp. 14531–14539 (2020)
6. Dhariwal, P., Nichol, A.: Diffusion models beat gans on image synthesis. Adv. Neural. Inf. Process. Syst. **34**, 8780–8794 (2021)
7. Dinh, L., Sohl-Dickstein, J., Bengio, S.: Density estimation using real nvp. arXiv preprint arXiv:1605.08803 (2016)
8. Dosovitskiy, A., et al.: An image is worth 16x16 words: Transformers for image recognition at scale. arXiv preprint arXiv:2010.11929 (2020)
9. Emami, H., Aliabadi, M.M., Dong, M., Chinnam, R.B.: Spa-gan: spatial attention gan for image-to-image translation. IEEE Trans. Multimedia **23**, 391–401 (2020)
10. Fei, B., et al.: Generative diffusion prior for unified image restoration and enhancement. In: Proceedings of the IEEE/CVF Conference on Computer Vision and Pattern Recognition, pp. 9935–9946 (2023)
11. Ganin, Y., Bartunov, S., Li, Y., Keller, E., Saliceti, S.: Computer-aided design as language. Adv. Neural. Inf. Process. Syst. **34**, 5885–5897 (2021)
12. Gao, R., et al.: Cat3d: create anything in 3d with multi-view diffusion models. arXiv preprint arXiv:2405.10314 (2024)
13. Gao, S., et al.: Implicit diffusion models for continuous super-resolution. In: Proceedings of the IEEE/CVF Conference on Computer Vision and Pattern Recognition, pp. 10021–10030 (2023)
14. Goodfellow, I., et al.: Generative adversarial networks. Commun. ACM **63**(11), 139–144 (2020)
15. Guo, H., Liu, S., Pan, H., Liu, Y., Tong, X., Guo, B.: Complexgen: cad reconstruction by b-rep chain complex generation. ACM Trans. Graph. (TOG) **41**(4), 1–18 (2022)
16. Han, W., Xiang, S., Liu, C., Wang, R., Feng, C.: Spare3d: a dataset for spatial reasoning on three-view line drawings. In: Proceedings of the IEEE/CVF Conference on Computer Vision and Pattern Recognition, pp. 14690–14699 (2020)
17. Ho, J., Jain, A., Abbeel, P.: Denoising diffusion probabilistic models. Adv. Neural. Inf. Process. Syst. **33**, 6840–6851 (2020)
18. Höllein, L., et al.: Viewdiff: 3d-consistent image generation with text-to-image models. In: Proceedings of the IEEE/CVF Conference on Computer Vision and Pattern Recognition, pp. 5043–5052 (2024)
19. Hu, W., Zheng, J., Zhang, Z., Yuan, X., Yin, J., Zhou, Z.: Plankassembly: robust 3d reconstruction from three orthographic views with learnt shape programs. In: Proceedings of the IEEE/CVF International Conference on Computer Vision, pp. 18495–18505 (2023)
20. Hudson, D.A., Zitnick, L.: Generative adversarial transformers. In: International conference on machine learning, pp. 4487–4499. PMLR (2021)
21. Isola, P., Zhu, J.Y., Zhou, T., Efros, A.A.: Image-to-image translation with conditional adversarial networks. In: Proceedings of the IEEE Conference on Computer Vision and Pattern Recognition, pp. 1125–1134 (2017)
22. Jayaraman, P.K., Lambourne, J.G., Desai, N., Willis, K., Sanghi, A., Morris, N.J.: Solidgen: an autoregressive model for direct b-rep synthesis. Trans. Mach. Learn. Res. (2022)
23. Jiang, Y., Chang, S., Wang, Z.: Transgan: two pure transformers can make one strong gan, and that can scale up. Adv. Neural. Inf. Process. Syst. **34**, 14745–14758 (2021)

24. Kawar, B., et al.: Imagic: text-based real image editing with diffusion models. In: Proceedings of the IEEE/CVF Conference on Computer Vision and Pattern Recognition, pp. 6007–6017 (2023)
25. Kingma, D.P., Ba, J.: Adam: A method for stochastic optimization. arXiv preprint arXiv:1412.6980 (2014)
26. Kingma, D.P., Welling, M.: Auto-encoding variational bayes. arXiv preprint arXiv:1312.6114 (2013)
27. Liu, M.Y., Huang, X., Mallya, A., Karras, T., Aila, T., Lehtinen, J., Kautz, J.: Few-shot unsupervised image-to-image translation. In: Proceedings of the IEEE/CVF International Conference on Computer Vision, pp. 10551–10560 (2019)
28. Liu, R., Wu, R., Van Hoorick, B., Tokmakov, P., Zakharov, S., Vondrick, C.: Zero-1-to-3: zero-shot one image to 3d object. In: Proceedings of the IEEE/CVF International Conference on Computer Vision, pp. 9298–9309 (2023)
29. Liu, Y., et al.: Syncdreamer: generating multiview-consistent images from a single-view image. In: The Twelfth International Conference on Learning Representations
30. Liu, Z., et al.: Swin transformer: Hierarchical vision transformer using shifted windows. In: Proceedings of the IEEE/CVF International Conference on Computer Vision, pp. 10012–10022 (2021)
31. Mirza, M., Osindero, S.: Conditional generative adversarial nets. arXiv preprint arXiv:1411.1784 (2014)
32. Van den Oord, A., Kalchbrenner, N., Espeholt, L., Vinyals, O., Graves, A., et al.: Conditional image generation with pixelcnn decoders. Adv. Neural Inform. Process. Syst. **29** (2016)
33. Para, W., et al.: Sketchgen: generating constrained cad sketches. Adv. Neural. Inf. Process. Syst. **34**, 5077–5088 (2021)
34. Park, T., Liu, M.Y., Wang, T.C., Zhu, J.Y.: Semantic image synthesis with spatially-adaptive normalization. In: Proceedings of the IEEE/CVF Conference on Computer Vision and Pattern Recognition, pp. 2337–2346 (2019)
35. Radford, A., Metz, L., Chintala, S.: Unsupervised representation learning with deep convolutional generative adversarial networks. arXiv preprint arXiv:1511.06434 (2015)
36. Ruan, D., Wang, D., Zheng, Y., Zheng, N., Zheng, M.: Gaussian context transformer. In: Proceedings of the IEEE/CVF Conference on Computer Vision and Pattern Recognition, pp. 15129–15138 (2021)
37. Seff, A., Zhou, W., Richardson, N., Adams, R.P.: Vitruvion: a generative model of parametric cad sketches. In: International Conference on Learning Representations (2021)
38. Sohl-Dickstein, J., Weiss, E., Maheswaranathan, N., Ganguli, S.: Deep unsupervised learning using nonequilibrium thermodynamics. In: International Conference on Machine Learning, pp. 2256–2265. PMLR (2015)
39. Sun, S., Wei, L., Xing, J., Jia, J., Tian, Q.: Sddm: score-decomposed diffusion models on manifolds for unpaired image-to-image translation. In: International Conference on Machine Learning, pp. 33115–33134. PMLR (2023)
40. Taigman, Y., Polyak, A., Wolf, L.: Unsupervised cross-domain image generation. In: International Conference on Learning Representations (2016)
41. Tang, H., Liu, H., Xu, D., Torr, P.H., Sebe, N.: Attentiongan: unpaired image-to-image translation using attention-guided generative adversarial networks. IEEE Trans. Neural Netw. Learn. Syst. **34**(4), 1972–1987 (2021)
42. Vaswani, A., et al.: Attention is all you need. Adv. Neural Inform. Process. Syst. **30** (2017)

43. Vougioukas, K., Petridis, S., Pantic, M.: Dino: a conditional energy-based gan for domain translation. In: International Conference on Learning Representations
44. Willis, K.D., Jayaraman, P.K., Lambourne, J.G., Chu, H., Pu, Y.: Engineering sketch generation for computer-aided design. In: Proceedings of the IEEE/CVF Conference on Computer Vision and Pattern Recognition, pp. 2105–2114 (2021)
45. Woo, S., Park, J., Lee, J.-Y., Kweon, I.S.: CBAM: convolutional block attention module. In: Ferrari, V., Hebert, M., Sminchisescu, C., Weiss, Y. (eds.) ECCV 2018. LNCS, vol. 11211, pp. 3–19. Springer, Cham (2018). https://doi.org/10.1007/978-3-030-01234-2_1
46. Wu, R., Xiao, C., Zheng, C.: Deepcad: a deep generative network for computer-aided design models. In: Proceedings of the IEEE/CVF International Conference on Computer Vision, pp. 6772–6782 (2021)
47. Xia, B., et al.: Diffir: efficient diffusion model for image restoration. In: Proceedings of the IEEE/CVF International Conference on Computer Vision, pp. 13095–13105 (2023)
48. Xu, S., Ma, Z., Huang, Y., Lee, H., Chai, J.: Cyclenet: rethinking cycle consistency in text-guided diffusion for image manipulation. Adv. Neural Inform. Process. Syst. **36** (2024)
49. Xu, X., Willis, K.D., Lambourne, J.G., Cheng, C.Y., Jayaraman, P.K., Furukawa, Y.: Skexgen: autoregressive generation of cad construction sequences with disentangled codebooks. In: International Conference on Machine Learning, pp. 24698–24724. PMLR (2022)
50. Yang, J., Cheng, Z., Duan, Y., Ji, P., Li, H.: Consistnet: enforcing 3d consistency for multi-view images diffusion. In: Proceedings of the IEEE/CVF Conference on Computer Vision and Pattern Recognition, pp. 7079–7088 (2024)
51. Yue, Z., Wang, J., Loy, C.C.: Resshift: efficient diffusion model for image super-resolution by residual shifting. Adv. Neural Inform. Process. Syst. **36** (2024)
52. Zhan, F., Zhang, J., Yu, Y., Wu, R., Lu, S.: Modulated contrast for versatile image synthesis. In: Proceedings of the IEEE/CVF Conference on Computer Vision and Pattern Recognition, pp. 18280–18290 (2022)
53. Zhang, B., et al.: Styleswin: Transformer-based gan for high-resolution image generation. In: Proceedings of the IEEE/CVF Conference on Computer Vision and Pattern Recognition, pp. 11304–11314 (2022)
54. Zhang, H., Goodfellow, I., Metaxas, D., Odena, A.: Self-attention generative adversarial networks. In: International Conference on Machine Learning, pp. 7354–7363. PMLR (2019)
55. Zhu, J.Y., Park, T., Isola, P., Efros, A.A.: Unpaired image-to-image translation using cycle-consistent adversarial networks. In: Proceedings of the IEEE International Conference on Computer Vision, pp. 2223–2232 (2017)

Hierarchical Feature Aggregation Network Based on Swin Transformer for Medical Image Segmentation

Hayato Iyoda[1], Yongqing Sun[2], and Xian-Hua Han[1(✉)]

[1] Graduate School of Artificial Intelligence and Science, Rikkyo University, 3-34-1 Nishi-Ikebukuro, Toshima-ku, Tokyo 171-8501, Japan
{24vr038t,hanxhua}@rikkyo.ac.jp
[2] College of Humanities and Sciences, Nihon University, 3-25-40 Sakurajosui Setagaya-Ku, Tokyo 156-8550, Japan

Abstract. Semantic segmentation plays a crucial role in computer-aided medical image analysis by achieving important and useful regions, which are vital for various diagnostic tasks. Recently, vision transformers (ViTs) have emerged as the leading approach in medical image segmentation, outperforming traditional convolutional neural networks (CNNs). The incorporation strategies of the ViTs for medical segmentation are dominated to leverage the widely used U-shape like architecture (U-Net) while replace the convolution blocks in both encoder and decoder paths using transformer blocks. It remains uncertain which components of the incorporated transformer block contribute most significantly to segmentation results in the medical field. This study presents a hierarchical feature aggregation method based on hierarchical Transformer features to enhance the performance of ViT-based architecture in data-constrained medical image segmentation. Specifically, our approach employs the hierarchical vision Transformer to configure the main encoder path for extracting multi-scale semantic features, and leverages several residual blocks to achieve local representation with detail spatial information. Then, we introduce a hierarchical feature aggregation module (HFAM) to serve as the decoder path for fusing multi-scale semantic features and residual spatial features. Compared with the existing transformer-based U-Net, the explored HFAM can not only effectively combine the diverse contexts but also potentially reduce the computational complexity. Experiments on 3 different medical image segmentation benchmarks have demonstrated our proposed method consistently outperformers the conventional U-Net, and various Transformer-based U-Net.

Keywords: Medical image segmentation · U-Net · Swin Transformer · Hierarchical feature aggregation

1 Introduction

Semantic segmentation is a fundamental process in computer-aided medical image analysis, serving the critical function of identifying and delineating regions of interest for various diagnostic tasks [17,20,21]. Medical segmentation, however, is often complicated due to variations in image modality, acquisition techniques, and inherent pathological or biological differences across patients. Such complexities introduce significant challenges to achieving accurate and reliable segmentation. Recently, the application of deep learning techniques has provided substantial advancements in addressing these challenges. Of particular significance is the introduction of the U-Net model [19], which has proven to be remarkably effective in medical image segmentation tasks. Thus, U-Net [19] and its numerous variants [6,7,11,13,14,16,30,32] have become the prevailing standard in many medical image segmentation tasks such as cardiac segmentation, organ segmentation, lesion segmentation and so on. The U-Net architecture is generally designed with a symmetric encoder-decoder structure with the dominated convolution components, and attempt to capture global context by creating large receptive fields with down-sampling and stacking multiple convolutional layers.

Despite the strong representational power of CNN-based U-net, they face challenges in establishing clear long-range dependencies because convolutional kernels have restricted receptive fields. This inherent limitation in the convolution operation makes it difficult to capture global semantic context [2], which is vital for dense prediction tasks such as segmentation. Motivated by the attention mechanism [22,25] in natural language processing, recent research addresses the limitations of CNNs by integrating attention into their architecture. For instance, Non-local neural networks [18] introduce a plug-and-play operator based on self-attention, allowing them to capture long-range dependencies within feature maps. However, this comes at the cost of significant memory and computational demands. Schlemper et al. [22] offer an alternative with the attention gate model, which enhances model sensitivity and prediction accuracy while introducing minimal computational overhead, making it easily adaptable to standard CNNs. In contrast, the Transformer architecture [5,25] is explicitly designed to handle long-range dependencies in sequence-to-sequence tasks, capturing relationships between any positions within a sequence. Recently, researchers have explored the application of Transformers in computer vision. The Vision Transformer (ViT) [5] was developed to tackle image recognition tasks by using 2D image patches with positional embeddings and pre-training on large datasets. ViT achieved performance comparable to CNN-based models. Furthermore, the Data-efficient Image Transformer (DeiT) [23] demonstrated that Transformers could be trained on mid-sized datasets, and its performance could be enhanced through distillation techniques. Additionally, the Swin Transformer [15], a hierarchical architecture, was later proposed as a vision backbone, achieving state-of-the-art results in image classification, object detection, and semantic segmentation. The successes of ViT, DeiT, and the Swin Transformer highlight the growing potential of Transformer models in computer vision applications.

In medical image segmentation filed, the incorporation of Transformers into U-Net architecture has extensively explored, and led to advancements in segmentation accuracy [1,24,26,27,31], especially in tasks that require precise delineation of complex structures. By incorporating Transformer blocks into the encoder and decoder paths, these Transformer-based U-Net enhances the ability to capture both local and global features. For example, Swin-UNet [10] uses Swin Transformer blocks within the U-Net structure to improve segmentation performance on 2D medical images by modeling long-range dependencies while maintaining spatial resolution through skip connections. Swin-Unet is a fully Transformer-based U-shaped architecture, incorporating the Swin Transformer block into all components: encoder, bottleneck, decoder, and skip connections. Despite the potential performance gain, the incorporation of the Transformer blocks into all components may cause high computational cost and memory usage, and possibly brings the overfitting problem especially for data-constrained medial image analysis tasks. Moreover, the impact of incorporating Transformer blocks into various components of the network on overall performance remains underexplored, with limited research addressing how these modifications influence segmentation accuracy and computational efficiency.

To handle the above issues, this study presents a novel hierarchical feature aggregation framework, leveraging hierarchical Transformer features to enhance the performance of Vision Transformer (ViT)-based architectures for medical image segmentation in data-limited scenarios. Our methodology centers on utilizing a hierarchical Vision Transformer as the primary encoder, which facilitates the extraction of multi-scale semantic features, crucial for capturing global context. Additionally, residual blocks are integrated to preserve fine-grained local representations and detailed spatial information, addressing the need for precise segmentation boundaries. To optimize feature fusion, we introduce a Hierarchical Feature Aggregation Module (HFAM) within the decoder, which effectively merges multi-scale semantic features with the residual spatial information. Compared to existing Transformer-based U-Net variants, the proposed HFAM not only efficiently combines rich contextual information but also demonstrates potential in reducing computational complexity. Extensive experiments conducted on three benchmark datasets for medical image segmentation consistently show that our approach surpasses both conventional U-Net architectures and several Transformer-based U-Net models, highlighting its efficacy and robustness in challenging segmentation tasks.

2 Related Work

CNN-Based Methods: Motivated by the great success of the development of deep learning, convolutional neural networks (CNNs) have widely applied for many medical image segmentation tasks. A pivotal work was the introduction of the U-Net architecture [19], specifically designed for biomedical image segmentation. The U-shaped architecture in U-Net [19], characterized by its encoder-decoder structure with skip connections, enabled both efficient feature extraction and precise localization, making it highly effective for segmentation tasks.

Due to its simplicity and strong performance, the U-Net framework has inspired numerous variants aimed at further enhancing its capabilities. Notable examples include Res-UNet [30], which incorporates residual connections to address gradient vanishing in deeper networks, Dense-UNet [14], which leverages dense connections to improve feature reuse and network efficiency, and U-Net++ [32], which refines the skip connections with nested architectures for better feature fusion. Additionally, UNet3+ [11] extends this design by introducing a full-scale skip connection mechanism, enabling a richer fusion of semantic and spatial features. U-Net and its variants have also been adapted for 3D medical image segmentation tasks, with architectures such as 3D U-Net [3] and V-Net [16] emerging to tackle volumetric data. These 3D models preserve the spatial coherence of medical images across slices, thereby improving segmentation accuracy in three-dimensional imaging modalities like CT and MRI. Overall, CNN-based methods, especially U-Net and its derivatives, have achieved remarkable success in medical image segmentation due to their powerful representation learning capabilities, adaptability to various tasks, and ability to handle both 2D and 3D medical imaging data. These advancements have significantly improved segmentation performance across a wide range of medical applications. However, all these methods generally employ the convolution layers as the dominated components, and have limited receptive fields to capture global context.

Vision Transformers: The Transformer model was originally introduced for machine translation tasks [25] and has since revolutionized the field of natural language processing (NLP). Transformer-based models have achieved state-of-the-art results across a wide range of NLP tasks [4], owing to their ability to capture long-range dependencies and model complex relationships through self-attention mechanisms. Inspired by the success of Transformers in NLP, researchers extended this architecture to the field of computer vision, leading to the development of the Vision Transformer (ViT) [5], marked a significant breakthrough in image recognition by offering an impressive balance between speed and accuracy, especially in large-scale tasks. Unlike CNN-based models, ViT relies on global self-attention, which allows it to model global context more effectively. However, a notable limitation of ViT is its reliance on large-scale datasets for pre-training. Unlike CNNs, which can be efficiently trained on smaller datasets, ViT requires extensive pre-training on datasets such as ImageNet to achieve competitive performance. To address this challenge, the Data-efficient Image Transformer (DeiT) [23] introduced several training techniques, including knowledge distillation, to enable ViT to perform well on mid-sized datasets, mitigating the need for vast amounts of data.

Building on ViT's foundations, a series of subsequent works [8,15,28] have further enhanced Transformer-based architectures for vision tasks. Among these, the Swin Transformer [15] stands out as a highly efficient and versatile model. The Swin Transformer introduces a hierarchical structure with shifted window-based attention, which enables the model to capture both local and global information in a computationally efficient manner. This design significantly reduces the complexity typically associated with full self-attention, making Swin

Transformer scalable and suitable for high-resolution inputs. As a result, it has achieved state-of-the-art performance across various computer vision tasks. Swin Transformer's ability to balance computational efficiency with high performance makes it an influential architecture in both research and practical applications within the vision domain. Most Transformers are originally proposed as the encoder in image classification tasks to extract different image representation, and require to configure a decoder to integrate the multi-scale encoder features for dense prediction tasks such as semantic image segmentation.

Transformers for Medical Image Segmentation: The success of ViT in traditional computer vision tasks has paved the way for a paradigm shift in medical image segmentation, and the integration of the Transformer block into the U-Net achitectires have increasingly been explored [1,24,26,27,31]. Among these advancements, TransUNet [2] represents the first framework to incorporate Transformers into medical image segmentation. It leverages the strengths of both CNNs and Transformers by combining the local feature extraction capabilities of CNNs with the global context modeling power of Transformers. Additionally, Valanarasu et al. [24] proposed the Gated Axial-Attention model (MedT), specifically designed to address the challenge of limited medical image data by incorporating attention mechanisms that are computationally more efficient and less data-intensive. Cao et al. [10] proposed Swin-Unet, the first pure Transformer-based U-shaped architecture for medical image segmentation. This model replaces traditional convolutional blocks with Swin Transformer layers, allowing for hierarchical and multiscale feature extraction while maintaining the U-Net's core encoder-decoder structure. However, the naive displacement of convolutional blocks with Swin Transformer blocks in both encoder and decoder paths may lead to structural redundancy and excessive computational overhead, without fully capitalizing on the strengths of Transformer encoding capability. Moreover, the impact of integrating Transformer blocks into different paths of the U-Net for medical image segmentation remains largely underexplored. Limited research has investigated how these architectural modifications affect key performance metrics, such as segmentation accuracy and computational efficiency.

3 Proposed Method

This study employs the Swin Transformer and simple ResBlocks to serve as dual branches of Encoder, and proposes a hierarchical feature aggregation module to server as the decoder for fusing various features learned in the Encoder. The overall framework is dubbed as hirachical feature aggregation network (HFANet), and the architecture is depicted in Fig. 1. Specifically, similar as the conventional U-Net, HFANet comprises three components: Encoder, decoder, and the skip connection bridging the interaction between the Encoder and Decoder. The Encoder path aims to incorporate Transformer blocks and convolution operation to extract both high-level semantic contexts and low-level detailed spatial structures. The first branch utilizes a Transformer architecture, initiating with window-based self-attention to model long-range dependencies and achieve

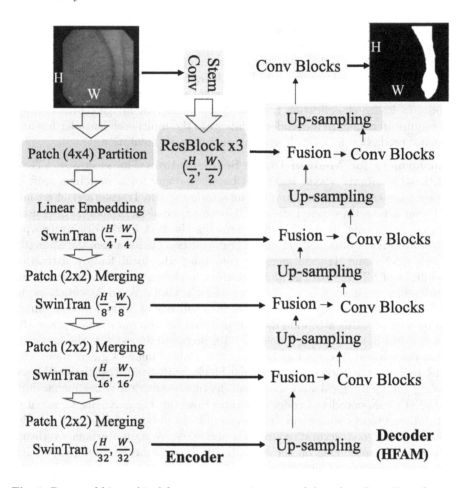

Fig. 1. Proposed hierarchical feature aggregation network based on Swin Transformer.

multi-scale semantic contexts. Concretely, we simply adopt the Swin Transformer proposed for generic vision task [15] to server as one branch of the Encoder, where the extracted feature even with the highest spatial resolution is reduced to $\frac{1}{4}$ of the input resolution, causing sptail detail lost for accurate segmentation. Thus, the second path employ the simple Resblocks taken from the stem convolution and the first stage of Resblock in the ResNet50 [9] to extract detailed spatial structure, which can provide the complementary information to the Transformer branch. Then, we transfer the multi-scale contexts from the Transformer branch and the detailed spatial structure from the Resblock branch to the Decoder path via skip connections, and investigate a hierarchical feature aggregation module (HFAM) as the decoder to fuse all features for final segmentation prediction. Particularly, the HFAM utilizes the simple operations such as the up-sampling, channel concatenation and convolutions to reduce the computational complexity. Next, we will give the detailed descriptions of the Encoder and Decoder of our proposed HFANet.

3.1 Encoder

The encoder architecture contains the main Swin Transformer branch and the complementary Resblock branch.

Transformer Branch: Transformer branch includes the 4 levels of Swin Transformer blocks, and each level has two blocks. The tokenized data with C-dimensional vectors and the reduced spatial resolution $\frac{1}{4}$ of the input data is firstly processed by two successive Swin Transformer blocks to extract the first level of semantic context $\mathbf{F}_1 \in \Re^{C \times W/4 \times H/4}$. These blocks facilitate representation learning while preserving both the dimensionality and spatial resolution of the features. Then, before pass $\mathbf{F}_1$ to the second level of Transformer block, a patch merging layer operates to downsample the input by a factor of 2, thereby reducing the number of tokens while simultaneously expanding the feature dimensionality to twice its original size as $\mathbf{F}_2 \in \Re^{2C \times W/8 \times H/8}$. This token reduction and feature enhancement process is iterated three times throughout the encoder, progressively refining the representation at each stage. Finally, we obtain the multi-scale semantic features as $\mathbf{F} = [\mathbf{F}_1, \mathbf{F}_2, \mathbf{F}_3, \mathbf{F}_4]$, all of which will be transferred to the Decoder for aggregation. Subsequently, we present the detailed explanations of the Swin Transformer block.

Each level of the Transformer branch contains two consecutive Swin transformer blocks. In contrast to the conventional multi-head self-attention (MSA) mechanism, the Swin Transformer block [15] is designed based on a shifted window paradigm, and comprises several components: a LayerNorm (LN) layer, a multi-head self-attention mechanism, a residual connection, and a two-layer multilayer perceptron (MLP) incorporating GELU activation. The first block leverages a window-based multi-head self-attention (WMSA) mechanism, while the second block employs a shifted window-based multi-head self-attention (SWMSA) mechanism. This successive Swin Transformer blocks with the above window partitioning strategy for $l-th$ level can be formulated as:

$$\hat{\mathbf{F}}_l^1 = WMSA(LN(\mathbf{F}_l)) + \mathbf{F}_l, \quad \mathbf{F}_l^1 = MLP(LN(\hat{\mathbf{F}}_l^1)) + \hat{\mathbf{F}}_l^1. \tag{1}$$

$$\hat{\mathbf{F}}_l^2 = SWMSA(LN(\mathbf{F}_l^1)) + \mathbf{F}_l^1, \quad \mathbf{F}_l^2 = MLP(LN(\hat{\mathbf{F}}_l^2)) + \hat{\mathbf{F}}_l^2, \tag{2}$$

where, $\hat{\mathbf{F}}_l^1$ and $\mathbf{F}_l^1$ denote the results of the WMSA and MLP module in the first block while $\hat{\mathbf{F}}_l^2$ and $\mathbf{F}_l^2$ refer to the output of the SWMSA and MLP module in the second block, respectively. The self-attention mechanism in the WMSA and SWMSA modules is calculated using the following equation:

$$Attention(\mathbf{Q}, \mathbf{K}, \mathbf{V}) = SoftMax(\frac{\mathbf{QK}^T}{\sqrt{d}})\mathbf{V}, \tag{3}$$

where, $\mathbf{Q}, \mathbf{K}, \mathbf{V} \in \Re^{S^2 \times d}$ refer to the query, key, and value matrices. d signifies the dimensionality of the query or key while S^2 represents the number of patches within a window.

Resblock Branch: Since the Transformer branch produce multi-scale features $\mathbf{F}$ even with the highest spatial resolution $W/4 \times H/4$, and thus may result in potential loss of detailed spatial structure. This study attempts to incorporate a simple convolution-based branch for compensating the lost structures in the Transformer branch. Specifically, we take the stem block with the Maxpool operation for reducing spatial decimation and the first layer of Resblock with three Bottleneck Residual structures to serve as the Resblock branch. This branch can produce the low-level feature $\mathbf{F}_0 \in \Re^{256 \times W/2 \times H/2}$ with more detailed spatial information, which is also skip connected to the decoder path for segmentation prediction.

3.2 Decoder

To effectively integrate the multi-scale features extracted by the encoder, the decoder implements a hierarchical fusion mechanism, dubbed as hierarchical feature aggregation module (HFAM), that replaces the computationally intensive Swin Transformer blocks with more efficient convolutional layers. This design choice not only enhances computational efficiency but also maintains the structural integrity of the feature representations. The HFAM involves an iterative process, where lower-resolution feature maps generated by the encoder are progressively upsampled by a factor of 2× and subsequently refined through convolutional operations. After each upsampling step, these refined features are fused with their corresponding encoder-derived feature maps that possess matching spatial dimensions. This hierarchical fusion strategy is executed recursively, enabling the gradual reconstruction of features at progressively higher resolutions. The process continues until the reconstructed feature map matches the input image's spatial resolution, thereby ensuring fidelity between the output and the original input in terms of both scale and detail. Specifically, given the lowest spatial resolution feature $\mathbf{F}_4$ extracted in the Encoder, we first employ a simple up-sampling operation and a point-wise convolution to double the spatial size and half the channel number, respectively. Then after concatenating with the feature $\mathbf{F}_3$ of the up one level, we further adopt two convolution layer to refine the fused feature. The above process can be formulated as:

$$\bar{\mathbf{F}}_3 = f_{conv}([\mathbf{F}_3, f_{up-Pw}(\mathbf{F}_4)]) \tag{4}$$

Then, $\bar{\mathbf{F}}_3$ follows the similar procedure as $\mathbf{F}_4$ for aggregating with $\mathbf{F}_2$. Finally, we achieve the fused feature $\bar{\mathbf{F}}_0$ by hierarchically aggregating all Encoder features, which is further up-sampled to the spatial resolution for producing segmentation output using a convolution block as the prediction head.

Table 1. Comparison with the state-of-the-art models.

Models	ClinicDB	BUSI	GLaS
U-Net [19]	90.66	72.27	87.99
MultiResUNet [12]	88.20	72.43	88.34
Swin-Unet [10]	90.69	76.06	86.45
UCTransnet [26]	**92.57**	75.56	87.17
SMESwin-Unet [29]	89.62	73.94	83.72
HFANet (Ours)	92.39	**84.33**	**91.70**

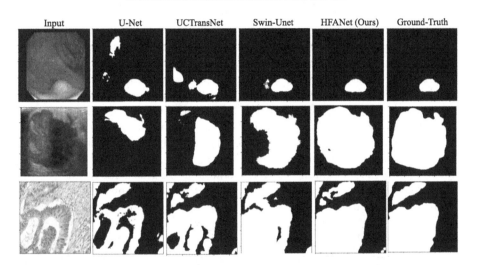

Fig. 2. Comparative qualitative results

4 Experiments

4.1 Datasets

To assess the performance of the proposed HFANet, we conducted a series of experiments utilizing three publicly available datasets, each representing diverse tasks and imaging modalities. The datasets employed in the evaluation BUSI [3], which includes 437 benign and 210 malignant breast ultrasound images, similar to those used in [13]; CVC-ClinicDB [4], a colonoscopy dataset with 612 images; and GlaS [20], for gland segmentation, consisting of 85 training and 80 test images. To ensure consistency, all images and their corresponding segmentation masks were resized to 224×224 pixels. For the GlaS dataset, we adhered to the predefined test split to evaluate the model. In contrast, for the remaining datasets, we randomly allocated 20% of the images for testing purposes. The remaining data was divided into 60% for training and 20% for validation.

Table 2. Ablation Study.

Encoder2	Decoder	ClinicDB	BUSI	GLaS
ResNet34	Swin	91.99	79.99	91.12
ResNet50	Swin	92.01	81.71	91.68
×	HFAM	92.15	82.79	**91.74**
ResNet50	HFAM	**92.39**	**84.33**	91.70

4.2 Implementation Details

The HFANet was implemented using PyTorch framework. To enhance data variability and improve model generalization, several data augmentation techniques, including horizontal and vertical flips as well as random rotations, were applied to the training dataset. Training was conducted on an Nvidia Geforce RTX 4090 GPU with 24 GB of memory, and the model's Encoder parameters were initialized using pre-trained weights from ImageNet, leveraging transfer learning to accelerate convergence and improve performance, while the Decoder parameters were randomly initialized. The model optimization was performed using the Adam optimizer, initialized with a learning rate of 10^{-5}, and dynamically adjusted throughout training using a cosine annealing scheduler. The models were trained over a total of 1000 epochs with the training process incorporating an early stopping mechanism. Specifically, an early stopping patience of 100 epochs was employed, meaning that training was halted if no improvement in performance was observed over 100 consecutive epochs, thus preventing overfitting. To optimize the model, we minimized a hybrid loss function that combined cross-entropy loss and Dice loss, a strategy designed to balance pixel-wise classification accuracy with segmentation overlap quality.

4.3 Comparisons with State-of-the-Art Methods

We conducted a comprehensive evaluation of The HFANet by comparing it against five representative models from U-Net-based architectures: U-Net [19], MultiResUNet [12], Swin-Unet [10], UCTransnet [26], and SMESwin-Unet [29]. These models were selected to represent key variants within the UNet family. The compared results in terms of Dice score, are summarized in Table 1, which presents the performance across various test datasets. It can be observed from Table 1 that our proposed model outperforms competing methods, achieving the best segmentation accuracy for the BUSI and GLaS datasets, and the second rank for the ClinicDB dataset in terms of the Dice Similarity Coefficient (DSC). The improvement in DSC compared to existing methods, such as U-Net [19] and Swin-Unet [10], is marginal for the ClinicDB dataset while our model demonstrates a substantial gain for BUSI and GLaS datasets. Concretely, our approach achieves an improvement of approximately 12% over U-Net and 8% over Swin-Unet for the BUSI dataset. Finally, we provide the visualizations of the segmentation results with several representative models including U-Net, UCTarnsNet and Swin-Unet in Fig. 2, and have demonstrated that our proposed HFANet achieves much better segmentations.

4.4 Ablation Study

We conducted an ablation study to systematically evaluate the impact of the proposed Encoder and Decoder components on all three datasets. Aa introduced above that our HFANet contains two Encoders: Swin Transformer branch and ResBlock branch eaxtected from the stem and first layers of the pretrained ResNet with The ImageNet dataset (Denoted as Encoder2), and one Decoder with the proposed HFAM. To verify the effectiveness of the HFAM for feature aggregation, we also employed the symmetric Swin Transformer blocks by replacing the patch merging with upsampling operation as the Decoder component. For Encoder, we removed the Resblock branch or utilized the layer from the ResNet34 and ResNet50, respectively. The compared results are manifested in Table 2, and manifested that the proposed HFAM and the incorporation the Resblock branch can improve the segmentation performance.

5 Conclusions

This paper introduced a novel approach for improving the performance of ViT-based architectures in medical image segmentation, especially when data is limited. The core of our method lies in a hierarchical feature aggregation strategy built upon hierarchical Transformer features. Our framework employed the hierarchical Vision Transformer as the primary encoder to capture multi-scale semantic information, while residual blocks are incorporated to preserve fine spatial details and local representations. To combine these features effectively, we proposed a hierarchical feature aggregation module (HFAM) that acts as the decoder, seamlessly merging the multi-scale semantic and spatial features. Compared to traditional Transformer-based U-Net models, the HFAM not only enhances context fusion but also holds the potential to reduce computational demands. Extensive experiments across three medical image segmentation datasets showed that our method consistently surpasses both standard U-Net and other Transformer-based U-Net models in performance.

References

1. Ailiang, L., Xu, J., Jinxing, L., Guangming, L.: Contrans: improving transformer with convolutional attention for medical image segmentation. In: Medical Image Computing and Computer Assisted Intervention (MICCAI), pp. 297–307 (2022)
2. Chen, J., et al.: Transunet: Transformers make strong encoders for medical image segmentation. CoRR (2021)
3. Cicek, O., Abdulkadir, A., Lienkamp, S., Brox, T., Ronneberger, O.: 3d u-net: learning dense volumetric segmentation from sparse annotation. Med. Image Comput. Comput.-Assisted Interv. (MICCAI) **9901**, 424–432 (2016)

4. Devlin, J., Chang, M.W., Lee, K., Toutanova, K.: Bert: pre-training of deep bidirectional transformers for language understanding. In: The 2019 Conference of the North American Chapter of the Association for Computational Linguistics: Human Language Technologies **1** (2019)
5. Dosovitskiy, A., et al.: An image is worth 16x16 words: Transformers for image recognition at scale. In: International Conference on Learning Representations (2021)
6. Isensee, F., Jaeger, P.F., Kohl, S.A.A., Petersen, J., Maier-Hein, K.H.: Nnu-net: a self-conguring method for deep learning-based biomedical image segmentation. Nat. Methods **18**(2), 203–211 (2021)
7. Gu, Z., et al.: Ce-net: context encoder network for 2d medical image segmentation. IEEE Trans. Med. Imaging **38**(10), 2281–2292 (2019)
8. Han, K., Xiao, A., Wu, E., Guo, J., Xu, C., Wang, Y.: Transformer in transformer. CoRR (2021)
9. He, K., Zhang, X., Ren, S., Sun, J.: Deep residual learning for image recognition. arXiv preprint arXiv:1512.03385 (2015)
10. Hu, C., et al.: Swin-unet: Unet-like pure transformer for medical image segmentation. ECCV Computer Vision Workshop, pp. 205–218 (2023)
11. Huang, H., et al.: Unet 3+: a full-scale connected unet for medical image segmentation. In: ICASSP, pp. 1055–1059 (2020)
12. Ibtehaz, N., Rahman, M.S.: Rethinking the u-net architecture for multimodal biomedical image segmentation. Neural Netw. **121** (2020)
13. Jin, Q., Meng, Z., Sun, C., Cui, H., Su, R.: Ra-unet: a hybrid deep attentionaware network to extract liver and tumor in ct scans. Front. Bioeng. Biotechnol. **8**, 1471 (2020)
14. Li, X., Chen, H., Qi, X., Dou, Q., Fu, C.W., Heng, P.A.: H-denseunet: hybrid densely connected unet for liver and tumor segmentation from ct volumes. IEEE Trans. Med. Imaging **37**(12), 2663–2674 (2018)
15. Liu, Z., et al.: Swin transformer: Hierarchical vision transformer using shifted windows. CoRR(2021)
16. Milletari, F., Navab, N., Ahmadi, S.A.: V-net: fully convolutional neural networks for volumetric medical image segmentation. In: Fourth International Conference on 3D Vision (3DV), pp. 565–571 (2016)
17. Naik, S., Doyle, S., Agner, S., Madabhushi, A., Feldman, M., Tomaszewski, J.: Automated gland and nuclei segmentation for grading of prostate and breast cancer histopathology. In:5th IEEE International Symposium in Biomedical Imaging: From Nano to Macro, pp. 284–287 (2008)
18. abd R. Girshick, X.W., Gupta, A., He, K.: Non-local neural networks. In: IEEE Conference on Computer Vision and Pattern Recognition, pp. 7794–7803 (2018)
19. Ronneberger, O., Fischer, P., Brox, T.: U-Net: convolutional networks for biomedical image segmentation. In: Navab, N., Hornegger, J., Wells, W.M., Frangi, A.F. (eds.) MICCAI 2015. LNCS, vol. 9351, pp. 234–241. Springer, Cham (2015). https://doi.org/10.1007/978-3-319-24574-4_28
20. Rouhi, R., Jafari, M., Kasaei, S., Keshavarzian, P.: Benign and malignant breast tumors classication based on region growing and cnn segmentation. Expert Syst. Appli. **42**(3), 990–1002 (2015)
21. Schindelin, J., Rueden, C.T., Hiner, M.C., Eliceiri, K.W.: The imagej ecosystem: an open platform for biomedical image analysis. Mol. Reprod. Dev. **82**, 518–529 (2015)
22. Schlemper, J., et al.: Attention gated networks: learning to leverage salient regions in medical images. Med. Image Anal. **53**, 197–207 (2019)

23. Touvron, H., Cord, M., Douze, M., Massa, F., Sablayrolles, A., Jegou, H.:Training data-efficient image transformers & distillation through attention. CoRR (2020)
24. Valanarasu, J.M.J., Oza, P., Hacihaliloglu, I., Patel, V.M.: Medical transformer: gated axial-attention for medical image segmentation. CoRR (2021)
25. Vaswani, A., et al.: Attention is all you need. Adv. Neural Inform. Process. Syst. **30** (2017)
26. Wang, H., Cao, P., Wang, J., Zaiane, O.: Uctransnet: rethinking the skip connections in u-net from a channel-wise perspective with transformer. In: AAAI Conference on Artificial Intelligence vol. 36(3), 2441–2449 (2022)
27. Wang, W., Chen, C., Ding, M., Li, J., Yu, H., Zha, S.: Transbts: Multimodal brain tumor segmentation using transformer. CoRR (2021)
28. Wang, W., et al.: Pyramid vision transformer: A versatile backbone for dense prediction without convolutions. CoRR (2021)
29. Wang, Z., Min, X., Shi, F., Jin, R., Nawrin, S., Yu, I., Nagatomi, R.: Smeswin unet: merging cnn and transformer for medical image segmentation. In: Medical Image Computing and Computer Assisted Intervention (MICCAI), pp. 517–526 (2022)
30. Xiao, X., Lian, S., Luo, Z., Li, S.: Weighted res-unet for high-quality retina vessel segmentation. In: 9th International Conference on Information Technology in Medicine and Education (ITME), pp. 327–331 (2018)
31. Zhang, Y., Liu, H., Hu, Q.: Transfuse: fusing transformers and cnns for medical image segmentation. CoRR (2021)
32. Zhou, Z., Rahman Siddiquee, M.M., Tajbakhsh, N., Liang, J.: UNet++: a nested U-Net architecture for medical image segmentation. In: Stoyanov, D., et al. (eds.) DLMIA/ML-CDS -2018. LNCS, vol. 11045, pp. 3–11. Springer, Cham (2018). https://doi.org/10.1007/978-3-030-00889-5_1

E2CANet: An Efficient and Effective Convolutional Attention Network for Semantic Segmentation

Yuerong Mu and Qiang Guo(✉)

School of Computer Science and Technology, Shandong University of Finance and Economics, Jinan, China
guoqiang@sdufe.edu.cn

Abstract. Many semantic segmentation methods employ various attention mechanisms to improve segmentation accuracy. However, as the accuracy of the model increases, the computational cost is relatively expensive, which is not favorable for some practical applications. To solve this problem, this paper presents an efficient and effective convolutional attention network (E2CANet), which is designed to achieve a good trade-off between segmentation accuracy and computational efficiency. E2CANet adopts an encoder-decoder architecture with skip connections to preserve details and semantic information. For the encoder, we use cheap convolutional operations to introduce two different attentions, i.e. global attention and multi-scale attention, which can significantly reduce the computational cost while highlighting important features and suppressing unnecessary ones. A lightweight All-MLP decoder, which only consists of six linear layers, is used to aggregate features from the encoder. The simple design of this decoder is also the key to reduce computational complexity. Extensive experiments are performed on ADE20K, Cityscapes, and COCO-stuff datasets. The proposed E2CANet delivers very competitive results on all datasets. Especially, E2CANet-Tiny (a lightweight version of E2CANet) achieves 41.92% mIoU on ADE20K dataset with less than 4.4M parameters, which demonstrates the efficiency and effectiveness of our method. Code is available at https://github.com/muyuerong/E2CANet.

Keywords: Semantic segmentation · Multi-scale attention · Global attention · Feature extraction · Lightweight All-MLP decoder

1 Introduction

Semantic segmentation is an important task of computer vision, aiming at labelling each pixel in an image as a corresponding semantic category. Different from image classification that identifies the categories of the whole image, semantic segmentation classifies each pixel in the image. It has a wide range of applications such as medical image segmentation [29], autonomous driving [37], saliency detection [19], human-machine interaction [25], and many other fields.

In recent years, attention mechanisms are introduced into semantic segmentation to focus on regions or features of interest. The attention mechanisms assign different weights to different input data for concentrating on the most relevant parts. Using attention mechanisms can help the segmentation model to better understand the input image and thus improve segmentation accuracy. In semantic segmentation, common attention mechanisms include spatial attention and channel attention. Channel attention pays attention to 'what' is meaningful given an input image. Different from channel attention, spatial attention focuses on 'where' is an informative part. Based on the above two attentions, some attention modules are designed to further improve segmentation accuracy. Large kernel attention (LKA) was proposed in VAN [14] to build channel attention and spatial attention. It combines the long-range dependence of self-attention and the advantage of large kernel convolution to make full use of contextual information. In RepLKNet [10], it can be found that using large kernel convolutions can significantly improve the effective receptive fields compared to increase the number of layers with smaller kernel size, and can even leverage more shape information. However, traditional large kernel convolutions are computationally expensive. To tackle this issue, Guo et al. [13] designed a multi-scale convolutional attention module, which introduces large kernels by using multi-branch lightweight strip convolutions. Besides, Woo et al. [36] proposed a convolutional block attention module, which sequentially applies channel attention and spatial attention. The module learns what and where to emphasize or suppress and refines intermediate features effectively.

Inspired by the impact of attention mechanisms on segmentation accuracy improvement, we propose an efficient and effective convolutional attention network (E2CANet) to achieve high segmentation accuracy and low computational cost. Our network adopts an encoder-decoder architecture with skip connections. The encoder uses a common hierarchical structure with four stages. Each stage contains a novel convolutional attention block, which introduces global attention and multi-scale attention by using cheap convolution operations. The former employs global average pooling and max pooling to obtain global context information, and the latter uses strip convolutions at different scales to capture multi-scale features. E2CANet sequentially utilizes these two attentions to focus on useful features and suppress useless ones. To further improve the efficiency of E2CANet, a lightweight All-MLP decoder is introduced, which aggregates the features from different stages and obtains the final segmentation results. Such an encoder-decoder design can allow our network with a good trade-off between segmentation accuracy and computational cost. Extensive experimental results on three public datasets demonstrate the advantages of E2CANet in terms of segmentation performance and the number of parameters.

In summary, our contributions are as follows:

- To increase the segmentation accuracy, we propose a convolutional attention block (CA block) that is composed of global attention and multi-scale attention. Based on the CA block, E2CANet is designed for tackling the task of semantic segmentation.

- To obtain a powerful representation and make E2CANet more efficient, we use a lightweight decoder without computationally complex modules to aggregate features of four stages.
- We conduct extensive experiments on the ADE20K, Cityscapes, and COCO-stuff datasets to validate the high segmentation accuracy and low computational complexity of the proposed E2CANet.

2 Related Work

In this section, we briefly review some common methods for extracting global and multi-scale features in semantic segmentation, as well as some representative attention mechanisms.

2.1 Global Feature Extraction

Global features can help semantic segmentation models better understand the overall context. FCN [24] is a significant advancement in the field of semantic segmentation, which defines a skip connection to combine deep semantic information with shallow detail information. However, it ignores the global information of an image. As a variant of FCN, ParseNet [23] obtains global features by using global average pooling. Besides, a context encoding module is introduced in EncNet [43] to capture global context information, which can significantly improve segmentation results with only a small additional computational cost compared to FCN. Unlike the above three networks, SENet [18] applies the global context to recalibrate the weights of different channels, but the global context information is not fully utilized. To obtain the global features, Wang et al. [34] proposed NLNet to model the long-range dependencies by using self-attention mechanism. It first computes the pairwise relations between the query and all positions to form the attention map, and then aggregates the features of all positions by weighted sum. However, global context information modeled by NLNet are almost the same for different positions within an image, which results in waste of calculations. Inspired by the global context modeling capabilities of NLNet, Cao et al. [4] designed a lightweight global context block in GCNet, which is able to efficiently models global context information.

2.2 Multi-scale Feature Extraction

Multi-scale features play an important role in the field of computer vision and image processing. To capture multi-scale contextual information, PSPNet [44] uses the pyramid pooling module to gather multi-scale information. Different from PSPNet, DeepLabV3 [6] designs the multi-scale dilated convolution to extract features at different scales. Both PSPNet and DeepLabV3 apply $n \times n$ square convolutions to extract features at different scales. However, using square convolutions, especially the kernel size being larger than seven, introduces a significantly increase in computational complexity. To solve this issue, Szegedy et al.

[31] used $1\times n$ and $n\times 1$ strip convolutions in parallel instead of $n\times n$ convolutions to reduce the number of parameters. Besides, HRNet [33] extracts multi-scale features by connecting high-to-low resolution convolutions in parallel. The convolutional multi-scale fusion module is utilized to integrate multi-scale feature hierarchies in HRFormer [42]. To capture different scale semantic dependencies, Jiao et al. [22] proposed a multi-scale dilated attention, which uses different dilation rates for different heads.

2.3 Attention Mechanism

In recent years, various attention mechanisms are used to adaptively select important features, resulting in the improvement of accuracy and efficiency [32] [15] [16]. Channel attention and spatial attention are employed in semantic segmentation to focus on the important information in the channel dimension and spatial dimension, respectively. A representative model of the channel attention is SENet, which designs a "Squeeze-and-Excitation" block (SE block) to adaptively recalibrate channel-wise feature responses by explicitly modeling interdependencies between channels. However, SE block adopts global average pooling to capture global information, which limits the modeling capability. To address this issue, a global second-order pooling block [12] was proposed to gather global features. Moreover, spatial attention was first presented in STN [21] to pay attention to the most relevant regions. A learnable spatial transformer module is introduced in STN, which can be inserted into existing convolutional architectures, performing spatial transformation on the input images. Woo et al. [36] proposed a convolutional block attention module (CBAM), which performs global average pooling and max pooling to adaptively learn channel and spatial attention weights. The experimental results in CBAM showed that combining channel and spatial attentions outperforms using only one attention independently.

3 Method

This section introduces the efficient and effective convolutional attention network for semantic segmentation. As illustrated in Fig. 1, E2CANet mainly contains two parts: (i) a convolutional attention encoder that is designed to better extract global features and multi-scale features; (ii) a lightweight All-MLP decoder that aims to aggregate features from the encoder and obtain the semantic segmentation mask. In the rest of this section, we detail the proposed encoder and decoder designs.

3.1 Convolutional Attention Encoder

The encoder of E2CANet adopts a common hierarchical structure to extract feature maps with high-level semantic information. A simple stem block is applied for capturing low-level features, followed by four stages to further extract global and multi-scale features as shown in Fig. 2. The first stage is composed of a

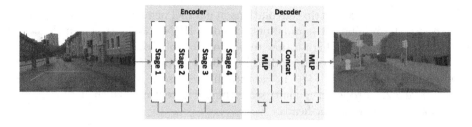

Fig. 1. The overall architecture of the proposed E2CANet

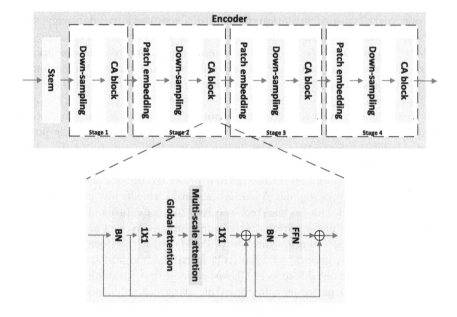

Fig. 2. The architecture of the proposed encoder. It is composed of a stem block, patch embedding blocks, down-sampling blocks, and convolutional attention blocks.

down-sampling block and a stack of convolutional attention blocks. The other three stages have a patch embedding block before each down-sampling block. The stem block stakes two 3 × 3 convolution layers. Each layer has a batch normalization, and a GELU activation is applied only for the first layer. The patch embedding block, containing a 3 × 3 convolution and a BN layer, is adopted to down-sample the spatial dimension by 4× and increase the channel dimension by 2×. As for the down-sampling block, we utilize a convolution with stride 2 and kernel size 3 × 3, followed by a batch normalization layer. To meet the needs of devices with less memory and less computational resources, we design a lightweight version named E2CANet-Tiny. Compared to E2CANet, it reduces the number of parameters by lowing the channel numbers and changing the number of building blocks. Detailed settings of E2CANet and E2CANet-Tiny are listed in Table 1.

Table 1. Detailed settings of different versions of the proposed E2CANet. C and L represent the numbers of channels and building blocks, respectively.

stage	E2CANet	E2CANet-Tiny
1	$C=64, L=2$	$C=32, L=3$
2	$C=128, L=2$	$C=64, L=3$
3	$C=320, L=4$	$C=160, L=5$
4	$C=512, L=2$	$C=256, L=2$

Convolutional Attention Block. The structural design of CA block is inspired by the transformer encoder in ViT. It is mainly composed of global attention, multi-scale attention, skip connections, and feed-forward network (FFN), which is shown in the bottom of Fig. 2. Skip connections preserve the details and semantic information of the input data, which can avoid detail information loss and improve the training speed. FFN is used to further extract features based on global attention and multi-scale attention, and enhance the ability of representation of our model. It consists of two 1×1 convolutions and a 3×3 convolution. Global attention and multi-scale attention are the main components of the CA block. In the following, we will detail these two attentions.

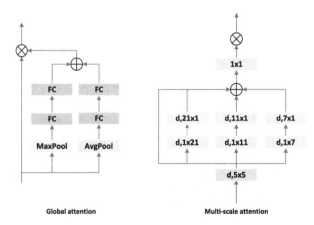

Fig. 3. The structures of global attention and multi-scale attention

In global attention, global average pooling and max pooling are commonly adopted to aggregate global information. Global average pooling computes the average of all pixel values in the feature maps, compressing the original two-dimensional feature maps into a one-dimensional feature vector. It can significantly reduce the number of parameters in the model while retaining the global information in feature maps. Different from global average pooling, global max pooling extracts the maximum value of all pixels, highlighting the most important features. Empirically, exploiting both types of pooling, which can extract richer high-level features, is more effective than using each independently. Therefore, we use global average pooling and max pooling simultaneously to obtain average-pooled and max-pooled features.

Fig. 3 (left) shows the structure of the global attention, which first performs the global average pooling and global max pooling on the input feature map to generate average-pooled and max-pooled features. Then features are forwarded to two fully connected layers to further extract the global context information. Finally, we utilize the element-wise summation to merge the attention map that is used as weights to reweigh the input. The global attention can be formulated as:

$$M_g(F) = MLP(GAP(F)) + MLP(MAX(F)),$$
$$F_g = M_g(F) \otimes F,$$

where F is the input feature map, GAP and MAX denote the global average pooling and max pooling, respectively. M_g is the global attention map, and F_g is the output of global attention. $\otimes$ represents the element-wise multiplication.

In multi-scale attention, a multi-scale convolutional architecture is designed to extract features at different scales, which is depicted in Fig. 3 (right). To aggregate local information, a 5 × 5 depth-wise convolution is used, followed by four branches. One of these branches is an identity connection. The other three branches employ large kernel convolutions with different sizes to capture features at different scales. Using large kernel convolutions can improve the segmentation accuracy due to the reason that it can enlarge the receptive fields. However, traditional $n \times n$ large kernel convolutions are costly, especially n is larger than seven. To tackle this issue, we utilize a depth-wise $1 \times n$ convolution and a $n \times 1$ depth-wise convolution to approximate the $n \times n$ depth-wise convolutions. Specifically, the kernel sizes of other three branches are set to 7, 11, and 21, respectively. Then, we obtain the attention weights by using the 1×1 convolution. The multi-scale attention weights are computed as:

$$F' = Conv_{5\times5}(F),$$
$$Conv_{n_i}(F) = Conv_{i\times1}(Conv_{1\times i}(F)),$$
$$M_m(F') = Conv_{1\times1}\big(F' + Conv_{n_7}(F') + Conv_{n_{11}}(F') + Conv_{n_{21}}(F')\big),$$
$$F_m = M_m(F') \otimes F,$$

where $Conv_{1\times1}$ and $Conv_{5\times5}$ represent depth-wise convolutions with kernel sizes of 1 and 5, $Conv_{n_i}$ and i denote strip convolutions and kernel sizes of strip convolutions, respectively. M_m is the multi-scale attention map, and F_m is the final output.

3.2 Lightweight All-MLP Decoder

The decoder is responsible for mapping the features extracted by the encoder to generate the segmentation mask. As shown in Fig. 1, we adopt a lightweight All-MLP decoder, which only consists of six linear layers. It first upsamples the features of different stages of the encoder to the same size as Stage 1. Then, a MLP layer is adopted to fuse the features and another MLP layer takes the fused feature to predict the segmentation masks. This lightweight decoder allows E2CANet to achieve a good trade-off between segmentation accuracy and computational cost.

4 Experiments

In this section, we evaluate the proposed E2CANet on ADE20K, Cityscapes, and COCO-stuff. We first summarize the datasets and the implementation details. Then, the contributions of each component are investigated in ablation studies on ADE20K. Finally, to verify the efficiency and effectiveness of E2CANet, we compare it with some state-of-the-art segmentation methods.

Datasets. ADE20K [45] is a large-scale dataset for scene parsing with more than 20, 000 images and 150 semantic tags of different categories, which covers a wide range of different scenes from indoor to outdoor, nature to urban. Each image has been labeled in detail to semantically classify each pixel in the image.

Cityscapes [9] is a semantic understanding image dataset on urban street scenes. It mainly contains street scenes from 50 different cities, with 5, 000 high-quality pixel-level annotated images of driving scenes in urban environments (2, 975 for train, 500 for validation, 1, 525 for test, with 19 categories) and 20, 000 roughly annotated images.

COCO-stuff [2], a large-scale dataset for scene understanding, is an extension of the COCO (Common Object in Context) dataset, which annotates new categories from images in the COCO dataset in order to provide a more comprehensive understanding of the scenes in the image. This dataset comprises over 200, 000 images and is labeled with 80 different object categories and 91 different pixel-level scene categories in each image. These pixel-level scene categories are marked as segmentation masks that can be used to train and evaluate semantic segmentation models.

We implement E2CANet based on MMSegmentation [8], which is an open source semantic segmentation toolbox. All the experiments are performed on the Pytorch [28] platform with a V100 GPU. The batch size is set to 8 for Cityscapes and 16 for other datasets. The iteration number is set to 160K for ADE20K and Cityscapes, and 80K for COCO-stuff. For all experiments, we use mean Intersection over Union (mIoU) and the number of parameters to serve as the evaluation metrics. For optimization, a learning rate of 0.01, a momentum of 0.9, and a weight decay of 0.5×10^{-3} are used in training.

4.1 Ablation Study

In this subsection, we conduct extensive ablation studies of our E2CANet on ADE20K to perform detail analysis of our proposed method in three aspects. We evaluate the benefits of using both global average pooling and max pooling, and examine the influence of different scale branches. Then, we verify the effectiveness of using global attention and multi-scale attention sequentially.

Selections of Pooling Methods in Global Attention. Table 2 shows the results of using three variants of the pooling methods in global attention, which contains global average pooling, global max pooling, and joint use of both poolings. From Table 2, it can be observed that using global average pooling performs slightly better than max pooling. We can also find using both types of pooling can improve mIoU by 1.2% compared to only using average pooling with no increase in the number of parameters. Therefore, we use both global average pooling and max pooling in E2CANet.

Table 2. Comparisons of different pooling methods used in global attention.

Methods	param (M)	mIoU (%)
AvgPool	4.4	40.7
MaxPool	4.4	40.6
MaxPool&AvgPool	4.4	41.9

Ablations on Multi-scale Attention Design. We perform ablation studies to verify the importance of three different scale branches and 1×1 convolution. $n \times n$ branch is composed of a depth-wise $1 \times n$ convolution and a $n \times 1$ depth wise convolution. In Table 3, it can be found that using three different scale branches is more efficient than using any two of them. The segmentation results are further improved by adding 1×1 convolution for channel mixing. The results demonstrate that branches at different scales and the 1×1 convolution are important for the final performance.

Usages of the Global and Multi-scale Attentions. In this ablation study, we compare four different ways of using the multi-scale (m.s.) attention and global (glo.) attention, i.e. m.s. attention only, glo. attention only, m.s.-glo. attention, and glo.-m.s. attention. The main difference between m.s.-glo. attention and glo.-m.s. attention is the attention order. The results are shown in Table 4. It can be observed that only using m.s. attention gains 1.2% mIoU over glo. attention. We can also find that using both attentions sequentially performs better than using them individually. This proves that both attentions play an important role in E2CANet. As each attention plays different roles, the order may affect the

Table 3. Ablation studies on multi-scale attention design. Br: branch.

7 × 7 Br	11 × 11 Br	21 × 21 Br	1 × 1 Conv	mIoU (%)
✗	✔	✔	✔	40.2
✔	✗	✔	✔	40.4
✔	✔	✗	✔	40.7
✔	✔	✔	✗	23.3
✔	✔	✔	✔	41.9

Table 4. Usages of the global and multi-scale attention.

Methods	mIoU (%)
m.s. attention	40.3
glo. attention	39.1
m.s.-glo. attention	40.9
glo.-m.s. attention	41.9

segmentation accuracy. In Table 4, the last two rows show that glo.-m.s. attention performs better than m.s.-glo. attention. Therefore, glo.-m.s. attention is adopted in our E2CANet.

4.2 Comparisons with State-of-the-Art Methods

In this subsection, to verify the efficiency and effectiveness of the proposed E2CANet, we compare it with several recent transformer-based segmentation methods include HRFormer [42], EfficientViT [3], Mask2Former [7], and SegFormer [39] and CNN-based state-of-the-art segmentation methods include ENet [27], SegNet [1], CGNet [38], MoblieNetV2 [30], BiSeNetV2 [41], PIDNet [40], DDRNet [26], RegSeg [11], and PSPNet [44] on ADE20K, Cityscapes, and COCO-Stuff datasets.

Comparisons with Transformer-Based Methods. We compare E2CANet-Tiny and E2CANet with the state-of-the-art transformer-based methods, and the results are shown in Table 5. SegFormer-B0 and EfficientViT-B0 are the lightweight models of SegFormer and EfficientViT for fast inference, and the number of parameters of E2CANet-Tiny are similar to them. From Table 5, we can find that E2CANet-Tiny gains 4.5% mIoU over SegFormer-B0 with the similar GFLOPs (floating point operations) on ADE20K and 3.7% mIoU over EfficientViT-B0 on Cityscapes. HRFormer-S and SegFormer-B1 are both the small scaled versions of HRFormer and SegFormer, which have the similar number of parameters to E2CANet. It can be also seen that E2CANet yields 2.7% mIoU improvement compared to HRFormer-S, and the GFLOPs of E2CANet are seven times smaller than the latter on COCO-Stuff. Besides, our proposed method surpasses SegFormer-B1 by 2.3% mIoU and EfficientViT-B1 by 0.5% with less computational cost on Cityscapes. Although the mIoU of Mask2Former is 2.8% higher than E2CANet on ADE20K, the number of parameters and GFLOPs of our method are far less. We visualize some segmentation results on Cityscapes dateset in Figs. 4 and 5. We observe that E2CANet-Tiny obtains finer segmentation results than SegFormer-B0 in Fig. 4. As shown in Fig. 5, E2CANet achieves improved results compared to SegFormer-B1. Moreover, it yields competitive performance in comparison to HRFormer-S, and slightly better than the latter in some details. Compared to E2CANet-Tiny, E2CANet achieves higher segmentation performance, which demonstrate that as the number of parameters increase, the segmentation accuracy increases.

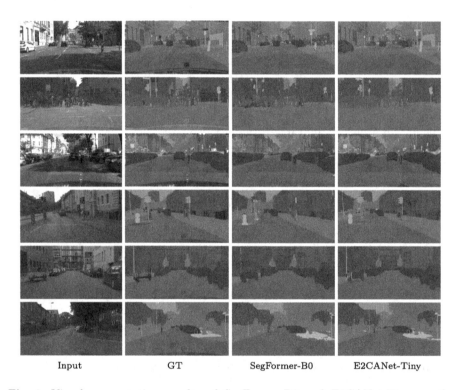

Input   GT   SegFormer-B0   E2CANet-Tiny

Fig. 4. Visual segmentation results of SegFormer-B0 and E2CANet-Tiny on the Cityscapes dataset

Comparisons with CNN-Based Methods. We further report the comparison results of the E2CANet-Tiny against to other CNN-based methods on Cityscapes. Table 6 shows that ENet and SegNet obtain the worst segmentation results. The mIoU of E2CANet-Tiny is about 9.2% higher than that of MobileNetV2 with a little increase of the number of the parameters. It can

Table 5. Comparisons with state-of-the-art transformer-based methods on ADE20K, COCO-stuff, and Cityscapes datasets.

Methods	params (M)	ADE20K		COCO-stuff		Cityscapes	
		GFLOPs	mIoU (%)	GFLOPs	mIoU (%)	GFLOPs	mIoU (%)
SegFormer-B0 [39]	3.8	8.4	37.4	8.4	35.6	125.5	76.2
EfficientViT-B0 [3]	7	-	-	-	-	-	75.7
E2CANet-Tiny	4.4	7.9	41.9	7.9	36.5	75.1	79.4
HRFormer-S [42]	13.5	109.5	44.0	109.5	37.9	835.7	80.0
SegFormer-B1 [39]	13.7	15.9	42.2	15.9	40.2	243.7	78.5
EfficientViT-B1 [3]	48	-	-	-	-	-	80.3
Mask2Former [7]	44	70.1	47.2	-	-	52.3	79.4
E2CANet	14.1	15.2	44.4	15.2	40.6	152.0	80.8

Table 6. Comparisons with state-of-the-art CNN-based methods on Cityscapes.

Methods	Params (M)	GFLOPs	mIoU (%)
ENet [27]	0.4	45.36	58.3
SegNet [1]	29.5	3.8	58.3
CGNet [38]	0.5	6	64.8
MoblieNetV2 [30]	2.1	9.1	70.2
BiSeNetV2 [41]	-	21.1	72.6
PSPNet [44]	65.6	286	78.4
PIDNet-S [40]	73.6	47.6	78.6
DDRNet-23-Slim [26]	5.7	36.3	77.4
RegSeg [11]	3.34	39.1	78.3
PEM-STDC1 [5]	17	92	78.3
E2CANet-Tiny	4.4	75.1	79.4

be found that our network yields 1% mIoU improvement compared to PSPNet with a dramatically decrease of the number of parameters. We also compare our method to three recent real-time semantic segmentation methods, e.g., PIDNet, DDRNet, and RegSeg. It can be find that our method achieves very competitive performance against these three methods. The results in Table 6 verify the efficiency and effectiveness of the E2CANet-Tiny.

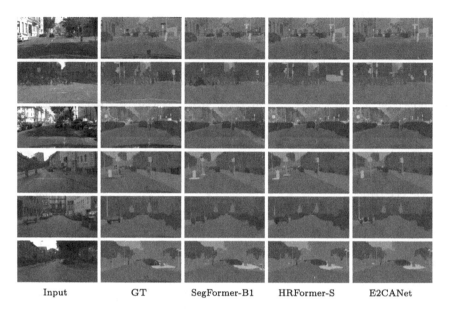

Fig. 5. Visual segmentation results of SegFormer-B1, HRFormer-S, and E2CANet on the Cityscapes dataset

5 Conclusion

In this paper, we present an efficient and effective convolutional attention network for semantic segmentation. The model introduces global attention and multi-scale attention by using cheap convolutional operations to highlight useful features and suppress unnecessary ones. Moreover, a lightweight All-MLP decoder is used to further reduce the computational complexity, which allows for a good trade-off between segmentation accuracy and computational cost. Extensive experiments on three commonly used datasets show that the proposed model achieves very competitive performance compares to some state-of-the-art methods.

In the future, we will explore the low rank approximation strategy as used in [35] [17] [20] to further improve the inference speed and reduce the number of parameters.

Acknowledgments.. This work was supported in part by the National Natural Science Foundation of China (61873145), in part by the Natural Science Foundation of Shandong Province for Excellent Young Scholars (ZR2017JL029), and in part by the Science and Technology Innovation Program for Distinguished Young Scholars of Shandong Province Higher Education Institutions (2019KJN045).

References

1. Badrinarayanan, V., Kendall, A., Cipolla, R.: Segnet: a deep convolutional encoder-decoder architecture for image segmentation. IEEE Trans. Pattern Analy. Mach. Intell. **39**(12), 2481–2495 (2017)
2. Caesar, H., Uijlings, J., Ferrari, V.: Coco-stuff: thing and stuff classes in context. In: Proceedings of the IEEE/CVF Conference on Computer Vision and Pattern Recognition, pp. 1209–1218 (2018)
3. Cai, H., Li, J., Hu, M., Gan, C., Han, S.: Efficientvit: lightweight multi-scale attention for high-resolution dense prediction. In: Proceedings of the IEEE/CVF International Conference on Computer Vision, pp. 17302–17313 (2023)
4. Cao, Y., Xu, J., Lin, S., Wei, F., Hu, H.: Gcnet: non-local networks meet squeeze-excitation networks and beyond. In: Proceedings of the IEEE/CVF International Conference on Computer Vision Workshops (2019)
5. Cavagnero, N., Rosi, G., Cuttano, C., Pistilli, F., Ciccone, M., Averta, G., Cermelli, F.: Pem: Prototype-based efficient maskformer for image segmentation. In: Proceedings of the IEEE/CVF Conference on Computer Vision and Pattern Recognition, pp. 15804–15813 (2024)
6. Chen, L.C., Papandreou, G., Schroff, F., Adam, H.: Rethinking atrous convolution for semantic image segmentation. In: Proceedings of the IEEE/CVF Conference on Computer Vision and Pattern Recognition (2017)
7. Cheng, B., Misra, I., Schwing, A.G., Kirillov, A., Girdhar, R.: Masked-attention mask transformer for universal image segmentation. In: Proceedings of the IEEE/CVF Conference on Computer Vision and Pattern Recognition, pp. 1290–1299 (2022)
8. Contributors M: MMSegmentation: Openmmlab semantic segmentation toolbox and benchmark. https://github.com/open-mmlab/mmsegmentation (2020)
9. Cordts, M., et al.: The cityscapes dataset for semantic urban scene understanding. In: Proceedings of the IEEE/CVF Conference on Computer Vision and Pattern Recognition, pp. 3213–3223 (2016)
10. Ding, X., Zhang, X., Han, J., Ding, G.: Scaling up your kernels to 31x31: revisiting large kernel design in cnns. In: Proceedings of the IEEE/CVF Conference on Computer Vision and Pattern Recognition, pp. 11963–11975 (2022)
11. Gao, R.: Rethinking dilated convolution for real-time semantic segmentation. In: Proceedings of the IEEE/CVF Conference on Computer Vision and Pattern Recognition, pp. 4675–4684 (2023)
12. Gao, Z., Xie, J., Wang, Q., Li, P.: Global second-order pooling convolutional networks. In: Proceedings of the IEEE/CVF Conference on Computer Vision and Pattern Recognition, pp. 3024–3033 (2019)
13. Guo, M.H., Lu, C.Z., Hou, Q., Liu, Z., Cheng, M.M., Hu, S.M.: Segnext: rethinking convolutional attention design for semantic segmentation. Adv. Neural Inform. Process. Syst. **35**, 1140–1156 (2022)
14. Guo, M.H., Lu, C.Z., Liu, Z.N., Cheng, M.M., Hu, S.M.: Visual attention network. Comput. Vis. Media **9**(4), 733–752 (2023)
15. Guo, M.-H., et al.: Attention mechanisms in computer vision: a survey. Comput. Vis. Media, 1–38 (2022). https://doi.org/10.1007/s41095-022-0271-y
16. Guo, Q., Fang, L., Wang, R., Zhang, C.: Multivariate time series forecasting using multiscale recurrent networks with scale attention and cross-scale guidance. IEEE Trans. Neural Netw. Learn. Syst. pp. 1–15 (2023)

17. Guo, Q., Zhang, Y., Qiu, S., Zhang, C.: Accelerating patch-based low-rank image restoration using kd-forest and lanczos approximation. Inform. Sci. **556**, 177–193 (2021)
18. Hu, J., Shen, L., Sun, G.: Squeeze-and-excitation networks. In: Proceedings of the IEEE/CVF Conference on Computer Vision and Pattern Recognition, pp. 7132–7141 (2018)
19. Hui, S., Guo, Q., Geng, X., Zhang, C.: Multi-guidance cnns for salient object detection. ACM Trans. Multimedia Comput. Commun. Appli. **19**(3), 1–19 (2023)
20. Jaderberg, M., Vedaldi, A., Zisserman, A.: Speeding up convolutional neural networks with low rank expansions. In: Proceedings of the British Machine Vision Conference, pp. 1–15 (2014)
21. Jaderberg, M., Simonyan, K., Zisserman, A., et al.: Spatial transformer networks. Adv. Neural Inform. Process. Syst. **28** (2015)
22. Jiao, J., et al.: Dilateformer: multi-scale dilated transformer for visual recognition. IEEE Trans. Multimedia **25**, 8906–8919 (2023)
23. Liu, W., Rabinovich, A., Berg, A.C.: Parsenet: Looking wider to see better. In: International Conference on Learning Representations (2016)
24. Long, J., Shelhamer, E., Darrell, T.: Fully convolutional networks for semantic segmentation. In: Proceedings of the IEEE Conference on Computer Vision and Pattern Recognition, pp. 3431–3440 (2015)
25. Oberweger, M., Wohlhart, P., Lepetit, V.: Hands deep in deep learning for hand pose estimation. In: Proceedings of the 20th Computer Vision Winter Workshop, pp. 21–30 (2015)
26. Pan, H., Hong, Y., Sun, W., Jia, Y.: Deep dual-resolution networks for real-time and accurate semantic segmentation of traffic scenes. IEEE Trans. Intell. Trans. Syst. **24**(3), 3448–3460 (2022)
27. Paszke, A., Chaurasia, A., Kim, S., Culurciello, E.: Enet: A deep neural network architecture for real-time semantic segmentation. arXiv preprint arXiv:1606.02147 (2016)
28. Paszke, A., et al.: Pytorch: an imperative style, high-performance deep learning library. Adv. Neural Inform. Process. Syst. **32** (2019)
29. Qiao, Q., Wang, W., Qu, M., Su, K., Jiang, B., Guo, Q.: Medical image segmentation via single-source domain generalization with random amplitude spectrum synthesis. In: Linguraru, M.G. et al. (eds.) MICCAI 2024. LNCS, vol. 15009, pp. 435-445. Springer (2024). https://doi.org/10.1007/978-3-031-72114-4_42
30. Sandler, M., Howard, A., Zhu, M., Zhmoginov, A., Chen, L.C.: Mobilenetv2: inverted residuals and linear bottlenecks. In: Proceedings of the IEEE/CVF Conference on Computer Vision and Pattern Recognition, pp. 4510–4520 (2018)
31. Szegedy, C., et al.: Going deeper with convolutions. In: Proceedings of the IEEE Conference on Computer Vision and Pattern Recognition, pp. 1–9 (2015)
32. Vaswani, A., et al.: Attention is all you need. Adv. Neural Inform. Process. Syst. **30** (2017)
33. Wang, J.: Deep high-resolution representation learning for visual recognition. IEEE Trans. Pattern Anal. Mach. Intell. **43**(10), 3349–3364 (2020)
34. Wang, X., Girshick, R., Gupta, A., He, K.: Non-local neural networks. In: Proceedings of the IEEE/CVF Conference on Computer Vision and Pattern Recognition, pp. 7794–7803 (2018)
35. Wen, W., Wu, C., Wang, Y., Chen, Y., Li, H.: Learning structured sparsity in deep neural networks. Adv. Neural Inform. Process. Syst. **29** (2016)

36. Woo, S., Park, J., Lee, J.-Y., Kweon, I.S.: CBAM: convolutional block attention module. In: Ferrari, V., Hebert, M., Sminchisescu, C., Weiss, Y. (eds.) ECCV 2018. LNCS, vol. 11211, pp. 3–19. Springer, Cham (2018). https://doi.org/10.1007/978-3-030-01234-2_1
37. Wu, J., Jiao, J., Yang, Q., Zha, Z.J., Chen, X.: Ground-aware point cloud semantic segmentation for autonomous driving. In: Proceedings of the 27th ACM International Conference on Multimedia, pp. 971–979 (2019)
38. Wu, T., Tang, S., Zhang, R., Cao, J., Zhang, Y.: Cgnet: a light-weight context guided network for semantic segmentation. IEEE Trans. Image Process. **30**, 1169–1179 (2020)
39. Xie, E., Wang, W., Yu, Z., Anandkumar, A., Alvarez, J.M., Luo, P.: Segformer: Simple and efficient design for semantic segmentation with transformers. Adv. Neural Inform. Process. Syst. **34**, 12077–12090 (2021)
40. Xu, J., Xiong, Z., Bhattacharyya, S.P.: Pidnet: a real-time semantic segmentation network inspired by pid controllers. In: Proceedings of the IEEE/CVF Conference on Computer Vision and Pattern Recognition, pp. 19529–19539 (2023)
41. Yu, C., Gao, C., Wang, J., Yu, G., Shen, C., Sang, N.: Bisenet v2: bilateral network with guided aggregation for real-time semantic segmentation. Inter. J. Comput. Vis. **129**, 3051–3068 (2021)
42. Yuan, Y., Fu, R., Huang, L., Lin, W., Zhang, C., Chen, X., Wang, J.: Hrformer: high-resolution vision transformer for dense predict. Adv. Neural. Inf. Process. Syst. **34**, 7281–7293 (2021)
43. Zhang, H., et al.: Context encoding for semantic segmentation. In: Proceedings of the IEEE/CVF Conference on Computer Vision and Pattern Recognition, pp. 7151–7160 (2018)
44. Zhao, H., Shi, J., Qi, X., Wang, X., Jia, J.: Pyramid scene parsing network. In: Proceedings of the IEEE/CVF Conference on Computer Vision and Pattern Recognition, pp. 2881–2890 (2017)
45. Zhou, B., Zhao, H., Puig, X., Fidler, S., Barriuso, A., Torralba, A.: Scene parsing through ade20k dataset. In: Proceedings of the IEEE/CVF Conference on Computer Vision and Pattern Recognition, pp. 633–641 (2017)

3-D Reconstruction from Consecutive Endoscopic Images Using Gaussian Splatting

Hung-Le Minh[1]([✉]), Duy-Van Truong[1], Huy-Xuan Manh[4], Viet-Hang Dao[2,3], Phuc-Binh Nguyen[2], Thanh-Tung Nguyen[2], and Hai Vu[1]

[1] Hanoi University of Science and Technology (HUST), Hanoi, Vietnam
hung.lm192890@sis.hust.edu.vn
[2] Institute of Gastroenterology and Hepatology (IGH), Hanoi, Vietnam
[3] Hanoi Medical University (HMU), Hanoi, Vietnam
[4] Koh Young Technology Vietnam, Hanoi, Vietnam

Abstract. Recent advancements in 3D reconstruction helped endoscopy doctors analyze the patients' gastrointestinal surfaces and abnormality detections. In this work, we expand this development further with a reconstruction method based on both classic techniques like structure from motion and recent advanced techniques like neural radiation fields and Gaussian splatting with new Gaussian encoding-decoding modules. In addition, an unique dataset was collected with some videos from daily endoscopy examinations. This development helped us achieve better reconstruction results and lower training time compared to existing methods.

Keywords: Gaussian Splatting · 3D Reconstruction · EndoScopy · Radiance Fields · Gastrointestinal Imaging

1 Introduction

Gastrointestinal diseases, including lesions in the esophagus, stomach, and duodenum appear quite popularly in digestive examinations. Several studies have shown that common lesions include gastritis (62.7%), gastric and duodenal ulcers (6.3%), reflux esophagus (41.3%), esophageal candidiasis (1.9%), and polyps (1.8%). Doctors usually use visual inspection with tools like tube camera or capsule camera to see the internal gastrointestinal surface, detecting abnormal like lesion and tumors. In these examinations, size is an important factor as it can be used to determine the stage of the tumors, thus give valuable information for doctors to provide treatments. However, the information in endoscopy data is usually presented as 2D RGB images that does not contribute much 3D information. Therefore, a computed-aided diagnostic or a CAD tool which is able to reconstruct 3D information from the video pipeline or an image sequences is needed.

3-D reconstruction from the consecutive image sequence is a conventional task in the computer vision field. Snavely et al.'s "Photo Tourism: Exploring Photo Collections in 3D" [11] introduced a ground-breaking approach to 3D reconstruction from photo collections of multiple viewpoints, laying the foundation for Structure-from-Motion (SfM). Since then, SfM has been used widely in different domains, and improved, notably with work of Schonberger et al. [10] to improve the traditional SfM (Shape-from-motion) pipeline. The authors address challenges in feature matching, camera model estimation, and point triangulation, proposing several improvements that advance the state-of-the-art in SfM. The revised SfM pipeline demonstrates superior performance, particularly in large and complex datasets, and has become the foundational reference in 3D reconstruction. With the advancement of deep learning and big data-based techniques, 3D reconstruction in endoscopy is promised to overcome challenges which made SfM and 3D computer vision methods alike produce sub-optimal results like requiring multiple viewpoints, light reflectance, narrow operating channels and tissue deformation.

Some studies about 3D reconstruction in other domains have yielded promising results. For example, the study of utilizing Gaussian Splatting [4], but the use of 3D reconstruction in endoscopy still faces several challenges, such as the inability to provide multiple angles and the deformation of damaged surfaces over time as doctors perform endoscopy. Additionally, due to the presence of numerous water bubbles in the gastrointestinal tract, the damaged areas may be overly illuminated, presenting a significant challenge, along with the lack of computational power and a large dataset for 3D in endoscopy. Therefore, in this study, we propose a solution that requires low computational resources, and can be integrated with endoscopy systems to reconstruct lesion surfaces efficiently. The main contribution is a new initialization method to eliminate input noise before training and to encode temporal changes of the damaged surfaces method. Based on HexPlane [2], it can reduce training time and resources. The experimental results show that the training time and memory required to train and run the deep learning model is much lighter than the based-line 3D reconstruction methods.

The remaining of this paper is organized as follows: Section 2 briefly reviews recent works on 3-D reconstructions from the endoscopic images sequence. Section 3 describes the proposed method. Section 4 shows the 3-D reconstruction results from consecutive images and evaluates in term of both quantitative and qualitative indexes. Finally, the conclusions and future works are given in Sect. 5.

2 Related Work

Recent advancements in depth estimation for monocular endoscopy have leveraged deep learning to overcome the challenges of accurately reconstructing 3D structures from 2D images. The paper "Deep learning-based depth estimation from a synthetic endoscopy image training set" [7] of Nicholas J. Durr and his colleges explores the use of synthetic datasets to train deep networks, allowing for robust depth estimation even in complex endoscopic environments. In additional, facing the challenges lack of in-depth information that makes training difficult, thus Xingtong Liu et al.'s "Dense Depth Estimation in Monocular Endoscopy With Self-Supervised Learning Methods" [6] introduces self-supervised approaches that eliminate the need for ground truth depth data, making the process more adaptable to real-world scenarios. Complementing these efforts, "A geometry-aware deep network for depth estimation in monocular endoscopy" [14] was introduced by Yongming Yang et al. integrates geometric constraints into the deep learning model, enhancing the accuracy of depth predictions by leveraging the inherent structure of the endoscopic scenes. One of the most successful studies on depth prediction presented by Beilei Cui et al. is "EndoDAC: Efficient Adapting Foundation Model for Self-Supervised Depth Estimation from Any Endoscopic Camera." [3] This method addresses challenges in self-supervised depth estimation from endoscopic images by incorporating Dynamic Vector-Based Low-Rank Adaptation (DV-LoRA) and Convolutional Neck blocks. EndoDAC autonomously estimates camera intrinsics, enabling it to work with various surgical video datasets without explicit camera information. The results show that EndoDAC outperforms existing methods in accuracy and efficiency, with fewer training epochs and reduced computational demands.

A different approach than the studies mentioned above to be able to reconstruct 3D scenes. The study [8] by Mildenhall et al. (2020) introduced a groundbreaking method where Neural Radiance Fields (NeRF) use neural networks to model the radiance field of a scene. This approach learns a continuous function to predict the density and color of points in 3D space, enabling the generation of new 2D images from different viewpoints. NeRF significantly enhances image quality and detail compared to previous methods by leveraging information from multiple views to construct a more accurate 3D representation of the scene (Mildenhall et al., 2020). Building on this, the works in [1] by Barron et al. (2021) extends NeRF's capabilities by introducing a multiscale representation to handle scenes with varying levels of resolution. Mip-NeRF improves the quality of rendered images by applying different layers of resolution during the learning process, enhancing detail and reducing artifacts in high-resolution areas. In [9], Pumarola et al. (2021) introduces D-NeRF, an extension of the Neural Radiance Fields (NeRF) framework designed to handle dynamic scenes. Unlike traditional NeRF, which is limited to static scenes, D-NeRF incorporates temporal information to model and reconstruct 3D scenes with moving objects or changing environments. This method enhances NeRF's capability to accurately represent and synthesize images of dynamic scenes by integrating dynamic changes into the 3D model. With the results that the above research brings as well as new approaches, there are a number of studies that have applied them in the medical field. The paper [12] by Yuehao Wang et al. (2022) introduces an adaptation of the NeRF framework specifically for endoscopic imaging. EndoNeRF addresses challenges like distortion, low resolution, and uneven lighting by modifying NeRF to better handle the unique characteristics of endoscopic data. It enables the accurate modeling and reconstruction of complex anatomical structures by predicting radiance and density at specific points in the scene. This allows for the creation of detailed 3D models from 2D endoscopic images, significantly improving reconstruction quality and offering potential benefits for medical applications such as diagnostics and surgical planning.

In recent years, the field of computer graphics has seen significant advancements in rendering techniques, particularly in the area of radiance field rendering. As researchers continue to push the boundaries of visual quality and computational efficiency, a new method has emerged that combines both speed and high fidelity. The paper [4] by Bernhard Kerbl et al. presents a novel approach for real-time radiance field rendering using a 3D Gaussian scene representation. This method achieves visual quality comparable to or surpassing previous state-of-the-art methods while requiring significantly shorter optimization times and delivering superior rendering speed. The approach leverages three key innovations: representing the scene with 3D Gaussians to avoid unnecessary computations in empty spaces, continuously optimizing the properties of the Gaussians including their anisotropic covariance, and developing a fast, visibility-aware rendering algorithm that supports real-time rendering at ≥ 30 fps at 1920×1080 resolution. The experimental results demonstrate that this method achieves state-of-the-art visual quality on various established datasets and provides the first real-time rendering solution for complex scenes with large depth complexity.

3 Proposed Method

The application of Structure from Motion (SfM) to generate input models for the Gaussian method has demonstrated significant improvements in both processing time and 3D reconstruction efficiency, as outlined in [4]. This approach enables the creation of three-dimensional structures from sequences of two-dimensional images, thereby enhancing the accuracy and effectiveness of the reconstruction process. However, when applied in the context of endoscopy, particularly in gastrointestinal endoscopy, several substantial challenges have emerged.

The environment within the gastrointestinal tract has unique characteristics, such as low light conditions and high levels of noise caused by unstable movements and interactions with soft tissues. Additionally, during endoscopic procedures, complex surgical instruments frequently enter the field of view, obstructing visibility and degrading the quality of the captured images. These factors increase the difficulty of accurately reconstructing 3D scenes from endoscopic images, leading to potential inaccuracies or gaps in the generated 3D models.

Recognizing these challenges, we have proposed several solutions to improve the reliability and efficiency of the 3D reconstruction process in endoscopic conditions. These solutions include optimizing image processing algorithms to minimize noise and enhance image recognition capabilities in low-light environments. Furthermore, we have focused on developing new approaches to better handle data processing when surgical instruments are present in the field of view. These advancements not only enhance the quality of the 3D models but also make the reconstruction process faster and more accurate, meeting the stringent demands of modern endoscopy (Fig. 1).

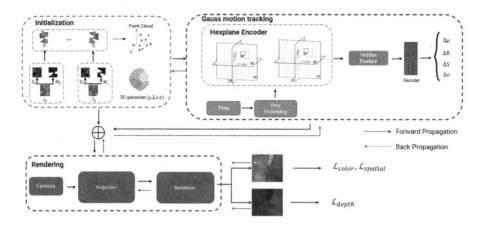

Fig. 1. The proposed method consists of three main modules. The initialization module helps to use mask image and depth image to convert 2D image to point cloud and remove noise. The motion encoding module helps to track the movement of tissues over time. The image rendering module helps to generate 2D depth image and 2D RGB color image after removing noise.

3.1 Initialization

The input data from SfM has several key limitations:

- **Sparse initialization points**: SfM often produces sparse initialization points, especially in areas with significant depth variation or rapid movement, leading to inaccurate models and requiring additional optimization to improve accuracy.
- **Training and rendering performance**: The sparsity of the initialization points can negatively impact the performance of model training and image rendering, particularly in complex scenes with significant depth variation.

To address the identified limitations, this study proposes a new initialization method designed to enhance the accuracy and efficiency of the 3D reconstruction process. Specifically, the system takes as input a sequence of images $\{I_i\}_{i=1}^{L}$ captured from a single camera, where L denotes the temporal length or number of frames in the sequence. In addition, based on the requirements set by the doctors to evaluate whether the input image contains noisy areas or not, we will then use the tool to create mask images $\{M_i\}_{i=1}^{L}$ to cover up the corresponding noisy areas. Instead of relying on traditional techniques such as SfM, we employ an advanced deep learning model [13] to predict depth maps $\{D_i\}_{i=1}^{L}$ for each image in the sequence.

This depth prediction process generates a depth map for each frame, representing the distance from points in the scene to the camera. With the obtained depth maps and the original input images, we combine them to reconstruct new images in the camera's space. This approach enables more accurate 3D space

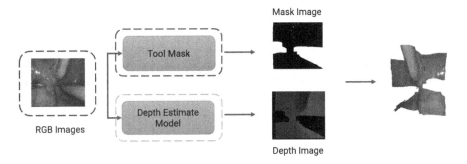

Fig. 2. The process of combining depth images and mask images to generate depth information and remove noise.

reconstruction, especially in complex environments such as endoscopy, where depth and shape variations are critical factors.

Based on the images obtained above, we propose a formula to initialize the point clouds.

$$P_i = K^{-1} T_i D_i (I_i \odot M_i) \tag{1}$$

where K, T_i, D_i, I_i, M_i are Extrinsic matrix, Intrinsic matrix, depth map, the image, the mask respectively.

Since the images are captured from a single camera, they may be limited and some areas might be obstructed by medical equipment, bubbles, etc. Thus, this study will integrate all point clouds to create a comprehensive initialization. Additionally, we want to create a common space that contains all the variations of the point clouds over time. So that when we optimize them, we can learn the most common features between the variations of the point clouds over time. This helps us to optimize better when combined with time embedding.

$$P = \{P_1, P_2, P_3, \ldots, P_T\} \tag{2}$$

3.2 Gaussian Motion Tracking

During surgery, tissues may move or change shape, requiring an accurate model to track and represent these changes. To address this issue, a deformation field $D(\mu, t)$ is needed to monitor the displacement of the attribute ΔG for each Gaussian distribution at time t. The deformed Gaussian distributions can be calculated as (Fig. 2):

$$G_t = G_0 + \Delta G \tag{3}$$

However, there are several challenges in practical implementation when using large neural networks to approximate the deformation field $D(\mu, t)$. First, the use of large neural networks can lead to slow inference speeds, making the computational process inefficient. Second, this model may not achieve optimal performance during training and inference, leading to subpar results in modeling surface dynamics.

Based on research [9], this study divides the deformation field into two lighter modules: D = F ∘ E. The E module is a disentangled voxel encoder that encodes the 4D inputs (the center of each Gaussian (x, y, z) and time t) into temporal latent features. The F module is a Gaussian transformation decoder that uses these latent features to compute Gaussian transformations. This approach improves inference speed and optimization while maintaining the accuracy of the model. The disentangled voxel encoder E is designed to effectively encode the 4D inputs.

Gaussian Encoding Module. This module is inspired by the research [2], which uses a multi-resolution HexPlane structure to represent the 4D voxel encoding. This means that the voxel encoding is split into six planes with corresponding vectors. The module E(μ, t) encodes the center of each Gaussian μ and the time t into temporal latent features, making the modeling of the temporal transformation of Gaussian distributions more efficient and accurate. After calculating the feature values of each plane and combining them, a latent feature is obtained. This process results in a complete and accurate latent feature from the 4D input coordinates. The combination formula is detailed below:

$$E = E_{XY} \otimes E_{ZT} \otimes v_1 + E_{XY} \otimes E_{ZT} \otimes v_2 + E_{XY} \otimes E_{ZT} \otimes v_3 \quad (4)$$

where $\otimes$ represents matrix multiplication, and $\mathbf{E}_{AB} \in \mathbb{R}^{A \times B}$ is a feature plane that has been learned from the training data. This matrix contains learned feature values and is used to represent important information of the data in the feature space $\mathbb{R}^{A \times B}$. This plane helps reduce the complexity of the data and focus on the most significant factors. $\mathbf{v}_i \in \mathbb{R}^D$ denotes the feature vector along the i-th axis. This vector contains feature values for a specific dimension in the input space. By using feature vectors along the axes, it is possible to analyze and represent data along different dimensions, allowing the model to capture the structure and characteristics of the data comprehensively.

Gaussian Decoding Module. This module plays a crucial role in adjusting and reconstructing the Gaussians over time. Designed to handle the variations of the Gaussian parameters from latent features, this decoder uses small multi-layer perceptrons (MLPs) to predict changes in position, rotation, scale, and opacity of each Gaussian. Specifically, this module is applied to predict changes in position ($\Delta\mu$), changes in rotation ($\Delta\mathbf{R}$), changes in scale ($\Delta\mathbf{S}$), and changes in opacity ($\Delta\mathbf{o}$). Each MLP aims to optimize the reconstruction of the Gaussians after they have been transformed, ensuring that the model can accurately reflect the dynamic changes in the surface over time. Therefore, the transformation of the Gaussian at time t is given by the following formula:

$$G_t = G_t + \Delta G = (\mu + \Delta\mu, \mathbf{R} + \Delta\mathbf{R}, \mathbf{S} + \Delta\mathbf{S}, \mathbf{o} + \Delta\mathbf{o}, \mathbf{SH}) \quad (5)$$

3.3 Optimization

Color Loss. In the process of scene reconstruction, ensuring that the colors of the reconstructed scene match the actual colors is crucial. The color loss function helps minimize the difference between the predicted colors and the actual colors, ensuring that the reconstructed image is realistic and accurate.

$$\mathcal{L}_{\text{color}} = \sum_{x \in \zeta} \left\| M(x) \left(\hat{C}(x) - C(x) \right) \right\|_1 \tag{6}$$

where M, $\hat{C}$, C and ζ are binary tool masks, predicted colors, real colors and 2D coordinate space.

Depth Loss. The depth loss function helps the model learn to accurately represent the distances between objects in the scene. This is essential for generating images with greater depth and accuracy. Additionally, this loss function utilizes soft constraints (sort constraints), which enable the model to focus on aligning spatial structures (such as shapes and relative distances between objects) rather than the absolute depth values.

$$\mathcal{L}_{\text{depth}} = 1 - \frac{\text{Cov}(M \odot \hat{D}, M \odot D)}{\sqrt{\text{Var}(M \odot \hat{D}) \text{Var}(M \odot D)}} \tag{7}$$

where M, $\hat{D}$, D and are binary tool masks, predicted depths, real depths.

Spatio-temporal Loss. To ensure the spatial and temporal smoothness of the reconstructed result, total variation (TV) functions are applied. This helps to avoid black/white regions in areas obscured by surgical instruments.

$$\mathcal{L}_{\text{spatial}} = \text{TV}(\hat{C}) + \text{TV}(\hat{D}^{-1}) \tag{8}$$

Total Loss. With the 3 loss functions mentioned above. They will be combined and create a synthetic loss function to ensure the quality of the image produced with the objectives each loss function brings.

$$\mathcal{L} = \lambda_1 \mathcal{L}_{\text{color}} + \lambda_2 \mathcal{L}_{\text{depth}} + \lambda_3 \mathcal{L}_{\text{spatial}} \tag{9}$$

where, the weights λ_1, λ_2, λ_3 play an important role in balancing the different loss functions. These weights help to adjust the influence of each loss function on the overall optimization of the model. But to get stable results, the color loss function will be focused on the most and the remaining two loss functions we test the settings to get the best results 4.

4 Experimental Results

4.1 Dataset and Evaluation Metrics

We performed experiments using two public datasets, including EndoNeRF [12] and another dataset. ENDONERF features two instances of in-vivo prostate surgery data recorded from a stereo camera at a single viewpoint, showcasing complex scenes with non-rigid deformations and instrument occlusions. In accordance with prior research [27], we divided the frame data from each scene into training and testing sets with a 7:1 ratio. Our method was assessed using several metrics such as PSNR, SSIM, and LPIPS. Additionally, we documented the training duration, inference speed (FPS, frames per second), and GPU memory consumption during training.

In addition, we collected three endoscopic, single viewpoint videos from IGH. On average, videos have a total of 430 frames, but we only select several frames according to the criteria related to lighting and angles to serve the reconstruction of the surface. The data was also divided into 8 parts, with 7 parts for training and 1 part for testing (Table 1).

Table 1. Overview of the datasets

Name Dataset	Total Frame
ENDONERF-cutting	155
ENDONERF-pulling	63
IGH dataset	430

4.2 Implementation Details

All our experiments are performed on a computer with an Intel core i7 12700k configuration and with a single RTX 3080 graphics card. Using the Adam [5] optimization function. The initial learning rate is initialized to 1.6×10^{-3}. We will perform the point cloud thickening without updating among the models for about 1000 iterations And the training process is performed for 6000 iterations

4.3 Results

We compared our proposed method with previous studies on surface structure reconstruction, including EndoNeRF and EndoSurf. The results are shown in Table 2 and Table 3 where we can see that the performance of EndoNeRF and EndoSurf achieved excellent results for temporally varying data but required substantial computational resources and lengthy training times. Our results show that our method achieved average result is 37.4895 (PSNR) with only 10 min of training on the EndoNeRF dataset, significantly faster than previous studies, and yielded improved outcomes. We have illustrated several scenes in EndoNeRF dataset which are reconstructed using the proposed method in Fig. 4 and Fig. 3. In both cases, they are reconstructed successfully even with the obstruction of some objects such as the endoscope tube and the cutting-device appeared in the endoscopy examination. In Fig. 5, we examine the proposed method to a practical image sequence. The 2-D original image sequence shows a stomach cancer region. The reconstructed images help examining doctor observe the abnormal region from different view-point (as given in 3-D visualization at following link http://surl.li/wedpmm) as well as clearly observe the shape of this abnormality. These results suggest that the proposed method has the potential to be applied in preoperative lesion assessment.

Table 2. Quantitative metrics of appearance (PSNR/SSIM/LPIPS)

Method	EndoNeRF [12]			EndoSurf [15]			Ours		
Metrics	PSNR↑	SSIM↑	LPIPS↓	PSNR↑	SSIM↑	LPIPS↓	PSNR↑	SSIM↑	LPIPS↓
ENDONERF-cutting	34.186	0.932	0.151	34.981	0.953	0.106	37.849	0.963	0.089
ENDONERF-pulling	34.212	0.938	0.161	35.004	0.956	0.120	37.13	0.930	0.004
Average	34.199	0.935	0.156	34.4925	0.9545	0.113	37.4895	0.9465	0.0465

Table 3. Comparison of GPU usage and training time across different methods.

Method	EndoNeRF [12]		EndoSurf [15]		Ours	
Metric	GPU↓	Time Training↓	GPU↓	Time Training↓	GPU↓	Time Training↓
ENDONERF-cutting	19 GB	7 h	19 GB	9 h	2 GB	10 min
ENDONERF-pulling	19 GB	7 h	19 GB	9 h	2 GB	10 min

Table 4. Weights for Loss Functions

λ_1	λ_2	λ_3	PSNR ↑
0.9	0.01	0.09	37.27
0.9	0.02	0.08	37.23
0.9	0.03	0.07	36.94
0.9	0.04	0.06	37.22
0.8	0.01	0.19	36.24
0.8	0.02	0.18	36.07
0.8	0.03	0.17	36.56
0.8	0.04	0.16	26.49

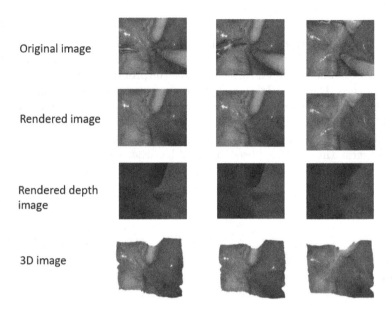

Fig. 3. Rendered images, depth images and 3D images are reconstructed from the original images from the dataset ENDONERF-pulling

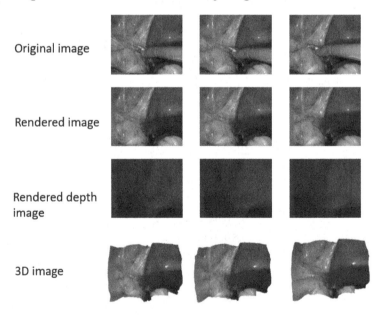

Fig. 4. Rendered images, depth images and 3D images are reconstructed from the original images from the dataset ENDONERF-cutting

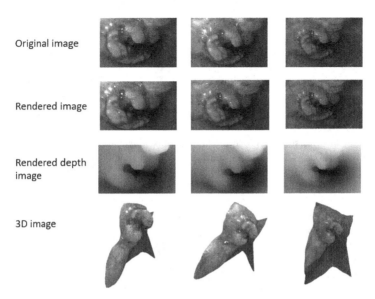

Fig. 5. Rendered images, depth images and 3D images are reconstructed from the original images from the IGH dataset (Please refer the reconstruction results in different views at following link http://surl.li/wedpmm)

5 Conclusion

In this paper, we present a real-time, high-quality framework for reconstructing dynamic surgical scenes. By leveraging Endo-Gaussian Initialization and Spatio-Temporal Gaussian Tracking, we effectively address the challenges of Gaussian initialization and tissue deformation. Extensive experiments demonstrate that our EndoGaussian achieves state-of-the-art reconstruction quality while significantly improving rendering speed. We believe that emerging Gaussian Splatting-based reconstruction techniques can open new pathways for better understanding robotic surgical scenes and support various clinical tasks, particularly in intraoperative applications.

References

1. Barron, J.T., Mildenhall, B., Tancik, M., Hedman, P., Martin-Brualla, R., Srinivasan, P.P.: Mip-NeRF: a multiscale representation for anti-aliasing neural radiance fields. In: Proceedings of the IEEE/CVF International Conference on Computer Vision (ICCV), pp. 5855–5864, October 2021
2. Cao, A., Johnson, J.: HexPlane: a fast representation for dynamic scenes. In: Proceedings of the IEEE/CVF Conference on Computer Vision and Pattern Recognition (CVPR), pp. 130–141, June 2023

3. Cui, B., Islam, M., Bai, L., Wang, A.C., Ren, H.: EndoDAC: efficient adapting foundation model for self-supervised depth estimation from any endoscopic camera. ArXiv abs/2405.08672 (2024). https://api.semanticscholar.org/CorpusID:269761398
4. Kerbl, B., Kopanas, G., Leimkühler, T., Drettakis, G.: 3D Gaussian splatting for real-time radiance field rendering (2023). https://arxiv.org/abs/2308.04079
5. Kingma, D.P.: Adam: a method for stochastic optimization. arXiv preprint arXiv:1412.6980 (2014)
6. Liu, X., et al.: Dense depth estimation in monocular endoscopy with self-supervised learning methods. IEEE Trans. Med. Imaging **39**(5), 1438–1447 (2020). https://doi.org/10.1109/TMI.2019.2950936
7. Mahmood, F., Durr, N.J.: Deep learning-based depth estimation from a synthetic endoscopy image training set. In: Angelini, E.D., Landman, B.A. (eds.) Medical Imaging 2018: Image Processing, vol. 10574, p. 1057421. International Society for Optics and Photonics, SPIE (2018). https://doi.org/10.1117/12.2293785
8. Mildenhall, B., Srinivasan, P.P., Tancik, M., Barron, J.T., Ramamoorthi, R., Ng, R.: NeRF: representing scenes as neural radiance fields for view synthesis. Commun. ACM **65**(1), 99–106 (2021). https://doi.org/10.1145/3503250
9. Pumarola, A., Corona, E., Pons-Moll, G., Moreno-Noguer, F.: D-NeRF: neural radiance fields for dynamic scenes. In: Proceedings of the IEEE/CVF Conference on Computer Vision and Pattern Recognition (CVPR), pp. 10318–10327, June 2021
10. Schonberger, J.L., Frahm, J.M.: Structure-from-motion revisited. In: Proceedings of the IEEE Conference on Computer Vision and Pattern Recognition, pp. 4104–4113 (2016)
11. Snavely, N., Seitz, S.M., Szeliski, R.: Photo tourism: exploring photo collections in 3D. ACM Trans. Graph. **25**(3), 835–846 (2006). https://doi.org/10.1145/1141911.1141964
12. Wang, Y., Long, Y., Fan, S.H., Dou, Q.: Neural rendering for stereo 3D reconstruction of deformable tissues in robotic surgery. In: Wang, L., Dou, Q., Fletcher, P.T., Speidel, S., Li, S. (eds.) Medical Image Computing and Computer Assisted Intervention - MICCAI 2022, pp. 431–441. Springer, Cham (2022). https://doi.org/10.1007/978-3-031-16449-1_41
13. Yang, L., Kang, B., Huang, Z., Xu, X., Feng, J., Zhao, H.: Depth anything: unleashing the power of large-scale unlabeled data. In: Proceedings of the IEEE/CVF Conference on Computer Vision and Pattern Recognition (CVPR), pp. 10371–10381, June 2024
14. Yang, Y., et al.: A geometry-aware deep network for depth estimation in monocular endoscopy. Eng. Appl. Artif. Intell. **122**, 105989 (2023). https://doi.org/10.1016/j.engappai.2023.105989. https://www.sciencedirect.com/science/article/pii/S0952197623001732
15. Zha, R., Cheng, X., Li, H., Harandi, M., Ge, Z.: EndoSurf: neural surface reconstruction of deformable tissues with stereo endoscope videos. In: Greenspan, H., et al. (eds.) MICCAI 2023. LNCS, vol. 14228, pp. 13–23. Springer, Cham (2023). https://doi.org/10.1007/978-3-031-43996-4_2. https://arxiv.org/abs/2307.11307

EMMA: EMotion Mixing Algorithm for Compound Expression Recognition Using Angle-Based Metric Learning

Riku Yamamoto[✉][iD] and Noriko Takemura[iD]

Kyushu Institute of Technology, 680-4 Kawazu, Iizuka 820-8502, Japan
yamamoto.riku878@mail.kyutech.jp, takemura@ai.kyutech.ac.jp

Abstract. Facial expression recognition (FER) is a key component in various AI-based systems and has been extensively studied. However, most FER research has focused on clear and simple basic emotions such as happiness and sadness, which are not suitable for real-world applications where numerous many ambiguous and complex emotions exist. Complex emotions are challenging to define and require ample data for each emotion to train a FER model. Moreover, due to their ambiguous nature, these emotions are difficult to annotate. Consequently, the difficulty in constructing comprehensive databases is a significant bottleneck in recognizing complex emotions. In this study, we propose complex emotion recognition method using only a database of basic emotions based through angle-base metric learning. This approach can mitigate the reduction in recognition accuracy caused by insufficient data and allows for the definition of new emotions in the future, unlike general FER tasks that require pre-definition of emotions.

Keywords: facial expression recognition · compound emotion · metric learning

1 Introduction

The development of facial expression recognition (FER) technology has enabled human-centric intelligent systems to provide adaptive services and support based on individual emotions and needs [19]. Specifically, it affects areas such as security [13,18], learning [17,21], and medical rehabilitation [25]. Facial expressions are a prominent expression of human emotions, which in fact, it is said that about 55% of human emotional communication is conveyed through facial expressions [2]. However, most previous studies have focused on Ekman's basic emotions [7], which are happiness, sadness, anger, surprise, fear, and disgust, and these emotions alone are insufficient to fully capture the diversity and complexity of human emotions. This is because human emotions go beyond these basic emotions and encompass more subtle and complex emotions that are important in real-world systems.

In recent years, FER has been dominated by methods based on convolutional neural networks (CNNs), which are type of neural networks, but while a CNN can achieve high recognition accuracy, it requires a large amount of data during training. It is possible to intentionally create facial expression data for basic emotions through acting; however, it is difficult to artificially create complex facial expressions because they are naturally expressed in daily life. In addition, complex emotions are challenging to define, and sufficient amount of data is required for each of the various emotions to be accurately recognized. Furthermore, complex emotions are often ambiguous, and it is not easy to annotate them correctly. The difficulty of constructing such a dataset is a bottleneck in complex emotion recognition.

By contrast, in the field of psychology, there is the idea that a wide variety of emotions can be expressed as compound emotions, which are combinations of basic emotions. Du et al. demonstrated that the Action Unit (AU) patterns of compound emotions reflect the Action Unit patterns of the basic emotions that constitute them [5,6]. AUs are facial muscle movements associated with specific emotions, serving as the basis for identifying these emotions [8]. This finding underscores the importance of basic emotions as fundamental building blocks in understanding more complex emotional states. Basic emotions are emotional categories that people generally tend to share. Explaining complex emotions on this basis greatly improves interpretability and allows for more detailed representations of subtle differences in human emotions. As an example, the complex emotion of nostalgia is expressed in terms of happy memories of the past (happiness) and the reality that these memories are in the past (sadness). In this way, it is possible to subdivide various emotions and analyze them in more detail.

In this study, we focus on the idea that complex emotions are expressed by combining basic emotions, and propose EMMA (EMotion Mixing Algorithm) for compound emotion recognition (CER) method using only the features related to basic emotions. In other words, we construct a CER method using only basic emotion data without considering compound emotion data. This method also solves the problem of constructing complex emotion datasets.

This method uses a CNN-based model and basic emotion data to train a feature extractor for basic emotions and a representative vector for the features of each basic emotion. In this training, we perform angle-based metric learning instead of the euclidean distance commonly used in general classification tasks. Angle-based metric learning does not consider the magnitude of vectors, thus preventing overfitting to factors such as the orientation of faces, occlusion, and gender in the training data. Therefore, it allows for the extraction of features that are more effectively separated by emotion. The compound emotion estimation is based on the compound emotion similarity calculated by linearly combining the basic emotion features of the image. In compound emotion estimation, each of compound emotion similarity is calculated as an unweighted average of the two basic emotion similarities that compose that.

The contributions of this study are as follows.

CER using only basic emotion data
This method does not require a dataset of compound emotions because it estimates compound emotions using only the features of the basic emotions. Conventional methods using compound emotion data have the problem that the amount of data for each emotion is not sufficient, and the bias in the amount of training data for each emotion significantly affects the accuracy of estimation.

An emotional space based on angles
Human facial images are diverse and include various elements. By introducing angle-based metric learning, it is expected to reduce the influence of elements other than expression information. Thus, it can lead to more isolated the features of basic emotion and improve the identification performance of compound emotions represented as linear combinations of basic emotions.

The ability to estimate a wide variety of emotions
Because the proposed method defines composite emotions as linear combinations of basic emotions, it is possible to freely set the composite emotions to be estimated. The estimated complex emotion need not be a human-definable emotion such as nostalgia or respect, as long as the proportion of the basic emotions that comprise the complex emotion is known. In other words, this method should be able to quantitatively express emotions that cannot be defined by humans, and hence advance our understanding of complex emotions in the field of psychology.

2 Related Work

2.1 Basic Emotion Recognition (BER)

The six emotions of surprise, fear, disgust, happiness, sadness, and anger are called Ekman's basic emotions and are the basis of current FER. As the name suggests, they are the basic elements of human emotional expression and are common to different cultures and societies around the world, and hence recognizing them plays an important role in understanding human emotions. Therefore, BER has been the subject of active research.

Among the various BER techniques, deep learning, which has contributed greatly to the field of image recognition in recent years, has also had a significant impact on the field of FER, improving recognition accuracy. Wang et al. proposed the Region Attention Network (RAN) as a model for FER that is robust to occlusion and pose changes in facial images [23]. RAN aggregates various numbers of domain features generated by the backbone CNN and embeds them in a compact fixed-length representation. The Distract your Attention Network (DAN) proposed by Zhengyao et al. extracts higher-order features of facial expressions from various regions of an expression image by convolution, and encodes these interactions to achieve a comprehensive understanding of facial expressions.

2.2 Complex Facial Emotion Recognition

As described above, deep learning has made a significant contribution to FER, but this method requires a large amount of annotated data for training. However, there is a lack of data for complex facial expressions, which poses a serious problem. This is due to the fact that complex emotions can be interpreted differently by different people, making it impossible to maintain consistent indices, and the annotation process becomes time-consuming as the number of labels increases [1]. Therefore, it is necessary to devise a way to achieve performance even with a small amount of data.

Through a multi-task learning approach that simultaneously performs Action Unit (AU) detection tasks, C-EXPR-NET [10] can efficiently learn from a small amount of complex facial expression data by exploiting the interaction of information obtained from each task. Because the middle layer of a CNN can be used as a feature extractor, there are also methods that pre-train on another task [4] and estimate complex emotions from the features. DLP-CNN [12] uses a model pre-trained with basic emotions as a feature extractor and classifies the features with a support vector machine to estimate compound emotions. Although this method uses features of the basic emotion to classify the compound emotion, the feature extractor for the basic emotions has not been designed with compound emotion estimation in mind.

2.3 Position of This Study

As mentioned in the previous section, conventional studies have improved the accuracy of complex FER by devising models. However, these methods require labeled data, which is a major barrier to their incorporation into actual systems, because new data must be prepared each time the emotion to be estimated changes. In this study, we propose a method for inference without using complex emotion data. This method significantly reduces the cost and time associated with data collection and labeling, allowing machine learning systems to be applied to real-world problem solving more quickly and efficiently.

3 Proposed Method

In this study, we propose EMotion Mixing Algorithm (EMMA) for CER that does not require data labeled with compound emotions. The specific procedure is as follows.

1) Pre-training on basic emotions

Through angle-based metric learning for basic emotions (six classes), we train a feature extractor that extracts features related to basic emotions from face images, and a representative vector for each basic emotion (basic EmoVec) in the feature space.

2) Compound emotion recognition (CER)

Using the Feature extractor and basic EmoVec trained in Step 1, we calculate similarity of each basic emotion in the input images. By calculating the average of these similarities, we obtain the similarity of compound emotion. The highest similarity is taken as the predicted value.

This method only performs pre-training for the basic emotions, it does not perform any new training for the compound emotions. Therefore, it is possible to estimate compound emotions using only the features of the basic emotions, without using compound-emotion data. The outline of the method is shown in Fig. 1, and the following sections describe each step in detail.

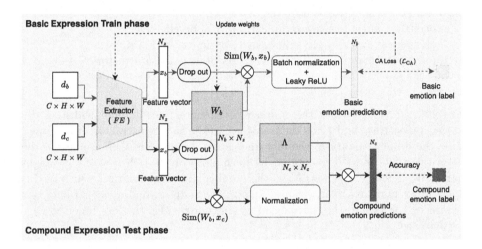

Fig. 1. Outline of the proposed method. The upper part shows the pre-training process for the basic emotions, and the lower part shows the estimation process of the compound emotions. In basic emotion train phase, Feature extractor and matrix of basic EmoVecs (W_b) are trained using the $\mathcal{L}_{CA}$ and basic emotion datasets; in compound emotion test phase, the similarity of the basic emotions is combined in the same ratio to define the compound emotion intensity and estimate it. Λ is the coupling coefficient of basic EmoVecs.

3.1 Pre-training on Basic Emotions

Feature extractors and basic EmoVecs are obtained through pre-training on basic emotions. For a single emotion, the feature vectors output by the feature extractor are distributed in close proximity to each other in the feature space, and a basic EmoVec determines the center of each emotion distribution. Therefore, to ensure that each emotion is clearly distinguished, the distance between basic EmoVecs should be large. To obtain such a feature extractor and basic EmoVecs, this study implements as angle-based metric learning, rather than pre-training as a general classification task using cross entropy loss ($\mathcal{L}_{CE}$).

In general, a model trained for a classification task can only classify the trained class, but the metric learning method should increase the separability between known and unknown emotions because it trains the transformations to the feature values. Angle-based metric learning is used in several methods proposed for the face recognition task, including sphereFace [14], arcFace [3], and cosFace [22]. This method is effective in terms of the scale invariance of the feature vectors for face images, which have many intra-class variations, such as lighting conditions and face angle. Specifically, unlike Euclidean distance, it does not consider the magnitude of the vectors, but focuses only on the direction, thus mitigating the increase in within-class variance due to these variations. Because facial expression images also have the characteristic of large intra-class variations, we expect that angle-based metric learning will be effective.

As a specific learning procedure, first, basic expression data $d_b \in \mathbb{Z}^{C \times H \times W}$ are input to the feature extractor $FE(\cdot)$ to obtain a feature vector $x_b \in N_z$.

$$x_b = FE(d_b) \tag{1}$$

where, C, H, and W are the number of channels, height, and width of the image, respectively, and FE is the transform of the feature extractor. In addition, N_z is the dimensionality of the feature vectors. Dropout is then applied to the feature vectors, and W_b is created by normalization in each row of the weight matrix $W_b' \in \mathbb{R}^{N_b \times N_z}$. Then, N_b is the number of basic emotions, which in this study is six (surprise, fear, disgust, happiness, sadness, and anger). Using this feature vector x_b and the normalized weight matrix W_b, the loss function $\mathcal{L}_{CA}$ is calculated (Eq. 2).

$$\mathcal{L}_{CA} = -\frac{1}{N} \frac{e^{\|x_b\| LReLU(\cos(w_{by}, x_b))}}{\sum_i e^{\|x_b\| LReLU(\cos(w_{bi}, x_b))}} \tag{2}$$

We design $\mathcal{L}_{CA}$ to be the loss based on the modified softmax loss [14] with the constrained by Leaky ReLU, which we call constrained angle loss in this paper. In Eq. 2, N is the number of training samples, $\cos(w_{bi}, x_b)$ is the cosine similarity between feature vector x_b and the i-th row vector w_{bi} of weight matrix W_b, and $LReLU$ is the Leaky ReLU function. In this loss function, if each row vector of the weight matrix W_b is regarded as a representative vector of each basic emotion (basic EmoVec), it can be interpreted that learning occurs so that the cosine similarity is large between feature vector x_b and the correct basic EmoVec w_{by}, and small between feature vector x_b and the other basic EmoVecs. The modified softmax loss minimizes the similarity with all emotions other than the correct emotion, but the minimum of the cosine is -1, resulting in a negative correlation. If the negative correlation is too strong, the within-class variance of each emotion in the feature space becomes too small, and the diversity of the feature vectors for one emotion may be lost. Because this study uses this feature vector to estimate compound emotions, it is desirable for the feature vector to be diverse to some extent. By contrast, if excessive restrictions such as setting the minimum value of the cosine similarity to 0 are applied to eliminate negative correlations, the classification of the training data itself becomes difficult. Therefore, Leaky ReLU is employed to moderate the negative correlation and achieve a balance between the diversity of feature vectors and separation performance. The accuracy of the Leaky ReLU parameters is evaluated in Sect. 4.3. The cosine similarity values are input to the batch normalization layer before being input to Leaky Relu to improve learning efficiency.

3.2 Compound Emotion Recognition (CER)

In compound emotion estimation, each of compound emotion similarity is calculated as an unweighted average of the two basic emotion similarities that compose that. The issue at hand is whether it is fair to combine basic emotions in a one-to-one ratio. Each emotion has a different distribution of similarity, which may prevent them from being evaluated on the same scale. Therefore, in this study, instead of using the similarity to basic emotions as is, we use the values after batch normalization layer. Batch normalization layer performs scaling by emotion similarity. This process is expected to enable fair treatment of each emotion.

Next, the specific CER process is described. First, compound expression data $d_c \in \mathbb{Z}^{C \times H \times W}$ are input to the feature extractor $FE(\cdot)$ trained in basic emotion to obtain a feature vector $x_c \in N_z$ (Eq. 3).

$$x_c = FE(d_c) \qquad (3)$$

The cosine similarity of each basic emotion is calculated by taking the product of matrix of basic EmoVecs $W_b \in \mathbb{R}^{N_z \times N_b}$ obtained by pre-training and the feature vector x_c (Eq. 4).

$$\mathrm{Sim}(W_b, x_c) = W_b^T x_c = \begin{bmatrix} w_{b0}^T x_c \\ w_{b1}^T x_c \\ \vdots \\ w_{bN_c}^T x_c \end{bmatrix} = \|x_c\| \begin{bmatrix} \cos(w_{b0}, x_c) \\ \cos(w_{b1}, x_c) \\ \vdots \\ \cos(w_{b1}, x_c) \end{bmatrix} \qquad (4)$$

where, each column of W_b are basic EmoVec and are normalized. That is, $\|w_{bi}\| = 1, (0 \leq i \leq N_b - 1, i \in \mathbb{Z})$. Finally, after inputting this similarity into Batch normalization layer $(BN(\cdot))$, a linear combination is performed to arrive at compound emotion. The largest compound emotion intensity created is then used as the predictive value y_c.

$$y_c = \mathrm{argmax}(\Lambda^T BN(\mathrm{Sim}(W_b, x_c))) \qquad (5)$$

where, $\Lambda \in \mathbb{R}^{N_b \times N_c}$ is a coupling matrix for combining the basic EmoVecs. In the case of a compound emotion that contains two basic emotions in a ratio of 1:1, the weight is 0.5 if the basic emotion constitutes the compound emotion and 0 otherwise.

4 Experiments and Discussion

To verify the usefulness of the framework proposed in this study for CER using only basic emotion data and the loss function $\mathcal{L}_{CA}$ for angle-base metric learning using Leaky ReLU, we conducted evaluation experiments using existing data sets on compound emotions.

4.1 Experimental Setup

Dataset. In this study, we use RAF-DB [11,12] as the dataset for basic and compound emotions. This dataset consists of natural facial images obtained from the Internet and includes diverse elements in terms of subject's age, gender, ethnicity, head pose, lighting conditions, occlusion (e.g., glasses or facial hair), and post-processing (various filters and special effects). They were also labeled by 40 crowd-sourced annotators. The data includes seven classes of basic emotions (including neutral) and eleven classes of compound emotions (composed of two basic emotions). The actual types of compound emotions and the number of data are shown in Table 2 below. Neutral represents the absence of emotion and is not considered to contribute much to the estimation of the compound emotion, so neutral images are excluded. In this dataset, the training data and test data are provided, but in this study, 20% of the training data were randomly selected to act as validation data. The input images were cropped to a square around the face area 224 × 224 pixels in size. However, in an experiment using DDRAMFN as the feature extractor has an image size of 112 × 112 pixels.

Hyperparameters. In this study, the batch size was 64 and the initial value of the learning rate was 0.001. Stochastic gradient descent (SGD) was used as the optimization method. In addition, early stopping was used to improve learning efficiency. Therefore, the number of training epochs varied from model to model. The patience of early stopping was set to 5, and learning was terminated when the value of the loss function at the time of validation rises five times in a row. Verification was performed at the end of each learning batch. Furthermore, negative slope of $\mathcal{L}_{CA}$ (Eq. 2), was set to 0.1.

4.2 Comparison with General Classification Methods

Table 1 presents the results obtained under various conditions in the CER model. These conditions include the type of backbone used as a feature extractor, the loss function, and predict method. The models used as feature extractors were ResNet50 [9], EffectNet B2 [20], DDAMFN [26], and DAN [24]. For the loss function, the cross entropy loss ($\mathcal{L}_{CE}$) used in general classification task models and the constrained angle loss ($\mathcal{L}_{CA}$) proposed in this study were compared and evaluated. As for the prediction method of compound emotion, we compared a general classification method, the EMMA method proposed in this study and transfer learning. General classification method is trained on a compound emotion dataset and directly predicts. EMMA uses models trained on basic emotion to predict compound emotions. Transfer learning is pre-trained on basic emotions and then undergoes additional training on the compound emotion dataset. Parameters are updated only for compound EmoVec. The transfer learning method is examined in detail in Sect. 5.

Furthermore, accuracy and unweighted average recall (UAR) evaluation metrics were used. Accuracy is the rate of correct answers out of all test data. UAR is the average recall of each class, which means it corresponds to the average of the diagonal elements of the confusion matrix. This indicator allows us to measure whether we are reasoning in a balanced manner.

Table 1. Result of CER. General classification (CLN) trained on the compound emotion dataset; EMMA is the proposed method and uses the basic emotion estimation model and no training data for compound emotions; Transfer is pre-trained on the basic emotion and then additionally trained on the compound emotion (this is described in Sect. 5).

FE	Loss	predict method					
		General CLN		EMMA(ours)		Transfer	
		Acc.[-]	UAR[-]	Acc.[-]	UAR[-]	Acc.[-]	UAR[-]
ResNet50 [9]	$\mathcal{L}_{CE}$	0.434	0.271	0.436	0.407	0.484	0.325
EffectNet B2 [20]	$\mathcal{L}_{CE}$	0.505	0.328	0.447	0.419	0.563	0.398
DAN [24]	$\mathcal{L}_{CE}$	0.606	0.440	0.539	0.502	0.621	0.495
DDRAMFN [26]	$\mathcal{L}_{CE}$	0.646	0.492	0.615	0.530	0.653	0.528
ResNet50	$\mathcal{L}_{CA}$	-		0.429	0.391	0.487	0.329
EffectNet B2	$\mathcal{L}_{CA}$	-		0.434	0.422	0.578	0.408
DAN	$\mathcal{L}_{CA}$	-		0.578	0.527	0.659	0.542
DDRAMFN	$\mathcal{L}_{CA}$	-		**0.617**	**0.559**	**0.659**	**0.559**

First, we compare the results of the general classification method (Table 1, red background), which uses $\mathcal{L}_{CE}$ and training data of compound emotions, and EMMA method (blue background), which uses $\mathcal{L}_{CA}$ and only the basic emotion data to construct the CER model. Despite not using a training dataset of compound emotions, EMMA achieves performance comparable to the general classification method across all models. In particular, UAR shows a significant improvement when compared to general method, which means that it can predict all emotions comprehensively. The confusion matrix of the proposed method is shown in Fig. 2 and the one when trained as a general classification task is shown in Fig. 3(b). Comparing these two confusion matrices, it can be seen that the proposed method has a high recognition rate without bias by class. We consider this to be a significant advantage, because it does not use composite emotion data with a disproportionate number of data.

In fact, to examine the relationship between the amount of data and recognition accuracy, the True Positive Rate (TPR) for each emotion is shown in Table 2. The numbers in parentheses in this table indicate the amount of training data for each emotion. The backbone was DAN. The results clearly show that the breakdown of the TPR is different between the EMMA and the general method. Specifically, the TPR of the general method is high for the happily surprised and sadly disgusted emotions, which have sufficient data, whereas it is low for the sadly surprised and disgustedly surprised emotions, which have insufficient data. In other words, the general method is highly dependent on data imbalance. By contrast, the proposed method does not have such a problem and can recognize the data in a relatively balanced manner.

Table 2. TPR for the EMMA and general classification method. The numbers in parentheses indicate the number of training data. Bold number is the higher TPR for the two methods.

Label	TPR[-]	
	EMMA(ours)	General classification
Happily Surprised	**0.81**	0.78 (438)
Happily Disgusted	**0.70**	0.40 (176)
Sadly Fearful	**0.64**	0.23 (85)
Sadly Angry	**0.27**	0.15 (110)
Sadly Surprised	**0.39**	0.11 (55)
Sadly Disgusted	0.67	**0.70** (483)
Fearfully Angry	0.58	**0.61** (98)
Fearfully Surprised	0.58	**0.73** (356)
Angrily Surprised	**0.37**	0.24 (116)
Angrily Disgusted	0.45	**0.72** (521)
Disgustedly Surprised	**0.34**	0.17 (91)

Next, comparing the performance obtained when using $\mathcal{L}_{CE}$ (green background) and when using $\mathcal{L}_{CA}$ (blue background) in this framework, we find that the proposed loss function improves UAR, indicating the effectiveness of angle-based metric learning. The confusion matrix in Fig. 3(c) uses Cross Entropy loss as the loss function. A comparison of this result with that of the proposed method (Fig. 2) shows that the overall trend is the same, but the actual recognition rate is improved by the proposed method. In other words, changing the loss function from Cross Entropy loss ($\mathcal{L}_c e$) to $\mathcal{L}_c a$ proposed in this study improves the overall recognition rate. In other words, changing the loss function from Cross Entropy loss ($\mathcal{L}_{ce}$) to Constrained Angle loss ($\mathcal{L}_{ca}$) proposed in this study improves the overall recognition rate.

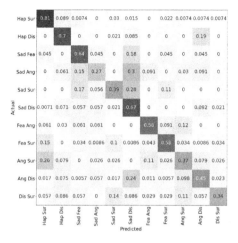

Fig. 2. Confusion matrix when predicting compound emotions using the proposed method. DAN is used as a feature extractor. This model trained basic emotional features using Constrained Angle loss.

4.3 Verification of the Leaky ReLU Gradient

In this section, we examine the effectiveness of Leaky ReLU in the loss function $\mathcal{L}_{CA}$ (Eq. 2) newly introduced in this study. Figure 4 shows the recognition accuracy of compound emotions when the hyperparameter of the Leaky ReLU function, negative slope (NS), was varied, and the shape of the cosine similarity in this case is shown in Fig. 5. The backbone used DAN, which resulted in the largest changes. When $NS = 0$, the result is consistent with the ReLU function, and when the cosine similarity is negative, the value is zero. When $NS = 1$, the input is directly output, so nothing is introduced (the cosine similarity ranges from -1 to 1). When $0 < NS < 1$, the value is slightly larger when the cosine similarity is negative. The results of Fig. 4 reveal that the highest accuracy was obtained when $NS = 0.1$ was used to constrain the results with Leaky ReLU.

Next, we consider the change in the distribution of feature vectors due to this constraint. The distribution of feature vectors for basic emotions was visualized using t-SNE, as shown in Fig. 6. This visualization was performed separately for three different groups: those employing $\mathcal{L}_{CE}$ as the loss function, those incorporating Leaky ReLU within $\mathcal{L}_{CA}$, and those not utilizing Leaky ReLU. Table 3 shows the intra-class variance for each emotion and inter-class in these three conditions. Comparing t-SNE of $\mathcal{L}_{CE}$ and $\mathcal{L}_{CA}$, we can see that $\mathcal{L}_{CA}$ has a better separation performance, especially for fear and disgust, because the distribution of $\mathcal{L}_{CA}$ is more coherent. Actually, Table 3 shows that the inter-class variance is smaller for $\mathcal{L}_{CA}$. By contrast, comparing the results with and without Leaky ReLU in $\mathcal{L}_{CA}$, we can see that the intra-class variance is larger and the feature vector diversity is higher when Leaky ReLU is applied. From these facts, the application of Leaky ReLU can improve the diversity within a class while providing separability between classes. Due to these characteristics, we consider

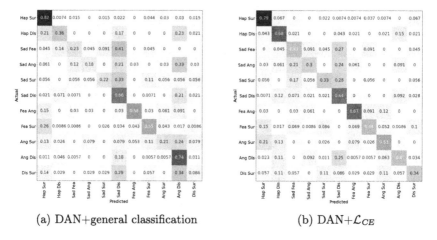

(a) DAN+general classification   (b) DAN+$\mathcal{L}_{CE}$

Fig. 3. Confusion matrix when predicting compound emotions. DAN is used as a feature extractor. (a) a model trained for the general classification task, i.e., using an 11-class compound emotion dataset, (b) a model trained basic emotional features using Cross Entropy loss.

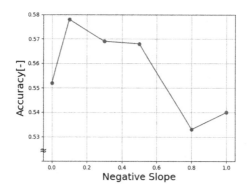

Fig. 4. Accuracy when changing NS in Leaky ReLU

that even a basic emotion feature extractor can capture the characteristics of compound emotions.

5 Compound Emotion Recognition Using Transfer Learning

When compound emotion data are available, the recognition accuracy of the proposed model trained on the basic emotion data should be improved if additional training on the compound emotion data is conducted. In this section, we use transfer learning to learn about compound emotions. Specifically, the parameters of the feature extractor (FE) pre-trained with the basic emotion shown in Fig. 1

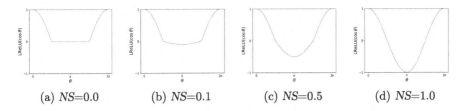

Fig. 5. Graph of $LReLU(\cos\theta)$. NS indicates the negative slope of Leaky ReLU. When $NS = 1$, it coincides with the general $\cos\theta$. The smaller the NS, the less training in the negative direction.

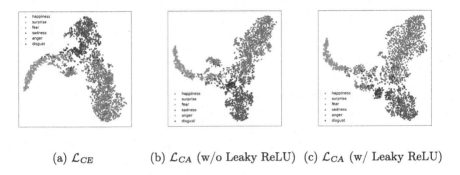

(a) $\mathcal{L}_{CE}$ (b) $\mathcal{L}_{CA}$ (w/o Leaky ReLU) (c) $\mathcal{L}_{CA}$ (w/ Leaky ReLU)

Fig. 6. t-SNE visualization of feature vector distribution. The feature extractor is DAN.

are fixed and matrix of basic EmoVecs (W_b) is excluded. Instead, we introduce the matrix of the representative vectors of the compound emotion (W_c) as the training parameter. This method makes it possible to estimate compound emotions using basic emotion features. The loss function is the constrained angle loss (NS = 1), which is the loss function excluding Leaky ReLU from constrained angle loss (Eq. 2). The reason for excluding Leaky ReLU is that, unlike basic emotion learning, diversity in the feature vectors is not required.

Next, the verification results are presented. The same backbone used in the previous section is used for the feature extractor. The results of the method using transfer learning are shown in Table 1, which reveals that the accuracy is improved by transfer learning. However, as with the conventional method (red background), the UAR is low relative to accuracy, and the results for resNet50 are instead significantly reduced by the transfer learning. This is also due to the imbalance in the data. By contrast, the DAN+$\mathcal{L}_{CA}$ results show that the UAR is improved by transfer learning, and this is a result not seen in the case of DAN+$\mathcal{L}_{CE}$. DDRAMFN has a similarly improved UAR. Common to all UAR-improved backbones is the use of complex attention mechanisms. We therefore infer that $\mathcal{L}_{CA}$ has a high affinity with this mechanism and contributes to the improvement of UAR through synergistic effects.

Table 3. Intra and inter-class variance of feature vector x_b transformed by t-SNE.

Label	Variance[-]		
	$\mathcal{L}_{CE}$	$\mathcal{L}_{CA}$ (w/o LReLU)	$\mathcal{L}_{CA}$ (w/ LReLU)
Surprise	639.9	603.9	618.3
Fear	481.3	214.7	468.5
Disgust	250.4	124.3	253.2
Happiness	552.7	268.5	549.3
Sadness	493.5	508.2	492.7
Anger	162.1	165.2	163.7
inter-class	433.8	327.2	353.1

6 Conclusion

In this study, we proposed a EMotion Mixing Algorithm (EMMA) for CER using basic emotional features. By introducing metric learning into the training of basic emotion features, it is possible to improve the performance of class separation in the feature space. The greatest advantage of this method is that it does not use compound emotion data for estimating compound emotions. It is difficult to annotate and collect data for complex emotions, including compound emotions, and large datasets are not available. The recognition accuracy of conventional methods is highly dependent on data size, making it difficult to recognize emotions that are difficult to collect. By contrast, the proposed method does not use compound emotion data at all, which avoids this problem.

In addition, after learning the features of basic emotions, we also tested a method for learning the additional features of compound emotions using transfer learning. Most CNN architectures are affected by data imbalance, but models that use complex attention mechanisms can solve this problem while improving recognition accuracy.

One of our future tasks is to examine more specifically what kind of dataset is appropriate for pre-learning basic emotions. For example, this study used a dataset consisting of natural facial expressions, but it will be necessary to verify whether the features learned using this dataset are sufficient to recognize natural complex emotions in other datasets such as those created by actors in a laboratory (e.g., CK+ [15] or JAFFE [16]). We also believe that there is room for improvement in the way the representative vectors of complex emotions are defined. In the method proposed in this study, they were defined by a simple linear combination, but the relationships among basic emotions are more complex, and a new definition that takes these into account is needed.

Acknowledgments. This work was supported by JSPS Grant-in-Aid for Scientific Research (B) Grant Number JP23H03486.

References

1. Bryant, D., Deng, S., Sephus, N., Xia, W., Perona, P.: Multi-dimensional, nuanced and subjective – measuring the perception of facial expressions. In: 2022 IEEE/CVF Conference on Computer Vision and Pattern Recognition (CVPR), pp. 20900–20909 (2022). https://doi.org/10.1109/CVPR52688.2022.02026
2. Darwin, C.: The Expression of Emotions in Man and Animals. Oxford University Press (1872)
3. Deng, J., Guo, J., Xue, N., Zafeiriou, S.: ArcFace: additive angular margin loss for deep face recognition. In: Proceedings of the IEEE/CVF Conference on Computer Vision and Pattern Recognition (CVPR), June 2019
4. Donahue, J., et al.: DeCAF: a deep convolutional activation feature for generic visual recognition. In: Xing, E.P., Jebara, T. (eds.) Proceedings of the 31st International Conference on Machine Learning. Proceedings of Machine Learning Research, vol. 32, pp. 647–655. PMLR, Bejing, China, 22–24 June 2014. https://proceedings.mlr.press/v32/donahue14.html
5. Du, S., Martinez, A.M.: Compound facial expressions of emotion: from basic research to clinical applications. Dialogues Clin. Neurosci. **17**(4), 443–455 (2015). https://doi.org/10.31887/dcns.2015.17.4/sdu
6. Du, S., Tao, Y., Martinez, A.M.: Compound facial expressions of emotion. Proc. Nat. Acad. Sci. **111**(15), E1454–E1462 (2014). https://doi.org/10.1073/pnas.1322355111. https://www.pnas.org/doi/abs/10.1073/pnas.1322355111
7. Ekman, P.: An argument for basic emotions. Cogn. Emot. **6**(3–4), 169–200 (1992)
8. Ekman, P., Friesen, W.V.: Facial action coding system. Environ. Psychol. Nonverbal Behav. (1978)
9. He, K., Zhang, X., Ren, S., Sun, J.: Deep residual learning for image recognition. In: 2016 IEEE Conference on Computer Vision and Pattern Recognition (CVPR), pp. 770–778 (2016). https://doi.org/10.1109/CVPR.2016.90
10. Kollias, D.: Multi-label compound expression recognition: C-expr database & network. In: Proceedings of the IEEE/CVF Conference on Computer Vision and Pattern Recognition (CVPR), pp. 5589–5598, June 2023
11. Li, S., Deng, W.: Reliable crowdsourcing and deep locality-preserving learning for unconstrained facial expression recognition. IEEE Trans. Image Process. **28**(1), 356–370 (2019)
12. Li, S., Deng, W., Du, J.: Reliable crowdsourcing and deep locality-preserving learning for expression recognition in the wild. In: 2017 IEEE Conference on Computer Vision and Pattern Recognition (CVPR), pp. 2584–2593. IEEE (2017)
13. Li, Z., Zhang, T., Jing, X., Wang, Y.: Facial expression-based analysis on emotion correlations, hotspots, and potential occurrence of urban crimes. Alex. Eng. J. **60**(1), 1411–1420 (2021)
14. Liu, W., Wen, Y., Yu, Z., Li, M., Raj, B., Song, L.: SphereFace: deep hypersphere embedding for face recognition. In: Proceedings of the IEEE Conference on Computer Vision and Pattern Recognition (CVPR), July 2017
15. Lucey, P., Cohn, J.F., Kanade, T., Saragih, J., Matthews, I.: The extended Cohn-Kanade dataset (CK+): a complete dataset for action unit and emotion-specified expression. In: 2010 IEEE Computer Society Conference on Computer Vision and Pattern Recognition - Workshops, pp. 94–101 (2010). https://doi.org/10.1109/CVPRW.2010.5543262
16. Lyons: The Japanese female facial expression (JAFFE) dataset, September 2019. https://doi.org/10.5281/zenodo.3451524

17. Mannepalli, K., Sastry, P.N., Suman, M.: A novel adaptive fractional deep belief networks for speaker emotion recognition. Alexandria Eng. J. **56**(4), 485–497 (2017). https://doi.org/10.1016/j.aej.2016.09.002. https://www.sciencedirect.com/science/article/pii/S1110016816302484
18. Nan, Y., Ju, J., Hua, Q., Zhang, H., Wang, B.: A-MobileNet: an approach of facial expression recognition. Alex. Eng. J. **61**(6), 4435–4444 (2022)
19. Sajjad, M., et al.: A comprehensive survey on deep facial expression recognition: challenges, applications, and future guidelines. Alexandria Eng. J. **68**, 817–840 (2023). https://doi.org/10.1016/j.aej.2023.01.017. https://www.sciencedirect.com/science/article/pii/S1110016823000327
20. Tan, M., Le, Q.: EfficientNet: rethinking model scaling for convolutional neural networks. In: Chaudhuri, K., Salakhutdinov, R. (eds.) Proceedings of the 36th International Conference on Machine Learning. Proceedings of Machine Learning Research, vol. 97, pp. 6105–6114. PMLR, 09–15 June 2019. https://proceedings.mlr.press/v97/tan19a.html
21. Tonguç, G., Ozaydın Ozkara, B.: Automatic recognition of student emotions from facial expressions during a lecture. Comput. Educ. **148**, 103797 (2020). https://doi.org/10.1016/j.compedu.2019.103797. https://www.sciencedirect.com/science/article/pii/S0360131519303471
22. Wang, H., et al.: CosFace: large margin cosine loss for deep face recognition. In: Proceedings of the IEEE Conference on Computer Vision and Pattern Recognition (CVPR), June 2018
23. Wang, K., Peng, X., Yang, J., Meng, D., Qiao, Y.: Region attention networks for pose and occlusion robust facial expression recognition. IEEE Trans. Image Process. **29**, 4057–4069 (2020). https://doi.org/10.1109/TIP.2019.2956143
24. Wen, Z., Lin, W., Wang, T., Xu, G.: Distract your attention: multi-head cross attention network for facial expression recognition. Biomimetics **8**(2) (2023). https://doi.org/10.3390/biomimetics8020199. https://www.mdpi.com/2313-7673/8/2/199
25. Yun, S.S., Choi, J., Park, S.K., Bong, G.Y.: Social skills training for children with autism spectrum disorder using a robotic behavioral intervention system. Autism Res. **10**(7), 1306–1323 (2017). https://doi.org/10.1002/aur.1778. https://onlinelibrary.wiley.com/doi/abs/10.1002/aur.1778
26. Zhang, S., Zhang, Y., Zhang, Y., Wang, Y., Song, Z.: A dual-direction attention mixed feature network for facial expression recognition. Electronics **12**(17) (2023). https://doi.org/10.3390/electronics12173595. https://www.mdpi.com/2079-9292/12/17/3595

LViTES: Leveraging Vision and Text for Enhancing Segmentation of Endoscopic Images

Thang La[1], Minh-Hanh Tran[1], Viet-Hang Dao[2,3], and Thanh-Hai Tran[1(✉)]

[1] School of Electrical and Electronic Engineering, Hanoi University of Science and Technology, Hanoi, Vietnam
hai.tranthithanh1@hust.edu.vn
[2] Hanoi Medical University Hospital, Hanoi, Vietnam
[3] Institute of Gastroenterology and Hepatology, Hanoi, Vietnam

Abstract Automatic lesion segmentation in endoscopic images is crucial for mitigating the risk of omissions during analysis, particularly for inexperienced physicians or in situations of medical overload. Traditional segmentation models predominantly rely on pixel-level labeled images, often neglecting auxiliary information such as physicians' diagnostic conclusions. This study proposes a novel approach to harness available lesion information—including segmentation regions, physician conclusions, and supplementary disease descriptions—to improve segmentation efficacy. Our method builds upon the successful integration of CNN and Vision Transformer architectures from the LViT model, originally designed for lung cancer lesion segmentation from X-ray images using dual inputs: images and text. We propose a new framework, namely called LViTES with four key advancements: 1) optimizing the LViT architecture to enhance image feature extraction by incorporating the EfficientNet backbone and integrating Cross-Attention, while also reducing model complexity and parameters; 2) addressing the scarcity of textual descriptions in current datasets by developing a module that generates text from segmentation masks based on attributes like shape, location, size, and quantity; 3) incorporating both image and text inputs during training while allowing adaptive prediction with only image inputs to align with typical use cases; and 4) evaluating model performance using both generated text and physician-provided descriptions. The effectiveness of our approach is validated on three types of lesions—gastric cancer, esophageal cancer (our self-collected datasets), and polyps (Kvasir-SEG dataset)—demonstrating superior performance compared to state-of-the-art methods.

Keywords: Segmentation · Endoscopic images · Deep learning · Transformer · CNN · LLM

1 Introduction

Digestive diseases are a significant global health concern, with millions affected worldwide [22]. The rising prevalence of these diseases underscores the urgent need for advanced diagnostic tools that can aid in early detection and accurate diagnosis. One promising solution is the automatic analysis and diagnosis of gastrointestinal conditions using artificial intelligence (AI).

Current methods for gastrointestinal endoscopy image segmentation predominantly rely on state-of-the-art convolutional neural networks (CNNs) or Transformers [19–21]. These techniques leverage the power of CNNs to accurately segment images, which is crucial for identifying abnormalities in the digestive tract. However, recent advancements in large language models (LLMs) have shown the potential to enhance image analysis by integrating language, thereby improving overall diagnostic accuracy [27].

Several models that combine language and image processing have been successfully applied to natural images. Notable examples include CLIP [16] and SAM [6], which have demonstrated significant improvements in various image analysis tasks. Despite their success in natural image domains, there has been limited research on applying these language-image models to medical imaging, particularly in the field of gastrointestinal endoscopy.

A few studies have explored the application of language and image models to chest X-rays [13], but there is currently no model specifically designed for endoscopic image analysis. The primary challenges include the scarcity of large databases and the lack of accompanying textual data. Addressing these challenges could pave the way for significant advancements in medical imaging analysis.

This paper presents an initial experimental study on the integration of language and images to enhance the segmentation quality of medical endoscopic images. We propose a novel approach LViTES (**L**everaging **Vi**sion and **T**ext for **E**ndoscopic **S**egmentation) that utilizes dual inputs: images and text. The text can be derived from doctors' conclusions or automatically generated from the ground truth masks. These modalities are processed through a CNN - Transformer (e.g. LViT) network architecture to extract and cross-interact features, ultimately improving segmentation outcomes.

Our main contributions are as follows: i) an original research combining text and images for gastrointestinal endoscopy image segmentation; ii) an integration of various textual generation methods to address the scarcity of medical documents; iii) a mechanism allowing training with both text and images, while enabling inference using only image inputs. Experimental evaluation across different datasets (gastric cancer, esophageal cancer, and polyps), demonstrated remarkable results. This pioneering study aims to open new avenues for the application of AI in medical imaging, ultimately contributing to better diagnostic tools and improved patient outcomes in the realm of digestive diseases.

2 Related Works

2.1 Semantic Segmentation of Medical Images

Semantic segmentation is a method in the field of computer vision and image processing that aims to classify each pixel in an image into different classes or labels. In this task, the Fully Convolutional Network (FCN) [10] is considered the first end-to-end pixel-to-pixel network to be published. Following this, models such as DeepLab [1] with three main components: Deep Convolutional Nets, Atrous Convolution, and Fully Connected CRFs have achieved impressive results in medical image segmentation and satellite image processing. Additionally, Mrdff [24,25] for CT whole heart segmentation were developed based on and improved upon the FCN. Among these, a pioneering model in the medical field is U-Net [17]. Subsequently, various U-Net variants have been developed to enhance performance and accuracy in different medical applications. Notably, UNet++ [28] utilizes more complex skip connections between the encoder and decoder layers to improve detail and accuracy in segmentation. Attention U-Net [15] incorporates attention mechanisms into the architecture, and Dense-UNet [7] is a variant combining U-Net and DenseNet. However, most methods face limitations with the amount of data and heavily depend on data quality, resulting in constrained model performance in generalization. Some recent approaches [8,11,12,23] employing semi-supervised learning methods have partially addressed the issue of data and label scarcity.

2.2 Text and Image Combination for Image Analysis

In recent years, vision-language models have gained importance and become a focal point of development. Among these, CLIP [16] stands as the first pioneering vision-language pre-trained (VLP) model, yielding impressive results in vision-and-language tasks and improving accuracy over previous pre-trained models based solely on images or text [18]. CLIP employs a neural network incorporating transformers, specifically utilizing the Vision Transformer (ViT) for image processing and a BERT-like transformer variant for text processing. The model is trained on a large dataset of images and text from the internet, combining image data with textual descriptions. Despite the advantages CLIP brings, it still faces challenges such as handling specific problem domains, computational efficiency, and model complexity.

Recent research has further explored the integration of image-text information to enhance performance in image segmentation tasks, with models like VLT [4], LAVT [26], and TransVG++ [2]. VLT can generate queries from textual descriptions to guide the model during image segmentation. Conversely, LAVT employs early fusion mechanisms and pixel-word attention to optimize image segmentation based on linguistic guidance. TransVG++ uses a Language Conditioned ViT trained to embed textual description information into the image processing workflow, leveraging attention mechanisms to synchronize information from images and text, thus enhancing object recognition accuracy.

However, these models are not specifically designed for medical images. Natural images and medical images exhibit numerous differences. Medical images of different body parts also have distinct characteristics, often lacking clear delineation and uniformity in size and shape. Consequently, directly applying computer vision models trained on natural images to medical image analysis poses challenges. Furthermore, textual descriptions of medical images differ significantly from those of natural images, resulting in weak correlations between text and images in the medical context.

To address these challenges, the LViT [9] model was developed, designed specifically for the medical field. It employs a hybrid CNN-Transformer architecture to retain both local and global features, utilizes only the Embedding layer to transform text features, reducing parameters and computational cost, and incorporates LV (LanguageVision) loss to supervise training of unlabeled images using direct text information. LViT was validated on chest X-ray images. Combining of language and vision for endoscopic images are underexplored.

3 Proposed Method

3.1 General Framework

In this paper, we reply upon LViT for endoscopic image segmentation with the training from both images and textual description. However, we have made several significant improvements based on the successful combination of CNN and Transformer in LViT, while also using the EfficientNet Backbone to image feature extraction. Additionally, a cross-attention mechanism has been added during the decoder stage to facilitate closer interaction between the features, thereby improving segmentation performance and synchronizing information between the different data streams. This combination not only enhances the model's analytical capabilities but also improves its ability to effectively handle lesion-related descriptions. Our proposed framework for lesion segmentation is illustrated in Fig. 1.

- At the training phase, it takes two inputs (the pair of original image and the corresponding mask annotated by the experts, and a textual description of the lesion). The textual description can come from two different sources: the medical report produced by doctors or automatically generated by our algorithm from the groundtruth mask. The textual stream goes through a text embedding to provide a text representation while the image is fed into the EfficientNet Backbone for feature extraction. Text features and image features at different layers interact in an attention model.
- At the inference phase, the model can take an image with or without a text description to generate a segmentation result.

In the following, we will detail each step of the framework.

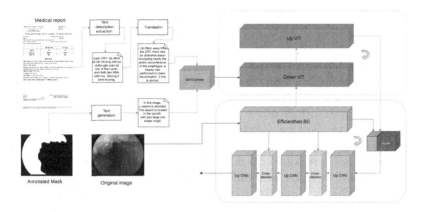

Fig. 1. Proposed framework LViTES for endoscopic image segmentation.

3.2 Text-Based Description Processing

Our disease description texts are generated from two sources: medical reports provided by the hospital and auto-generated texts from masks, as shown in Fig. 1.

Text Extraction. As we mentioned, there is no dataset of endoscopic images which contains both binary masks and the corresponding textual description about the lesion. To test our method, we have collected a dataset which contain two lesion categories: gastric cancer, esophageal cancer. The textual descriptions of the lesion are stored in structured pdf format. We develop a script to acronymize the patients identity. We then extract the conclusion about the lesion part. The top-left of Fig. 1 illustrates a medical report for a given patient. We observe that during an endoscopic examination, the doctor reports all relevant issues of the digestive system, such as those involving the esophagus, stomach, duodenal bulb, duodenum, etc. However, if a lesion is identified as Gastric Cancer or Esophageal Cancer, we focus exclusively on extracting information related to the that lesion.

Text Translation. As the collected texts are in Vietnamese or may be any other languages, therefore, we translated them into English for text encoding. In our work, our self-collected dataset contain textual description in Vietnamese. We use the VinAI Translate model [14] to translate text in Vietnamese into English. In Fig. 1, we illustrate an example where the Vietnamese text about Esophageal cancer is extracted as "Cách CRT~18-28cm có tổn thương loét sùi chiếm gần toàn bộ chu vi thực quản, sinh thiết làm MBH kiểm tra. Đường Z bình thường." It is translated into English as "~18-28cm away from the CRT, there was an ulcerative lesion occupying nearly the entire circumference of the esophagus, a biopsy was performed to check the ectropion. Zline is normal."

Mask-Based Text Description Generation. The textual description of the lesion is not always available. For example, for KVasir-SEG dataset, only groundtruth mask is provided. To address the lack of textual description, which is a common problem in practical situation, we propose a novel approach to generate text descriptions from the image data itself. This approach mitigates the dependency on textual annotations provided by physicians, making it particularly useful in situations where such annotations are scarce or unavailable.

Our proposed method includes following steps. First, we detect contours from the mask image using Canny. Then we apply connected component analysis to determine the legion regions in terms of number, location, and shape. For location, we divide the image into different regions (center, top-middle, bottom-middle, left-middle, right-middle, top-left, top-right, bottom-left, bottom-right, left, right, top, bottom) and determine the location of each wound within these regions. For shape description, we calculate characteristics of the wounds such as area, perimeter, roundness, shape, and size. Finally, based on the number, location, and characteristics of the wounds, we create a natural language description of the image. Figure 2 illustrated the main steps for textual description generation from the ground truth image.

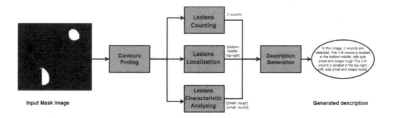

Fig. 2. Main steps for textual description generation from the groundtruth image.

3.3 Model Architecture

The input of our model in the training phase composes of two components: an original image and its groundtruth and the corresponding textual description (real of generated). Text embedding is computed using the well-know pre-trained BERT [3] while image pass through an EfficientNet-B0 for extracting visual features. To capture the correlation between visual and textual features, both T_{embed} and I_{embed} go through a down ViT then an up ViT. The output $Z_{reconstruct}$ from the up ViT and the visual feature continue to pass through a Pixel Level Attention Module (PLAM) before entering to a up sampling module for generating the segmentation map.

After the image and text features have been extracted through their respective branches, they pass through double Vision Transformer (ViT) blocks to optimize and combine the information before entering the decoder stage. The Vision Transformer block is completely inspired from the LViT model [9]. However, it is noted that our architecture LViTES makes some differences. First, for the image embedding, LViT utilized fours down CNNs which contains Conv, BatchNorm (BN), and ReLU activation layers for image embedding. Instead, LViTES employed EfficientNet-B0 as explained in the previous subsection. Second, we keep only one down ViT and one up ViT in the ViT U-block that capture interaction between text and image embeddings while the original LViT contains four down and four up ViTs. We reduce the number of ViT blocks to avoid overfitting and enhance the accuracy of the model as presented in our experiments. We detail our architecture LViTES as follows:

- **Down ViT Block**:
  - The image embedding $\mathbf{I}_{\text{embed}}$ and text embedding $\mathbf{T}_{\text{embed}}$ are processed simultaneously through Transformer layers to capture relationships between image patches and text features.
  - $\mathbf{I}_{\text{embed}}$ is passed through PatchEmbedding, while $\mathbf{T}_{\text{embed}}$ is aligned using a CTBN block (Conv, BatchNorm, ReLU), and both are merged:

    $$\mathbf{x}_{\text{merged}} = \text{PatchEmbedding}(\mathbf{I}_{\text{embed}}) + \text{CTBN}(\mathbf{T}_{\text{embed}})$$

  - The merged feature is passed through a Transformer's Multi-Headed Self-Attention (MHSA) and MLP layers:

    $$\mathbf{x}' = \text{MHSA}(\text{LN}(\mathbf{x}_{\text{merged}})) + \mathbf{x}_{\text{merged}}$$

    $$\mathbf{Z} = \text{MLP}(\text{LN}(\mathbf{x}')) + \mathbf{x}'$$

  - We briefly write the above calculation process as follows:

    $$\mathbf{Z} = f^1_{\text{ViT}}(\mathbf{I}_{\text{embed}}, \mathbf{T}_{\text{embed}})$$

- **Up ViT Block (Reconstruction)**:
  - The output $\mathbf{Z}$ from the first block is reconstructed by the second ViT block. Gives output the same size as the visual feature:

    $$\mathbf{Z}_{\text{reconstruct}} = f^2_{\text{ViT}}(\mathbf{Z})$$

  - The second block introduces a CTBN update:

    $$\mathbf{x}' = \mathbf{x} + \text{CTBN}(\mathbf{x})$$

- **Pixel-Level Attention Module (PLAM)**:
  - PLAM merges local image features with semantic text features. It uses Global Average Pooling (GAP) and Global Max Pooling (GMP) on the input X. For our model, X represents the sum of $\mathbf{Z}_{reconstruct}$ from the up ViT and the visual feature:

    $$\mathbf{z}_{\text{GAP}} = \text{GAP}(\text{RELU}(\text{Conv2D}(X))), \quad \mathbf{z}_{\text{GMP}} = \text{GMP}(\text{RELU}(\text{Conv2D}(X)))$$

  - The combined output:

    $$\mathbf{z}_{\text{concat}} = \text{CONCATENATE}(\mathbf{z}_{\text{GAP}}, \mathbf{z}_{\text{GMP}}, \mathbf{z}_{\text{GAP}} + \mathbf{z}_{\text{GMP}})$$

- This is passed through an MLP and scaled:

$$y = X \times \text{MLP}(\mathbf{z}_{\text{concat}})$$

- **Decoder:**
  - $\mathbf{Z}_{\text{reconstruct}}$, the initial image feature $\mathbf{I}_{\text{embed}}$, and PLAM outputs are combined.
  - The decoder applies convolutional layers, upsampling, and cross-attention between image and text features for final output generation.

In this work, we inherit the loss function design from the LViT model, which consists of a combination of **Dice loss** and **Cross-Entropy loss**. The loss function is used to guide the model in learning accurate segmentation by minimizing errors in both labeled and unlabeled data. To further improve the robustness of the model and mitigate overfitting, we incorporate an **L2 regularization** mechanism. The L2 regularization term penalizes large weights by adding the squared magnitude of weights to the loss function. This helps reduce the complexity of the model and prevents it from overfitting to the training data.

4 Experiments

4.1 Datasets and Implementation Details

In this study, we utilize two datasets: Kvasir-SEG [5] and IGHEndoLesion-Image-Report, which are divided into training, validation, and test sets in a ratio of 7:1:2, as shown in Table 1.

Kvasir-SEG: This is a widely used dataset for tasks related to the evaluation of models for detecting and segmenting lesions in the form of polyps, comprising 1,000 images with resolutions ranging from 332×487 to 1920×1072. This dataset provides only ground-truth image, not textual description. We must apply text generation module to train the segmentation model.

IGHEndoLesion-Image-Report: This dataset consists of 2,667 pairs of image reports with a resolution of 1280×995, where 1,383 pairs are related to esophageal cancer and 1,284 pairs pertain to gastric cancer. Each patient may have more than one pair of image reports. Each image is accompanied by a detailed report regarding the patient's condition, provided by physicians who directly conducted the examinations. These reports contain critical information about the diagnosis and severity of lesions. Each report includes three main sections of descriptive information: characteristics at the cancer site; detailed descriptions of the lesions such as size, shape, and location; and annotations regarding the patient's condition. Specifically, reports concerning gastric cancer have an average length of 232 characters per report and describe features such as gastric fluid, mucosa, curvature, antrum, and cardia. Reports related to esophageal cancer have an average length of 218 characters per report and describe features such as veins, esophageal lumen, Z-line (transition zone of the mucosa), and mucosa. The images and medical reports are provided by the Gastroenterology and Hepatology Institute of Hanoi Medical University.

Table 1. Summary of experimented datasets

Dataset	IGHEndo	Lesion-Image-Report	Kvasir-SEG
Lesion Type	Gastric Cancer	Esophageal Cancer	Polyp
Train set	898	968	700
Val set	128	138	100
Test set	258	277	200
Total	1284	1383	1000

The proposed approach is implemented using PyTorch. The main server parameters are as follows: the operating system is Ubuntu 20.04.6 LTS, the CPU is an Intel(R) Xeon(R) Gold 6130, the GPU is a single NVIDIA RTX 4090, and the memory capacity is 128 GB. The initial learning rate is set to 3×10^{-4} for all datasets. We also use an early stopping mechanism, which halts training if the model performance does not improve after 40 epochs. The default batch size is 16 for both datasets, and all images are resized to 224×224. Additionally, the cosine learning rate schedule is applied during training.

4.2 Experimental Results

We evaluated our model on three datasets: Gastric Cancer, Esophageal Cancer, and Kvasir-SEG. For the first two, we compared it to the original LViT model using text generated by Mask-based text description generation (G) and text from medical records (D). For Kvasir-SEG, we compared our results to state-of-the-art methods and conducted additional experiments to assess the effects of not using text (W) and only using text during training (IW). It is noted that for Kvasir-SEG dataset, we generated textual description about the polyps.

Table 2. Experimental results on the Gastric Cancer dataset.

Method	mIoU	DSC	Recall	Precision
LViT-D	0.7200	0.8125	0.8418	0.9050
LViT-G	0.7437	0.8318	0.8754	0.8906
LViTES-D	0.7618	0.8829	0.9114	0.9309
LViTES-G	**0.7855**	**0.8941**	**0.9127**	**0.9449**

Table 3. Experimental results on Esophageal Cancer dataset.

Method	mIoU	DSC	Recall	Precision
LViT-D	0.6707	0.7828	0.8746	0.8835
LViT-G	0.7006	0.8068	0.8725	0.9024
LViTES-D	0.7579	0.8596	0.9198	0.9411
LViTES-G	**0.7634**	**0.8702**	**0.9288**	**0.9497**

Table 4. Experimental results on the Kvasir-SEG dataset.

Method	mIoU	DSC	Recall	Precision
U-Net	0.7472	0.8264	0.8504	0.8703
U-Net++	0.7420	0.8228	0.8437	0.8607
ResU-Net++	0.5341	0.6453	0.6964	0.7080
HarDNet-MSEG	0.7459	0.8260	0.8485	0.8652
ColonSegNet	0.6980	0.7920	0.8193	0.8432
UACANet	0.7692	0.8502	0.8799	0.8706
UNeXt	0.6284	0.7318	0.7840	0.7656
TransNetR	0.8016	0.8706	0.8843	0.9073
LViT-G	0.8040	0.8779	0.9039	0.9172
LViTES-G (Our)	**0.8642**	**0.9306**	**0.9271**	**0.9363**

Table 5. Other experimental results on the Kvasir dataset.

Method	mIoU	DSC	Recall	Precision
LViTES-W	0.8125	0.8994	0.8872	0.8997
LViTES-IW	0.8311	0.9061	0.8988	0.9124
LViTES-G	**0.8642**	**0.9306**	**0.9271**	**0.9363**

Table 2 and Table 3 show the results obtained by our model, LViTES, compared to the original LViT in two scenarios: using textual descriptions provided by doctors and using textual descriptions generated by our modules. It is shown that in both scenarios, our proposed model, LViTES, outperformed LViT. Specifically, the generated textual descriptions provided slightly better performance than those provided by doctors. This can be explained by the fact that when textual descriptions from doctors are unavailable, the generated text becomes highly significant. Table 4 compares our model, LViTES-G, which uses generated textual descriptions, and demonstrates that it outperformed all existing state-of-the-art (SOTA) methods such as U-Net, U-Net++, ResU-Net++, HarDNet-MSEG, ColoSegNet, UACANet, UNext, and TransNetR. LViTES-G achieved a mIoU that is 6.02% higher than the original LViT-G.

Table 6. Qualitative results on the Kvasir-SEG dataset.

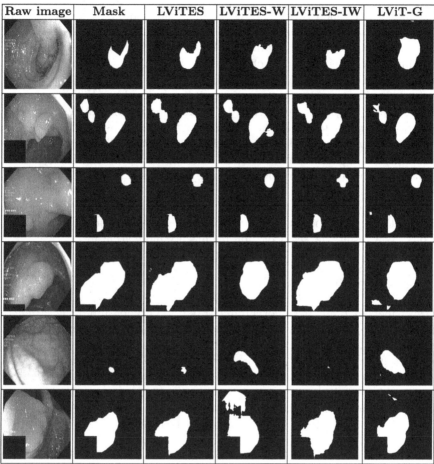

Table 7. Qualitative results on the Gastric Cancer and the Esophageal Cancer datasets.

Table 5 presents the results of our ablation study, and shows the effects of modality on segmentation performance. LViTES-W is the model we trained using only images, without text, while LViTES-IW is the model we trained using only text, without images. We compared these with the final model, LViTES-G, which was trained using both text and images. We observed that using both modalities for training increased the mIoU of LViTES-G by 3.31% and 5.17% compared to LViTES-IW and LViTES-W, respectively.

Table 6 visualizes some segmentation results obtained by LViTES on Kvasir-SEG, compared against the original LViT. It is noted that LViTES produces segmentation maps that best fit the ground truth, whereas LViT sometimes produces larger or smaller lesions than the ground truth. Table 7 illustrates examples comparing results using doctors' textual descriptions against generated textual descriptions. Due to the additional information in the doctors' textual descriptions, the segmentation results are sometimes more biased. In contrast, the generated descriptions are based on ground truth images, helping to focus more on the region of interest by describing the shape, location, and number of lesions. As a result, the segmentation regions are more accurate.

5 Conclusion

In this paper, we presented a novel framework, LViTES, for improving segmentation of endoscopic images by leveraging both visual and textual information. Our approach builds upon the LViT model but introduces significant enhancements, such as utilizing the EfficientNet backbone for image feature extraction and incorporating a cross-attention mechanism during the decoding process. Additionally, we addressed the lack of textual data by proposing a novel method for generating textual descriptions directly from lesion masks, which enhances the model's ability to generalize even when physician-provided reports are unavailable. We evaluated LViTES on three datasets, including gastric cancer, esophageal cancer, and polyps (Kvasir-SEG), and demonstrated that our model outperforms state-of-the-art methods in terms of segmentation accuracy. Specifically, LViTES showed superior performance in both mIoU and Dice Score metrics compared to the original LViT by around 4% on gastric cancer and esophageal cancer. On Kvasir-SEG dataset, the mIoU of LViTES is higher than TransNetR and the original VLiT by 6%, and outperformed the U-NET by 11.70%. Overall, LViTES provides a significant step forward in integrating multimodal data for medical image analysis. Future work may explore expanding the scope of this framework to other medical imaging tasks and further improving the quality of generated textual descriptions to boost segmentation performance.

Acknowledgments. This research is funded by Vietnam Ministry of Science and Technology under grant number KC-4.0-17/19-25.

References

1. Chen, L.C., Papandreou, G., Kokkinos, I., Murphy, K., Yuille, A.L.: DeepLab: semantic image segmentation with deep convolutional nets, atrous convolution, and fully connected CRFs. IEEE TPAMI **40**(4), 834–848 (2017)
2. Deng, J., et al.: TransVG++: end-to-end visual grounding with language conditioned vision transformer. IEEE Trans. Pattern Anal. Mach. Intell. (2023)
3. Devlin, J.: BERT: pre-training of deep bidirectional transformers for language understanding. arXiv preprint arXiv:1810.04805 (2018)
4. Ding, H., Liu, C., Wang, S., Jiang, X.: Vision-language transformer and query generation for referring segmentation. In: Proceedings of the IEEE/CVF International Conference on Computer Vision, pp. 16321–16330 (2021)
5. Jha, D., et al.: Kvasir-SEG: a segmented polyp dataset. In: Ro, Y.M., et al. (eds.) MMM 2020. LNCS, vol. 11962, pp. 451–462. Springer, Cham (2020). https://doi.org/10.1007/978-3-030-37734-2_37
6. Kirillov, A., et al.: Segment anything. In: Proceedings of the IEEE/CVF International Conference on Computer Vision, pp. 4015–4026 (2023)
7. Li, X., Chen, H., Qi, X., Dou, Q., Fu, C.W., Heng, P.A.: H-DenseUNet: hybrid densely connected UNet for liver and tumor segmentation from CT volumes. IEEE Trans. Med. Imaging **37**(12), 2663–2674 (2018)
8. Li, Y., et al.: GT U-net: a U-Net like group transformer network for tooth root segmentation. In: Lian, C., Cao, X., Rekik, I., Xu, X., Yan, P. (eds.) MLMI 2021. LNCS, vol. 12966, pp. 386–395. Springer, Cham (2021). https://doi.org/10.1007/978-3-030-87589-3_40
9. Li, Z., et al.: LViT: language meets vision transformer in medical image segmentation. IEEE Trans. Med. Imaging (2023)
10. Long, J., Shelhamer, E., Darrell, T.: Fully convolutional networks for semantic segmentation. In: CVPR, pp. 3431–3440 (2015)
11. Luo, X., Chen, J., Song, T., Wang, G.: Semi-supervised medical image segmentation through dual-task consistency. In: Proceedings of the AAAI Conference on Artificial Intelligence, vol. 35, pp. 8801–8809 (2021)
12. Luo, X., Hu, M., Song, T., Wang, G., Zhang, S.: Semi-supervised medical image segmentation via cross teaching between CNN and transformer. In: International Conference on Medical Imaging with Deep Learning, pp. 820–833. PMLR (2022)
13. Malaviya, N., Rahevar, M., Virani, A., Ganatra, A., Bhuva, K.: LViT: vision transformer for lung cancer detection. In: 2023 International Conference on Artificial Intelligence and Smart Communication (AISC), pp. 93–98. IEEE (2023)
14. Nguyen, T.H., et al.: A Vietnamese-English neural machine translation system. In: Annual Conference of the International Speech Communication Association (was Eurospeech) 2022, pp. 5543–5544. International Speech Communication Association (ISCA) (2022)
15. Oktay, O., et al.: Attention U-Net: learning where to look for the pancreas. arXiv preprint arXiv:1804.03999 (2018)
16. Radford, A., et al.: Learning transferable visual models from natural language supervision. In: International Conference on Machine Learning, pp. 8748–8763. PMLR (2021)
17. Ronneberger, O., Fischer, P., Brox, T.: U-Net: convolutional networks for biomedical image segmentation. In: Navab, N., Hornegger, J., Wells, W.M., Frangi, A.F. (eds.) MICCAI 2015. LNCS, vol. 9351, pp. 234–241. Springer, Cham (2015). https://doi.org/10.1007/978-3-319-24574-4_28

18. Shen, S., et al.: How much can clip benefit vision-and-language tasks? arXiv preprint arXiv:2107.06383 (2021)
19. Tang, S., et al.: Transformer-based multi-task learning for classification and segmentation of gastrointestinal tract endoscopic images. Comput. Biol. Med. **157**, 106723 (2023)
20. Tran, T.H., et al.: DCS-UNet: dual-path framework for segmentation of reflux esophagitis lesions from endoscopic images with U-Net-based segmentation and color/texture analysis. Vietnam J. Comput. Sci. **10**(02), 217–242 (2023)
21. Wang, S., et al.: Multi-scale context-guided deep network for automated lesion segmentation with endoscopy images of gastrointestinal tract. IEEE J. Biomed. Health Inform. **25**(2), 514–525 (2020)
22. Wang, Y., et al.: Global burden of digestive diseases: a systematic analysis of the global burden of diseases study, 1990 to 2019. Gastroenterology **165**(3), 773–783 (2023)
23. Wu, Y., et al.: Mutual consistency learning for semi-supervised medical image segmentation. Med. Image Anal. **81**, 102530 (2022)
24. Xu, F., et al.: A multi-resolution deep forest framework with hybrid feature fusion for CT whole heart segmentation. In: International Conference on Bioinformatics and Biomedicine (BIBM), pp. 1119–1124. IEEE (2021)
25. Xu, F., et al.: MRDFF: a deep forest based framework for CT whole heart segmentation. Methods **208**, 48–58 (2022)
26. Yang, Z., Wang, J., Tang, Y., Chen, K., Zhao, H., Torr, P.H.: LAVT: language-aware vision transformer for referring image segmentation. In: Proceedings of the IEEE/CVF Conference on Computer Vision and Pattern Recognition, pp. 18155–18165 (2022)
27. Zhang, J., Huang, J., Jin, S., Lu, S.: Vision-language models for vision tasks: a survey. IEEE Trans. Pattern Anal. Mach. Intell. (2024)
28. Zhou, Z., Rahman Siddiquee, M.M., Tajbakhsh, N., Liang, J.: UNet++: a nested U-Net architecture for medical image segmentation. In: Stoyanov, D., et al. (eds.) DLMIA/ML-CDS -2018. LNCS, vol. 11045, pp. 3–11. Springer, Cham (2018). https://doi.org/10.1007/978-3-030-00889-5_1

RichMediaGAI

GameIR: A Large-Scale Synthesized Ground-Truth Dataset for Image Restoration over Gaming Content

Lebin Zhou[1,2(✉)], Kun Han[2], Nam Ling[1], Wei Wang[2], and Wei Jiang[2]

[1] Santa Clara University, Santa Clara, USA
{lzhou,nling}@scu.edu
[2] Futurewei Technologies Inc., San Jose, USA
{lzhou,khan,rickweiwang,wjiang}@futurewei.com

Abstract. Image restoration methods like super-resolution and image synthesis have been successfully used in commercial cloud gaming products like NVIDIA's Deep Learning Super Sampling (DLSS). However, restoration over gaming content is not well studied by the general public. The discrepancy is mainly caused by the lack of ground-truth gaming training data that match the test cases. Due to the unique characteristics of gaming content, e.g., sharp and clear low-resolution (LR) images, the common approach of generating pseudo training data by degrading the original high-resolution (HR) images results in inferior restoration performance. In this work, we develop GameIR, a large-scale high-quality computer-synthesized ground-truth dataset to fill in the blanks, targeting at two different applications. The first is super-resolution with deferred rendering, to support the gaming solution of rendering and transferring LR images only and restoring HR images on the client side. We provide 19200 LR-HR paired ground-truth frames coming from 640 videos rendered at 720p and 1440p for this task. The second is novel view synthesis (NVS), to support the multiview gaming solution of rendering and transferring part of the multiview frames and generating the remaining frames on the client side. This task has 57,600 HR frames from 960 videos of 160 scenes with 6 camera views (with associated camera intrinsic and extrinsic parameters). In addition to the RGB frames, the GBuffers during the deferred rendering stage (i.e., segmentation maps, and depth maps) are also provided, which can be used to help restoration. Furthermore, we evaluate several SOTA super-resolution algorithms and NeRF-based NVS algorithms over our dataset, which demonstrates the effectiveness of our ground-truth GameIR data in improving restoration performance for gaming content. Also, we test the method of incorporating the GBuffers as additional input information for helping super-resolution and NVS. We release our dataset and models to the general public to facilitate research on restoration methods over gaming content.

Keywords: computer-synthesized dataset · super-resolution · novel view synthesis · cloud gaming

1 Introduction

Modern cloud gaming has become increasingly popular with an expected global market share value reaching over $12 billion by 2025. By streaming frames rendered on remote servers to users' devices, cloud gaming benefits both users and game developers. Users can play a large library of games on any device, without requiring expensive hardware, and game developers can optimize games for known server-side hardware, without dealing with heterogeneity of client-side devices. In recent years, Generative AI (GAI) technologies, such as GAN and diffusion models, are transforming the gaming industry by enabling fast and accessible high-quality content creation. GAI makes it possible for anyone to build and design games without professional artistic and technical knowledge, further empowering immeasurable market growth.

Being the next-generation game changer, cloud gaming poses tremendous challenges for data compression and transmission. Most current solutions rely on heavy server-side computation and network delivery, where the client device is merely used for display. It is difficult for a client to enjoy high-quality gaming if the bandwidth is limited, even with a powerful client device. To avoid input delay and over-consuming bandwidth, high-quality frames need to be heavily compressed with extremely low latency. Traditional codecs like H.264/H.265/H.266 [17,18] or recent neural video coding [31] targeting natural videos cannot resolve this transmission bottleneck.

Generative methods like GAN, when applied to super-resolution and image rendering and synthesis, can largely alleviate the transmission issues. Server-side computation and transmission can be reduced by leveraging the computation power of client devices. For example, the server can render low-resolution (LR) frames to transfer, and high-resolution (HR) frames can be computed on the client side. In multiview (e.g., immersive VR) gaming, the server can render part of the frames or views to transfer, and the remaining frames or views can be computed by client devices. NVIDIA's Deep Learning Super Sampling (DLSS) technology [37–39] has commercialized this idea, demonstrating the great potential of optimizing the gaming experience by leveraging bandwidth conditions and computation power of client devices.

The key factor of the success of DLSS is the large-scale ground-truth LR-HR paired data or multiview gaming data used for training that matches the test scenarios. In comparison, the research community uses pseudo training data for many restoration tasks [2,25,41,51]. For example, for super-resolution, the LR data is generated from the HR data by downsampling and adding degradation-like noises, blurs, and compression artifacts. Such pseudo training data does not match the real gaming data. For example, as shown in Fig. 3, true LR gaming frames are high-quality, sharp, and clear without noises or blurs, different from generated pseudo LR data. Also, there are unnatural visual effects and object movements, but with little motion blur, different from captured natural videos. As a result, we have to resort to ground-truth gaming data for effective training. Unfortunately, it is non-trivial to obtain such ground truth, which requires technical skills, labor, and computation using graphics engines.

In this paper, we provide GameIR, a large-scale computer-synthesized ground-truth dataset to facilitate the research of restoration methods over the gaming content. We aim to bring the success of commercial-level DLSS to the public research community so that AI-empowered cloud gaming solutions using image restoration techniques can be more effectively investigated in the field. Our contributions can be summarized as follows.

- We develop a large-scale, high-quality, computer-synthesized ground-truth dataset aiming at two different applications: super-resolution with deferred rendering to support the gaming solution of transferring LR images and restoring HR images on the client side, and novel view synthesis (NVS) to support the gaming solution of transferring part of the multiview frames and generating the remaining frames on the client side. For super-resolution, the GameIR-SR dataset contains 19,200 LR-HR paired ground-truth frames derived from 640 videos rendered at 720p and 1440p using the open-source CARLA simulator with the Unreal Engine. In addition to the LR-HR paired RGB images, the additional GBuffers during the deferred rendering stage (*i.e.*, segmentation maps, and depth maps) are also provided. An example is shown in Fig. 2. For NVS, the GameIR-NVS dataset contains 57,600 HR frames at 1440p from 960 videos of 160 scenes with 6 camera views (with associated camera intrinsic and extrinsic parameters). Besides multiview RGB frames, the dataset also includes the corresponding segmentation maps and depth maps. An example is illustrated in Fig. 4.
- We evaluate over our dataset several existing SOTA algorithms. For super-resolution, we test Anime4K [14] that is designed for anime/cartoon content, RealESRGAN [46] that learns real-world degradations for general images, and AdaCode [30] that uses learned generative codebook priors. For NVS, we test Instant-NGP [36], NeRFacto [43], DSNeRF [9] and PyNeRF [45], which represent the latest NeRF-based NVS methods. We aim to provide a baseline to understand how current methods perform over real gaming data so that improved solutions can be further studied.
- We further evaluate how models can benefit from the additional GBuffers. For super-resolution, GBuffers are either used as additional inputs by concatenating with the RGB frames or used as generative conditions during the conditional restoration process by feature modulation. For NVS, the depth map is used to assist NeRF-based models by providing additional geometry information for the scene.

We release our GameIR dataset and the evaluated models. To the best of our knowledge, this is the first large-scale video set providing ground-truth computer-synthesized LR-HR paired or multiview frames with associated GBuffer data at the scene level. Our dataset can help to advance the research of restoration methods over gaming content for the general public.

2 Related Works

2.1 Super-Resolution: Methods and Datasets

The pioneering work of SRCNN [10,11] has inspired extensive research on deep-learning-based single image super-resolution (SISR). Earlier works [20,23,27,48,55,58] assumed that LR images were generated by ideal degradation models. To handle the complex real degradations consisting of unknown factors like blurring and noises, blind SISR [33,56] has become the research focus, which can be categorized as explicit and implicit methods. Explicit methods [15,32,52,56] explicitly model and estimate the degradation process, and perform reconstruction based on the estimated degradation model. However, real-world degradation is too complicated to model accurately through simple combinations of multiple degradations. In comparison, implicit methods [14,26,30,46,48,53] automatically learn and adapt to various degradation conditions based on LR training data distribution. Although implicit methods have achieved large improvements over real-world images, their performance is highly limited by the training degradations, making them difficult to generalize to out-of-distribution images. Previous methods mitigate this issue by increasing the variety of training degradation types and scales. However, such a training strategy does not work well for gaming content. Real-rendered LR images are clear, sharp, and without blur or artifacts, which are quite different from pseudo LR images generated by applying degradations. As a result, SISR models need to be trained on real LR-HR paired gaming data to learn true degenerative features and improve their performance.

Existing datasets for SISR mainly consist of HR images. Commonly used datasets include DIV2K [1], DIV8K [16], Flickr2K [27], and Flickr-Faces-HQ (FFHQ) [21]. Other popular datasets for general vision tasks, such as ImageNet [8] and COCO [28], are also used for SISR. LR images for these datasets are generated by applying degradations to the HR images. To improve the generalization of models when applied to real-world scenarios, complex degradations have been employed, such as multiple simulated degradations [13] and BSRGAN generated degradations [53]. However, the gap between the simulated and real degradations still exists. The problem is especially prominent for gaming images due to their unique characteristics different from natural images.

There are some datasets providing real-world ground-truth image pairs, *e.g.*, City100 [5], RealSR [4], and DRealSR [49], by using two calibrated devices with varying focal lengths to directly capture LR-HR image pairs. However, due to the expensive process, scale and content diversity is usually highly limited. Also, time synchronization and pixel-level alignment still remain challenging.

2.2 Novel View Synthesis: Methods and Datasets

NVS aims to generate novel view images by integrating image data from multiple camera perspectives. Recently, methods based on Neural Radiance Fields (NeRF) [35] have shown great performance over a large variety of scenes. NeFR++ [54] builds upon NeRF with improved representations and volume rendering. Mip-NeRF [3] uses conical frustums rendering to reduce aliasing and enhance applicability to multiscale and high-resolution scenes. Instant-NGP [36] uses hash tables and multi-resolution grids to speed up training and inference. DSNeRF [9] leverages depth information for supervision to improve performance. 3D Gaussian Splatting (3DGS) [22] uses Gaussian functions for real-time, high-quality rendering. PyNeRF [45] enhances the rendering speed and quality by training models across various spatial grid resolutions.

NVS datasets are generally divided into synthetic and real-world. Most synthetic datasets are at the object level. Blender [35], Objaverse [7], and D-NeRF [40] are typical synthetic datasets, which contain 3D CAD models with varied textures and geometries without real-world noises or non-ideal conditions.

Earlier real-world datasets were developed for multi-view stereo tasks, such as Tanks and Temples [24] and DTU [19], offering limited scene variety. ScanNet [6] contains 3D scans and RGB-D video data, but with motion blur and narrow field-of-view. Later datasets featuring outward-facing and forward-facing scenes have limited diversity in general. For instance, LLFF [34] provides 24 cellphone-captured forward-facing scenes. Mip-NeRF 360 [3] provides 9 indoor and outdoor scenes with uniform distance around central subjects. Mill 19 [44] provides 2 industrial and open-space scenes. BlendedMVS [50] offers multi-view images and depth maps but with limited scenes. Recently, large-scale scene-level real-world datasets have emerged. For example, RealEstate10K [59] offers diverse indoor scenes through real estate videos, but with low-resolution and inconsistent quality. Replica [42] provides high-quality data including RGB images, depth maps, and semantic annotations, but is limited to indoor environments only. The most recent DL3DV-10K [29] significantly enriched the real-world scene collection by providing 10,510 videos captured from 65 types of scene locations, with different levels of reflection, transparency, and lighting conditions.

In comparison, there is a lack of large-scale scene-level synthetic datasets for NVS research over synthetic gaming data. Similar to the super-resolution task, due to the unique characteristics of gaming content, *e.g.*, unnatural object motion with limited motion blur, NVS methods need to be trained and evaluated over scene-level synthetic datasets to assess their effectiveness for gaming data.

3 Our Dataset

In this work, we develop the GameIR dataset, which is a large-scale synthetic scene-level dataset to facilitate image restoration research for cloud gaming solutions. GameIR provides ground-truth LR-HR pairs and synchronized multiview video frames to support both super-resolution and NVS tasks.

3.1 Acquisition Environment and Settings

GameIR was collected using CARLA [12], an autonomous driving simulator developed based on the UE4 game engine. CARLA provided 8 towns: Town01, Town02, Town03, Town04, Town05, Town06, Town07, and Town08. Each town

Fig. 1. Representative views of 8 different towns.

has a distinct style and environment, including various simulation entities such as weather, roads, buildings, vehicles, pedestrians, and vegetation. Figure 1 gives example views of these towns. For each town, we collected two types of scenes: the static autonomous driving scene where there were no other moving vehicles in the scene; and the dynamic autonomous driving scene, where there were other moving vehicles in the scene. Data were collected by controlling an agent vehicle driving in different towns with different camera setups. There were many spawn points for driving agent vehicles in each town, and after initializing the agent vehicle, we set it to autonomous driving mode. Different cameras were initialized and attached to the agent vehicle, which dynamically recorded the surrounding data as the vehicle drove through. During data collection, we set the CARLA simulation to synchronous mode, which prevented discrepancies between the camera capture and storage to avoid frame drops in the stored camera photos.

3.2 GameIR-SR: Dataset for Super-Resolution

For both static and dynamic autonomous driving scenes, we randomly selected 20 spawn points in each of the 8 towns, totaling 320 scenes for the GameIR-SR dataset. To provide ground truth for super-resolution, we placed one set of HR cameras and one set of LR cameras at the front of the agent vehicle, each set capturing synchronized RGB images, segmentation maps, and depth maps with 1920×1440 resolution and 960×720 resolution, respectively.

Each video in the GameIR-SR dataset is 2-s long at 30 fps, totaling 60 frames. During capture, the GBuffer data from the deferred rendering phase, as well as the cameras's intrinsic parameters and extrinsic 6-DoF parameters, were also collected synchronously. Figure 2 gives an example of the GameIR-SR dataset. Finally, GameIR-SR has 19200 LR-HR paired ground-truth frames from 320 LR-HR paired videos, along with the corresponding GBuffers and camera parameters. The ground-truth LR frames are clear and sharp, different from the pseudo LR images generated by degrading the HR images, as illustrated in Fig. 3. Such

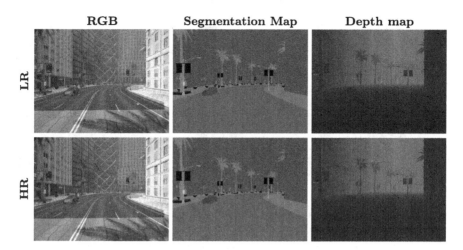

Fig. 2. An example of GameIR-SR dataset, consisting of paired LR-HR (LR top row, HR bottom row) RGB images with associated GBuffer information.

ground-truth LR-HR pairs can better serve as training data for super-resolution methods targeting at gaming content, where the real degradation features can be learned to improve the models' performance.

3.3 GameIR-NVS: Dataset for Novel View Synthesis

For NVS, we only collected data with static autonomous driving scenes. We randomly selected 20 spawn points in each of the 8 towns and recorded 160 scenes in total for the GameIR-NVS dataset. To provide ground truth for NVS, we placed 6 sets of cameras in 6 directions around the agent vehicle: front view, left 60° view, right 60° view, left 120° view, right 120° view, and back view. Each set captured the RGB images, semantic segmentation maps, and depth maps at the resolution of 1920 × 1440 when the vehicle drove through different parts of the towns. Adjacent cameras have some overlapping field-of-view. For each scene, the video is 2-s long at 30 fps, totaling 60 frames. Figure 4 gives an example of the GameIR-NVS dataset. The camera intrinsic parameters and the 6-DoF camera extrinsic parameters for each frame are also recorded. Finally, the GameIR-NVS dataset comprises 960 videos from 160 scenes, totaling 57,600 HR frames. These 360-degree scene-level synthetic data are suitable for training and evaluating NVS methods over gaming content. In addition, the associated depth maps and segmentation maps can be leveraged by NVS algorithms to further improve the generation performance.

4 Evaluation of Super-Resolution over GameIR-SR

Our evaluation has 3 progressive stages: test pretrained SISR models, test finetuned models using the GameIR-SR training set, and test modified SISR models

(a) True LR (b) Pseudo LR

Fig. 3. True LR input versus pseudo LR input. The pseudo LR input is downsampled from the HR input with added noise and blur as commonly used degradation [55,58].

using the GBuffer information as additional inputs or as generative conditions, trained over the GameIR-SR training set.

Tested SISR Methods. We evaluated 3 methods: Anime4K [14] that is specially designed and trained for anime/cartoon images, Real-ESRGAN [46] that handles diverse real-world image degradations, and Adacode [30] that employs a learned codebook-based visual representation as learned generative priors. These methods have their unique strengths in different application areas and give SOTA performance over both anime/cartoon and real-world images. By evaluating their pretrained, finetuned, and modified improved models, we can obtain a good assessment of how current SISR methods perform over gaming content.

Implementation Details. We used data from Town01-04 and Town06-08 for training, and data from Town05 for testing. This ensures that training and test data have distinct styles and town structures. We used published source code by Anime4K [14], Real-ESRGAN [46], and Adacode [30]. During training or finetuning, we followed the original methodologies and hyperparameters described in each method's seminal papers. We used eight V100 GPUs for training and a single V100 GPU for testing.

Evaluation Metrics. We evaluated PSNR, SSIM, FID, and LPIPS. PSNR and SSIM focus on pixel-level distortions. LPIPS highlights local image quality. FID measures the distance between the distributions of generated and real images, reflecting the overall perceptual quality. These metrics are widely used for restoration tasks to provide complementary perspectives on image quality.

4.1 Performance Without GBuffer

Without using GBuffer, we tested the pretrained models and finetuned models with the GameIR-SR training set. Table 1 gives the quantitative performance comparison. Figure 5 gives the qualitative comparison of example results. From the results, we can see that, after finetuning over the GameIR-SR training set,

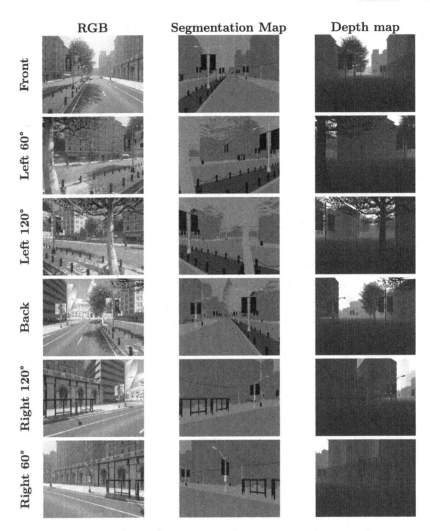

Fig. 4. An example of our GameIR-NVS dataset, consisting of 6 views of RGB images with associated segmentation maps and depth maps. From the top to bottom, the sequence gives views of a 360° rotation, starting from the front.

compared to the pretrained models, all three tested methods have large performance improvements across all metrics. As clearly shown in Fig. 5, the finetuned models are capable of restoring intricate image details more accurately than the pretrained models. The finetuned models generate images with enhanced sharpness and clarity, especially in regions with rich textures. These results demonstrate the importance of training over ground-truth LR-HR paired data, where models can better learn real degradation features that match the inference stage.

To further prove our point, we also used BSRGAN [53] as the degradation model to generate pseudo LR training data and finetuned the tested pretrained

Table 1. Performance of pretrained (PT) and finetuned (FT) super-resolution methods, using pseudo-LR training data or ground-truth (GT) LR training data: PSNR and SSIM, the higher the better; LPIPS and FID, the lower the better.

	PSNR↑			SSIM↑			FID↓			LPIPS↓		
	PT	FT(pseudo)	FT(GT)	PT	FT(pseudo)	FT(GT)	PT	FT(pseudo)	FT(GT)	PT	FT(pseudo)	FT(GT)
Anime4k	30.9274	30.5923	**31.2974**	0.9008	0.8874	**0.9057**	4.7365	6.7408	**4.6172**	0.0866	0.1069	**0.0788**
AdaCode	28.6791	27.9637	**29.5762**	0.8382	0.8146	**0.8741**	14.8789	17.6222	**11.4608**	0.0884	0.0899	**0.0451**
Real-ESRGAN	29.1421	29.2743	**30.2517**	0.8639	0.8623	**0.8916**	17.7232	14.7961	**8.3153**	0.0905	0.0706	**0.0379**

models using such pseudo pairs. The results are also shown in Table 1, which clearly show that such pseudo training data can not give any performance gain. That is, improvements can not be obtained by simply increasing pseudo training data. Ground-truth LR-HR gaming data are necessary for effective training.

4.2 Performance with Additional GBuffer

We modified the Real-ESRGAN [46] to make use of GBuffers as additional information to help SISR. There are two common ways to include additional input information into a model: as additional inputs or as generation conditions.

Table 2. Performance of using different GBuffer information: segmentation map (S), depth map (D), or both (SD), by different methods: as additional input (concat), or as generative condition (SFT).

Input Channels	PSNR↑	SSIM↑	FID↓	LPIPS↓
RGB (pretrained models)	30.1798	0.9040	10.0374	0.0535
w/ S(concat)	30.1065	0.9026	**9.9637**	0.0518
w/ D(concat)	**30.2236**	0.9038	10.2756	0.0499
w/ SD(concat)	30.1639	0.9033	9.9164	0.0524
w/ S(SFT)	30.0750	0.9025	10.2182	0.0519
w/ D(SFT)	30.1049	0.9020	9.5882	**0.0492**
w/ SD(SFT)	30.1996	**0.9052**	9.8511	0.0507

GBuffer as Additional Input. The GBuffers provide two types of information: segmentation maps and depth maps. We experimented with adding segmentation map or depth map alone, or adding both of them together. Specifically, we concatenated either the segmentation map or depth Map, or both of them with the RGB image to form a 4-channel or 5-channel input and adjusted the input layer of Real-ESRGAN accordingly. We followed the two-phase training procedure of the original Real-ESRGAN [46]. In phase 1, Real-ESRNet with RRDBNet was finetuned as the backbone using the GameIR-SR training data. In phase 2, we continued tuning with GAN loss, where Real-ESRNet served as the generator tuned together with a UNet discriminator.

Fig. 5. Qualitative comparison of pretrained (PT) and finetuned (FT) super-resolution methods. Finetuning with ground-truth LR-HR paired data can restore intricate details more accurately with better clarity.

GBuffer as Generation Condition. The GBuffer information can serve as generation conditions during reconstruction. Following the two-phase training process of Real-ESRGAN, in phase 1, we added a control network (CtrlNet) [57] to Real-ESRNet, feeding the segmentation map, depth map, or both of them into CtrlNet respectively. Then, we used a Spatial Feature Transform (SFT) module [47] to fuse the output of the Real-ESRNet and CtrlNet in reconstructing the final image output. CtrlNet comprises upsample layers and ResBlocks, which extract multi-scale features from segmentation maps or depth maps, or both. In the SFT module, the scale and shift layers use outputs from CtrlNet to perform spatial feature transformations on the feature maps. The scale layer dynamically adjusts the scales of the output feature maps by Real-ESRNet. The shift layer adjusts the baseline activation levels of the Real-ESRNet outputs, tuning feature activation by the conditional information.

Table 2 gives the performance comparison of different ways of using GBuffer inputs. From the results, by using both the segmentation map and depth map as generative conditions, we can both improve distortion and perceptual quality across all metrics. Note that due to modifications of the models, no pretrained model is used here. Better performance can be expected in the future if larger-scale and more diverse gaming data can be available for learning a general pretrained model for general gaming content.

5 Evaluation of NVS over GameIR-NVS

We evaluated SOTA NeRF-based NVS algorithms over the GameIR-NVS dataset. For each town, we randomly selected 5 out of the 20 multiview videos with different spawn sites. For each selected video, we randomly sampled a subset of multiview frames to train NeRF-based models and then rendered the remaining frames to compare with the ground-truth frames for evaluation.

We evaluated 4 latest methods: Instant-NGP [36] that targets fast computation, NeRFacto [43] that improves efficiency in complex scenes, PyNeRF [45] that improves both speed and quality by training models across different spatial grid resolutions, and DSNeRF [9] that uses depth maps to improve performance.

We used the NeRFStudio implementation [43] of the tested methods. The default provided hyperparameters were used. The training and test were done on a single V100 graphics card.

We evaluated PSNR, SSIM, LPIPS, and FID to measure both distortion and visual quality. In addition, we evaluated normalized root mean square error (NRMSE) that measured the quality of generated depth maps.

5.1 Performance Without GBuffer

We tested 2 cases: using only the front view for both training and testing or using all 6 views (360°). In this experiment, for each video, we randomly extracted 10% of the frames as the test set, with the remainder for training. Table 3 gives the performance comparison.

The results show that PyNeRF outperforms the other methods across all metrics on our GameIR-NVS dataset. Also, metrics obtained using only front views are better than those using 360° views. We attribute this to the front view's single-directional nature as opposed to the comprehensive perspective provided by 360° views, simplifying the model's task by lowering the complexity involved in handling variations in lighting and perspective differences.

Table 3. Performance of different NVS methods, using front views or 360° training data: PSNR and SSIM, the higher the better; LPIPS and FID, the lower the better.

	PSNR↑		SSIM↑		FID↓		LPIPS↓	
	front	360°	front	360°	front	360°	front	360°
Instant-NGP	30.9481	27.7004	0.9373	0.9089	30.8528	44.7195	0.1095	0.1703
NeRFacto	26.6457	26.7710	0.8539	0.8774	36.1718	43.0672	0.1206	0.1648
PyNeRF	**36.1329**	**32.0542**	**0.9612**	**0.9357**	**20.2392**	**36.3312**	**0.0758**	**0.1301**

5.2 Performance with Additional GBuffer

Motivated by the finding of DSNeRF [9] that supervising training with depth maps enhances the model's understanding of the scene, allowing it to render high-quality images with fewer training views, in this experiment, we aim to test how depth maps from GBuffer can help NVS with reduced training views. Specifically, we used NeRFactor and Depth-NeRFactor, where Depth-NeRFactor extended NeRFactor by incorporating depth maps as inputs and employed depth loss for supervised training. Instead of using 10% frames for testing in the previous experiments, here we used 10% randomly sampled frames for training, and the remaining 90% for evaluation. Table 4 gives performance metrics, and Fig. 6 gives qualitative results.

As shown in Table 4, using depth maps improves the NVS performance across all metrics. This demonstrates that depth maps can enrich the available geometric information even with limited views, facilitating faster learning and a deeper understanding of the scene's 3D structure. Furthermore, Fig. 6 illustrates that depth maps enhance the realism of the generated RGB images, reduce distortions, and result in images with more detailed and textured representations. Results also clearly show that training with depth maps can significantly reduce depth error, and the depth maps generated by Depth-NeRFactor are very close to the ground truth. Such improvement is quite meaningful for gaming scenarios, as accurately generated depth maps provide precise three-dimensional geometric information.

Table 4. Performance with 10% training views: PSNR and SSIM, the higher the better; LPIPS, FID, and Depth Error, the lower the better.

	PSNR↑		SSIM↑		FID↓		LPIPS↓		Depth Err%↓	
	front	360°	front	360°	front	360°	front	360°	front	360°
NeRFacto	22.9597	23.3168	0.8133	0.8276	60.3013	40.8520	0.1654	0.1830	32.8053	30.5997
Depth-NeRFacto	**25.3482**	**24.4990**	**0.8425**	**0.8306**	**29.3219**	**28.7427**	**0.1249**	**0.1736**	**5.4212**	**7.2538**

Fig. 6. Qualitative comparison of NeRFacto and Depth-NeRFacto. The RGB (left) and depth map (right) generated by Depth-NeRFacto better match the ground truth compared to NeRFacto.

6 Conclusion

In this paper, we proposed GameIR, a large-scale synthetic dataset specifically designed for image restoration in gaming content. This dataset comprises two subsets aiming at two tasks: the GameIR-SR dataset for super-resolution and the GameIR-NVS dataset for NVS. GameIR-SR contains ground-truth LR and HR image pairs, and GameIR-NVS contains multiview videos with associated camera parameters. The corresponding GBuffers from deferred rendering are also provided for both datasets. We evaluated several SOTA algorithms for super-resolution and for NVS on our dataset to establish a baseline assessment for subsequent research on real gaming data. Additionally, we explored methods of utilizing GBuffers as supplementary information to help the super-resolution and NVS tasks. Our results demonstrated that GBuffers can provide enriched contextual information to improve performance.

This paper is our first attempt to provide ground-truth gaming data to facilitate public research on image restoration methods over the gaming content. In the future, we will continue to enrich our data collection by increasing the diversity of the data content, such as collecting different types of gaming data with different styles.

References

1. Agustsson, E., Timofte, R.: NTIRE 2017 challenge on single image super-resolution: dataset and study. In: The IEEE Conference on Computer Vision and Pattern Recognition (CVPR) Workshops, July 2017
2. Al-Mekhlafi, H., Liu, S.: Single image super-resolution: a comprehensive review and recent insight. Front. Comput. Sci. (181702) (2024)
3. Barron, J.T., Mildenhall, B., Tancik, M., Hedman, P., Martin-Brualla, R., Srinivasan, P.P.: Mip-NeRF: a multiscale representation for anti-aliasing neural radiance fields. In: Proceedings of the IEEE/CVF International Conference on Computer Vision, pp. 5855–5864 (2021)
4. Cai, J., Zeng, H., Yong, H., Cao, Z., Zhang, L.: Toward real-world single image super-resolution: a new benchmark and a new model. In: Proceedings of the IEEE/CVF International Conference on Computer Vision, pp. 3086–3095 (2019)
5. Chen, C., Xiong, Z., Tian, X., Zha, Z.J., Wu, F.: Camera lens super-resolution. In: Proceedings of the IEEE/CVF Conference on Computer Vision and Pattern Recognition, pp. 1652–1660 (2019)
6. Dai, A., Chang, A.X., Savva, M., Halber, M., Funkhouser, T., Nießner, M.: ScanNet: richly-annotated 3D reconstructions of indoor scenes. In: Proceedings of the IEEE Conference on Computer Vision and Pattern Recognition, pp. 5828–5839 (2017)
7. Deitke, M., et al.: Objaverse-XL: a universe of 10m+ 3D object. arXiv preprint arXiv:2307.05663 (2023)
8. Deng, J., Dong, W., Socher, R., Li, L.J., Li, K., Fei-Fei, L.: ImageNet: a large-scale hierarchical image database. In: 2009 IEEE Conference on Computer Vision and Pattern Recognition, pp. 248–255. IEEE (2009)
9. Deng1, K., Liu, A., Zhu, J., Ramanan, D.: Depth-supervised NeRF: fewer views and faster training for free. In: CVPR (2022)
10. Dong, C., Loy, C.C., He, K., Tang, X.: Learning a deep convolutional network for image super-resolution. In: Fleet, D., Pajdla, T., Schiele, B., Tuytelaars, T. (eds.) ECCV 2014. LNCS, vol. 8692, pp. 184–199. Springer, Cham (2014). https://doi.org/10.1007/978-3-319-10593-2_13
11. Dong, C., Loy, C.C., He, K., Tang, X.: Image super-resolution using deep convolutional networks. IEEE Trans. Pattern Anal. Mach. Intell. **38**(2), 295–307 (2015)
12. Dosovitskiy, A., Ros, G., Codevilla, F., Lopez, A., Koltun, V.: CARLA: an open urban driving simulator. In: Proceedings of the 1st Annual Conference on Robot Learning, pp. 1–16 (2017)
13. Elad, M., Feuer, A.: Restoration of a single superresolution image from several blurred, noisy, and undersampled measured images. IEEE Trans. Image Process. **6**(12), 1646–1658 (1997)
14. Feng, A.: Anime4k: a high-quality real time upscaler for anime video (2019). https://github.com/bloc97/Anime4K
15. Gu, J., Lu, H., Zuo, W., Dong, C.: Blind super-resolution with iterative kernel correction. In: Proceedings of the IEEE/CVF Conference on Computer Vision and Pattern Recognition, pp. 1604–1613 (2019)
16. Gu, S., Lugmayr, A., Danelljan, M., Fritsche, M., Lamour, J., Timofte, R.: DIV8K: diverse 8K resolution image dataset. In: 2019 IEEE/CVF International Conference on Computer Vision Workshop (ICCVW), pp. 3512–3516 (2019). https://doi.org/10.1109/ICCVW.2019.00435

17. International Telecommunication Union-Telecommunication (ITU-T) and International Standards Organization/Int/Electrotechnical Commission (ISO/IEC JTC 1): High efficiency video coding, rec. ITU-T H.265 and ISO/IEC 23008-2 (2019)
18. ITU-T and ISO: Versatile video coding, rec. ITU-T H.266 and ISO/IEC 23090-3 (2020)
19. Jensen, R., Dahl, A., Vogiatzis, G., Tola, E., Aanæs, H.: Large scale multi-view stereopsis evaluation. In: CVPR (2014)
20. Johnson, J., Alahi, A., Fei-Fei, L.: Perceptual losses for real-time style transfer and super-resolution. In: Leibe, B., Matas, J., Sebe, N., Welling, M. (eds.) ECCV 2016. LNCS, vol. 9906, pp. 694–711. Springer, Cham (2016). https://doi.org/10.1007/978-3-319-46475-6_43
21. Karras, T., Aila, T., Laine, S., Lehtinen, J.: Progressive growing of GANs for improved quality, stability, and variation. arXiv preprint arXiv:1710.10196 (2017)
22. Kerbl, B., Kopanas, G., Leimkuhler, T., Drettakis, G.: 3D Gaussian splatting for real-time radiance field rendering. ACM Trans. Graph. (ToG) **42**, 1–14 (2023)
23. Kim, J., Lee, J.K., Lee, K.M.: Accurate image super-resolution using very deep convolutional networks. In: Proceedings of the IEEE Conference on Computer Vision and Pattern Recognition, pp. 1646–1654 (2016)
24. Knapitsch, A., Park, J., Zhou, Q.Y., Koltun, V.: Tanks and temples: benchmarking large-scale scene reconstruction. ACM Trans. Graph. (ToG) **36**(4), 1–13 (2017)
25. Lepcha, D., Goyal, B., Dogra, A., Goyal, V.: Image super-resolution: a comprehensive review, recent trends, challenges and applications. Inf. Fusion **91**, 230–260 (2023)
26. Liang, J., Cao, J., Sun, G., Zhang, K., Van-Gool, L., Timofte, R.: SwinIR: image restoration using swin transformer. arXiv preprint arXiv:2108.10257 (2021)
27. Lim, B., Son, S., Kim, H., Nah, S., Mu Lee, K.: Enhanced deep residual networks for single image super-resolution. In: Proceedings of the IEEE Conference on Computer Vision and Pattern Recognition Workshops, pp. 136–144 (2017)
28. Lin, T.-Y., et al.: Microsoft COCO: common objects in context. In: Fleet, D., Pajdla, T., Schiele, B., Tuytelaars, T. (eds.) ECCV 2014. LNCS, vol. 8693, pp. 740–755. Springer, Cham (2014). https://doi.org/10.1007/978-3-319-10602-1_48
29. Ling, L., et al.: A large-scale scene dataset for deep learning-based 3D vision. In: CVPR (2024)
30. Liu, K., Jiang, Y., Choi, I., Gu, J.: Learning image-adaptive codebooks for class-agnostic image restoration. In: ICCV (2023)
31. Lu, G., Ouyang, W., Xu, D., Zhang, X., Cai, C., Gao, Z.: DVC: an end-to-end deep video compression framework. In: CVPR, pp. 11006–11015 (2019)
32. Maeda, S.: Image super-resolution with deep dictionary. In: Avidan, S., Brostow, G., Cissé, M., Farinella, G.M., Hassner, T. (eds.) ECCV 2022. LNCS, vol. 13679, pp. 464–480. Springer, Cham (2022). https://doi.org/10.1007/978-3-031-19800-7_27
33. Michaeli, T., Irani, M.: Nonparametric blind super-resolution. In: Proceedings of the IEEE International Conference on Computer Vision, pp. 945–952 (2013)
34. Mildenhall, B., et al.: Local light field fusion: practical view synthesis with prescriptive sampling guidelines. ACM Trans. Graph. (TOG) (2019)
35. Mildenhall, B., Srinivasan, P.P., Tancik, M., Barron, J.T., Ramamoorthi, R., Ng, R.: NeRF: representing scenes as neural radiance fields for view synthesis. Commun. ACM **65**(1), 99–106 (2021)
36. Müller, T., Evans, A., Schied, C., Keller, A.: Instant neural graphics primitives with a multiresolution hash encoding. ACM Trans. Graph. **41**(4), 102:1–102:15 (2022). https://doi.org/10.1145/3528223.3530127

37. NIVIDIA: Introducing Nvidia DLSS 3. https://www.nvidia.com/en-us/geforce/news/dlss3-ai-powered-neural-graphics-innovations/
38. NIVIDIA: NVIDIA announces DLSS 3.5 with ray reconstruction, boosting RT quality with an AI-trained denoiser. EuroGamer. https://www.eurogamer.net/digitalfoundry-2023-nvidia-announces-dlss-35-with-ray-reconstruction-boosting-rt-quality-with-an-ai-trained-denoiser
39. NIVIDIA: truly next-gen: adding deep learning to games and graphics. https://www.gdcvault.com/play/1026184/Truly-Next-Gen-Adding-Deep
40. Pumarola, A., Corona, E., Pons-Moll, G., Moreno-Noguer, F.: D-NeRF: neural radiance fields for dynamic scenes. arXiv preprint arXiv:2011.13961 (2020)
41. Rota, C., Buzzelli, M., Bianco, S., et al.: Video restoration based on deep learning: a comprehensive survey. Artif. Intell. Rev. **56**, 5317–5364 (2023)
42. Straub, J., et al.: The replica dataset: a digital replica of indoor spaces. arXiv preprint arXiv:1906.05797 (2019)
43. Tancik, M., et al.: Nerfstudio: a modular framework for neural radiance field development. In: ACM SIGGRAPH 2023 Conference Proceedings, SIGGRAPH 2023 (2023)
44. Turki, H., Ramanan, D., Satyanarayanan, M.: Mega-NeRF: scalable construction of large-scale nerfs for virtual fly-throughs. In: Proceedings of the IEEE/CVF Conference on Computer Vision and Pattern Recognition (CVPR), pp. 12922–12931, June 2022
45. Turki, H., Zollhöfer, M., Richardt, C., Ramanan, D.: PyNeRF: pyramidal neural radiance fields. In: Advances in Neural Information Processing Systems, vol. 36 (2024)
46. Wang, X., Xie, L., Dong, C., Shan, Y.: Real-ESRGAN: training real-world blind super-resolution with pure synthetic data. In: International Conference on Computer Vision Workshops (ICCVW)
47. Wang, X., Yu, K., Dong, C., Loy, C.C.: Recovering realistic texture in image super-resolution by deep spatial feature transform. In: Proceedings of the IEEE Conference on Computer Vision and Pattern Recognition, pp. 606–615 (2018)
48. Wang, X., et al.: ESRGAN: enhanced super-resolution generative adversarial networks. In: Leal-Taixé, L., Roth, S. (eds.) ECCV 2018. LNCS, vol. 11133, pp. 63–79. Springer, Cham (2019). https://doi.org/10.1007/978-3-030-11021-5_5
49. Wei, P.: Component divide-and-conquer for real-world image super-resolution. In: Vedaldi, A., Bischof, H., Brox, T., Frahm, J.-M. (eds.) ECCV 2020. LNCS, vol. 12353, pp. 101–117. Springer, Cham (2020). https://doi.org/10.1007/978-3-030-58598-3_7
50. Yao, Y., et al.: BlendedMVS: a large-scale dataset for generalized multi-view stereo networks. In: Proceedings of the IEEE/CVF Conference on Computer Vision and Pattern Recognition, pp. 1790–1799 (2020)
51. Zhai, L., Wang, Y., Cui, S., Zhou, Y.: A comprehensive review of deep-learning-based real world image restoration. IEEE Access (2023)
52. Zhang, K., Gool, L.V., Timofte, R.: Deep unfolding network for image super-resolution. In: Proceedings of the IEEE/CVF Conference on Computer Vision and Pattern Recognition, pp. 3217–3226 (2020)
53. Zhang, K., Liang, J., Van Gool, L., Timofte, R.: Designing a practical degradation model for deep blind image super-resolution. In: IEEE International Conference on Computer Vision, pp. 4791–4800 (2021)
54. Zhang, K., Riegler, G., Snavely, N., Koltun, V.: NeRF++: analyzing and improving neural radiance fields. arXiv preprint arXiv:2010.07492 (2020)

55. Zhang, K., Zuo, W., Gu, S., Zhang, L.: Learning deep CNN denoiser prior for image restoration. In: Proceedings of the IEEE Conference on Computer Vision and Pattern Recognition, pp. 3929–3938 (2017)
56. Zhang, K., Zuo, W., Zhang, L.: Learning a single convolutional super-resolution network for multiple degradations. In: Proceedings of the IEEE Conference on Computer Vision and Pattern Recognition, pp. 3262–3271 (2018)
57. Zhang, L., Rao, A., Agrawalau, M.: Adding conditional control to text-to-image diffusion models. ArXiv, arXiv:2302.05543 (2023)
58. Zhang, Y., Tian, Y., Kong, Y., Zhong, B., Fu, Y.: Residual dense network for image super-resolution. In: Proceedings of the IEEE Conference on Computer Vision and Pattern Recognition, pp. 2472–2481 (2018)
59. Zhou, T., Tucker, R., Flynn, J., Fyffe, G., Snavely, N.: Stereo magnification: learning view synthesis using multiplane images. arXiv preprint arXiv:1805.09817 (2018)

Vector Logo Image Synthesis Using Differentiable Renderer

Ryuta Yamakura and Keiji Yanai

The University of Electro-Communications, Tokyo, Japan
{yamakura-r,yanai}@mm.inf.uec.ac.jp

Abstract. Vector images are widely used in the design field, particularly for logos and icons, due to their scalable properties. Consequently, the flexible and high-quality creation of such images is expected to support creative activities. In this study, we leverage a recently proposed differentiable renderer and the strong raster image generation capabilities of Stable Diffusion to generate vector-format logo images. This is achieved through optimizing vector parameters based on losses calculated from text prompts and shape images. Additionally, we address the self-intersection issue, a common challenge in vector image generation through optimization methods, by introducing a new technique called Radiation Loss. This approach explicitly monitors control points to enhance the quality of the output. While this method successfully generates logo images that maintain the input text and shape, challenges remain, including the persistence of unnecessary paths and difficulty in controlling the output entirely by text prompts. The experimental results showed the effectiveness of the proposed methods.

Keywords: vector graphics · differentiable renderer · logo image

1 Introduction

Vector images, which represent visuals through parametric mathematical primitives, are widely used in design fields such as fonts and icons due to their inherent advantages. These advantages include the ability to scale without losing image quality and typically smaller data sizes compared to equivalent raster images. Vector graphics are often employed for various visual elements, with logo creation being especially challenging. This is because logos need to embody artistic expression while also aligning with a specific visual identity.

Moreover, creating vector images is a time-consuming and labor-intensive process that requires not only proficiency in vector graphics but also the use of specialized tools. The ability to intuitively and flexibly generate vector images using text and image inputs as guides could significantly support artistic activities, streamlining the design process and minimizing unnecessary iterations.

This paper focuses on the generation of vector-format logo images. It explores methods for producing a diverse range of vector images that incorporate both visual identity and artistic design by utilizing shape-indicating images and text prompts to define the content.

Currently, vector graphics generation using generative models is primarily achieved through two methods: language-based approaches, which rely on models trained with explicit supervision of vector graphics command sequences, and image-based approaches, which combine pre-trained Text-to-Image models with image vectorization. The former approach is often limited to specific domains, such as logos and fonts, due to the challenges of collecting large, high-quality vector graphics datasets and the fact that vector graphics do not always have a unique representation.

In contrast, the latter approach benefits from recent advances in image generation models, as it can draw upon the knowledge of models trained on large datasets of raster images. Moreover, with the advent of differentiable renderers [6], raster images can now be vectorized through a straightforward optimization process using loss functions, as illustrated in Fig. 1, without the need for complex vector transformation steps.

This paper adopts the latter approach, leveraging the capabilities of Text-to-Image models to generate diverse and stable vector graphics. Our contributions in this paper are summarized as follows:

- We propose a method for generating vector-format logo images by combining a differentiable renderer with the raster image generation capabilities of Stable Diffusion.
- We introduce Radiation Loss, a novel loss function that addresses the self-intersection problem in vector image generation, improving the quality and structure of the output.
- We demonstrate the successful generation of logo images that preserve both input text and shapes through extensive experiments using text prompts and shape images.

2 Related Work

2.1 Differentiable Renderer

Traditionally, the rendering process for vector graphics has been unidirectional, and vectorization of raster images required special methods [3,15] that involve tracing edges. However, vectorization by these methods is unrelated to the original vector metrics and the vector graphics generated have a dramatically different structure, making it impossible to apply raster-based algorithms to vector graphics. Li et al. [6] addressed the problems of such vectorization methods by using two methods for pixel prefiltering based on the differentiable 3D renderer [7]: analytical prefiltering and multisampling antialiasing. They proposed a renderer that can automatically compute gradients for vector parameters.

2.2 Image Generation with Differentiable Renderer

With the advent of differentiable renderers, several methods were proposed to optimize vector parameters by directly applying raster-based loss functions and machine learning methods to vector content, bypassing the limitations of the need to collect data for high-quality, large-scale vector content.

CLIPDraw [2] uses a large model of learned image-text relationships published by OpenAI, CLIP [10], to compute the similarity between an input text and an output image. The method optimizes the parameters of Bezier curves via a differentiable renderer, and was the basis for similar frameworks that followed.

CLIPDraw uses CLIP's image and text encoders to optimize the input text and output image similarity (CLIP Loss) for each iteration as a loss function to obtain a vector image output that matches the text. However, CLIPDraw uses Bezier curves as lines because the main purpose of CLIPDraw is to draw with vector paths, and the poor representation remains as an issue.

StyleCLIPDraw [14] used the VGG16 model to condition images with auxiliary style loss, and extended it to transfer styles to images generated by CLIPDraw.

In addition, VectorFusion [5] uses a trained text-to-image (T2I) model, Stable Diffusion [11]. Score Distillation Sampling (SDS) [9], a method that distills the model so that its output is transformed into an arbitrary parameter space, such as 3D or vector parameters, tailored for vector graphics. By using it as a loss function, we were able to generate more consistent vector graphics. The SDS Loss used in VectorFusion is also used in this study. Compared to CLIPDraw, VectorFusion has greatly improved drawing capability by using a robust output space of diffusion models. However, it does not have the ability to maintain the shape of the input image because it only receives text as input.

3 Method

3.1 Outline

A schematic diagram of the method is shown in Fig. 1. In this method, a set of Bezier curve parameters (color, opacity, and control point coordinates), which is generated from the input image I, is rendered by using a differentiable renderer $\mathcal{R}$ with gradient generation enabled and $\hat{I}_i$ is obtained in each cycle, i. After enhancement using cropping and perspective transformation, the result is input into a stable diffusion model.

In this process, the Tone Loss, Radiation Loss, and SDS Loss are calculated and used to optimize the Bezier curve parameters through the gradient descent method.

3.2 Representation Format and Initialization

The primitives that make up the vector image are restricted to closed cubic Bezier curves, based on the concept of CLIPDraw [2]. Bezier curves are defined

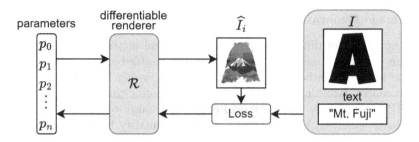

Fig. 1. Outline

by a set of control points, and by connecting multiple Bezier curves, a wide variety of shapes can be approximated. This allows for a simple implementation and evaluation without losing expressive power.

In this method, a vector graphic $\hat{I} = \{P_0, ..., P_{N-1}\}$ is formed using N closed Bezier curve paths. Each path consists of M cubic Bezier curve segments, and the four control points p_m that make up each segment define $P_n = \{p_{0,0}, p_{0,1}, p_{0,2}, ..., p_{M-1,0}, p_{M-1,1}, p_{M-1,2}\}$. Note that a path always passes through the starting control point of a segment, and the ending control point of a segment is shared with the starting control point of the next segment to ensure connectivity. Furthermore, each P_n possesses a single color, and vector generation is performed by individually optimizing the control point coordinates and colors.

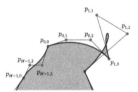

Fig. 2. Relationship of path and control points

The vector path is initialized by arranging all control points in a circular layout, following the idea of LIVE [8]. This method is expected to prevent path self-intersection issues in advance.

Additionally, to improve the convergence speed of the Tone Loss, the center coordinates of the path are pre-set within the range of the input image I. An example of this initialization is shown in Fig. 3.

3.3 Self Crossing Problem

LIVE [8] points out that some vector paths self-intersect as a result of the path initialization and optimization process, leading to the generation of detrimental artifacts and improper topology. As shown in the upper left corner of Fig. 4,

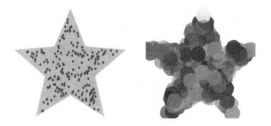

Fig. 3. Example of initialization. The center position of the Bezier curve is randomly obtained according to the input image (left figure) and initialized to a circular shape (right figure).

paths with self-intersecting problems may generate artifacts when scaling beyond the default rendering size and additional paths may be generated to cover the artifacts. To address this problem, the study assumes that all vector primitives are cubic Bezier curves, and if the control points of a path are A, B, C, D in that order, the angle between $\overrightarrow{AB}$ and $\overrightarrow{CD}$ is The Xing Loss is expressed by the following equation:

$$D_1 = \mathbb{I}\left(\overrightarrow{AB} \times \overrightarrow{CD}\right), D_2 = \frac{\overrightarrow{AB} \cdot \overrightarrow{CD}}{\left\|\overrightarrow{AB}\right\|\left\|\overrightarrow{CD}\right\|} \tag{1}$$

$$\mathcal{L}_{xing} = D_1(\text{ReLU}(-D_2)) + (1 - D_1)(\text{ReLU}(D_2)), \tag{2}$$

where $\times$ is the outer product, $\cdot$ is the inner product, and $\mathbb{I}(\cdot)$ is the sign function.

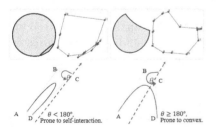

Fig. 4. Self crossing problem. (cited from [8])

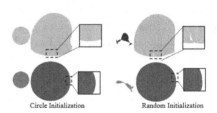

Fig. 5. Example where the path intersects between segments due to the initialization method. (cited from [8])

3.4 Loss Function

Figure 6 shows the derivation flow of the loss functions. Since logo images possess the appearance of a specific shape as a unique identity that preserves artistic graphics and identity, SDS Loss and Tone Preserving Loss are introduced to control the content by text and shape, respectively, and to restrict the shape to the region of the image where the input is located. In addition, Radiation Loss is introduced to solve the self-intersection problem, in which control points are optimized into uninterpretable shapes during the optimization process, a problem inherent in vector image generation using optimization methods.

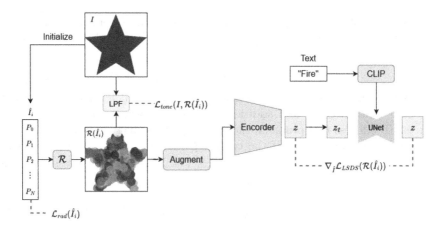

Fig. 6. Loss Function

Tone Preserving Loss. Tone Preserving Loss is the loss proposed in Word-As-Image [4], which applies a low-pass filter to the rendered vector image and the image of the shape being referenced and takes the Euclidean distance between the images. This loss preserves the local tones of the image and limits the adjusted image from deviating too much from the input image.

$$\mathcal{L}_{tone} = ||LPF(\mathcal{R}(I)) - LPF(\mathcal{R}(\hat{I}_i))||_2^2 \qquad (3)$$

where $\mathcal{R}(\cdot)$ is the renderer and P and $\hat{P}$ are the set of parameters that make up the vector image.

This method also introduces $\mathcal{L}_{tone}$ to restrict the shape to that of the input image. However, since Word-As-Image is a method intended to manipulate monotone font shapes, applying Eq. (3) directly or applying a gray scale transformation to a rendered image will optimize the color value to be close to the mask value (usually 0), resulting in blackening the corresponding parts of the generated image. and the corresponding areas of the generated image will be blackened.

Therefore, Equation (5) applied to Tone Loss is used to create an image showing the region of the path from the difference with the background using an image rendered with a random noise background image, bg, according to the following equation:

$$\mathcal{R}(\hat{I}_i)_{binary} = \text{Sigmoid}\left(\alpha(\mathcal{R}(\hat{I}_i) - bg)^2)\right), \tag{4}$$

where α is the scaling factor and $\alpha = 10^6$.

$$\mathcal{L}_{tone} = ||LPF(\mathcal{R}(I)) - LPF(\mathcal{R}(\hat{I}_i)_{binary})||_2^2 \tag{5}$$

Score Distillation Sampling Loss. Score Distillation Sampling (SDS) Loss is the loss function proposed in DreamFusion [9] using the sampling results when the Jacobian term in the diffusion model is omitted. More intuitively, it can be said to be a loss that optimizes the parameter θ to fit the conditional text prompt by minimizing.

At each iteration $t \in 1, 2, ..., T$, a randomly rendered image x is generated and then noise is added to this image, accompanied by the terms $\epsilon \sim \mathcal{N}(0, I)$, α_t, σ_t that control the noise schedule, $x_t = alpha_t x + \sigma_t \epsilon$ is formed. The noisy image is then passed to the Imagen [13] pre-trained UNet [12] model, which outputs a prediction of the noise ϵ. The SDS Loss is defined by the following equation:

$$\nabla_\theta \mathcal{L}_{SDS} = \mathbb{E}_{t,\epsilon}\left[w(t)\big(\hat{\epsilon}_\phi(x_t, t, y) - \epsilon\big)\frac{\partial x}{\partial \theta}\right], \tag{6}$$

where $\hat{\epsilon}_\phi$ is the UNet-based denoising network, y is the conditional text prompt, θ is the NeRF parameter, and $w(t)$ is a constant multiplier that depends on α_t.

While SDS Loss was utilized for the 3D object generation task in the proposed DreamFusion, VectorFusion [5] utilized SDS Loss for the vector graphics generation task. VectorFusion is defined similarly to DreamFusion with the raster images generated by Stable Diffusion as vectorized vector images or randomly initialized vector images as initial values, defined by the following formula:

$$\nabla_{\hat{I}} \mathcal{L}_{LSDS} = \mathbb{E}_{t,\epsilon}\left[w(t)\big(\hat{\epsilon}_\phi(\alpha_t z_t + \sigma_t \epsilon, y) - \epsilon\big)\frac{\partial z}{\partial z_{aug}}\frac{\partial x_{aug}}{\partial \theta}\right], \tag{7}$$

where x_{aug} is the enhancement by perspective transformation and cropping as shown in CLIPDraw [2], z is the encoding applying the stable diffusion pre-trained encoder $\mathcal{E}$, $z = \mathcal{E}(x_{aug})$. Similarly in this paper, SDS Loss is used to effectively utilize the T2I feature of Stable Diffusion as well as VectorFusion for the vector graphics generation task.

Radiation Loss. Although the Xing Loss (Eq. (2)), which solves the self-intersection problem, fully prevents self-intersections for single cubic Bezier curves, it is difficult to prevent segment intersections for actual Bezier curves with connected segments, and there is room for improvement in this respect. In this regard, LIVE prevents segment intersections by initializing the paths circularly. However, this method, which does not allow area-based initialization based on the target image and has a large variation of control points during the optimization process, is likely to cause segment intersections even if the same initialization is used. Figure 5 shows an example of segment intersections due to the path initialization method.

Therefore, we propose Radiation Loss, which is an extension of Xing Loss, taking into account the positional relationship with the segments before and after.

The simplest condition necessary for the entire connected path to not intersect is that all control points are in order toward one of the rotation directions. That is, with c as the median of the starting control point for each segment, for all control points comprising the path P_n, $\angle p_n c p_{n+1} < \angle p_n c p_{n+2}$ must be fulfilled. Figure 7 shows an overview of Radiation Loss. Radiation Loss is therefore defined as in the following equation:

$$\mathcal{L}_{rad} = \sum_n \text{ReLU}(\angle p_n c p_{n+1} - \angle p_n c p_{n+2}) \tag{8}$$

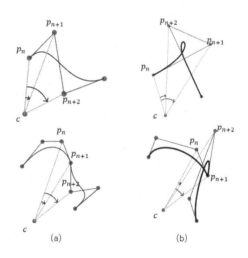

Fig. 7. Overview of Radiation Loss. (a) Example of sequential placement with respect to the direction of rotation, (b) Example of crossed paths as a result of placement against the direction of rotation.

Total Loss. The above three losses are weighted and added together to define Total Loss.

$$\mathcal{L}_{total} = \lambda_{tone}\mathcal{L}_{tone} + \lambda_{rad}\mathcal{L}_{rad} + \lambda_{LSDS}\mathcal{L}_{LSDS} \tag{9}$$

Note that λ_{tone}, λ_{rad}, and λ_{LSDS} are weights to adjust each loss.

3.5 Path Elimination

Since SDS Loss is calculated using rasterized images in this method, path reduction and transparency occur so that Loss becomes smaller. These path ambiguities are not a problem for the raster image after rendering, but for the vector image, they have a negative impact on the data size and rendering process. Therefore, a more concise vector image is generated by removing the paths considering their impact on the rendered image during the optimization process. With a threshold value of τ, P_n is removed when the conditions represented in Eq. (11) are satisfied.

$$\hat{I}_{i,n} = \mathcal{R}(\{P_n, ..., p_N\}) - \mathcal{R}(\{P_{n+1}, ..., p_N\}) \tag{10}$$

$$\frac{\sum_{x,y} \text{alpha}(\hat{I}_{i,n})}{w \times h} < \tau, \tag{11}$$

where $\mathcal{R}$ is the renderer, alpha($\hat{I}_{i,n}$) is the image opacity, and w and h are the output image sizes. The expression 10 represents how much P_n is visible in the raster image after rendering, and in the experiment $\tau = 5.0 \times 10^{-4}$.

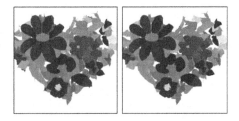

Fig. 8. Example before path deletion (left) and with path deletion applied (right). In this example, the number of Bezier curves has decreased from 200 to 134.

4 Experiments

4.1 Experimental Settings

The size of the input/output image is 600×600. The pre-diffusion model enhancement crops $\mathcal{R}(\hat{I}_i)$ to 512 × 512. By default, the number of Bezier curves is 200, the number of segments is 6, the number of parameter updates i is 1000, and the loss weights are $\lambda_{tone} = 200$, $\lambda_{rad} = 1$, $\lambda_{LSDS} = 1$, and for Tone loss the kernel size of the low-pass filter used was set to 101 and σ to 30. These values were set empirically. Also, the path was removed in case of $i = 800$. Prompt engineering was used for the input with reference to VectorFusion [5] and "a logo of {concept}. minimal flat 2d vector. lineal color. trending on artstation."

4.2 Results

The experimental results of the proposed method are shown in Fig. 9. It can be seen that the structure shown in the text is generated in the shape of the input image. For example, in the case of the input "Mt. Fuji", a structure like Mt. Fuji appears as the largest structure, while a path is generated that does not seem to make much sense and fills the input image area. In addition, as in the "rabbit" example, the color scheme is far from what is intuitively imagined from the input text.

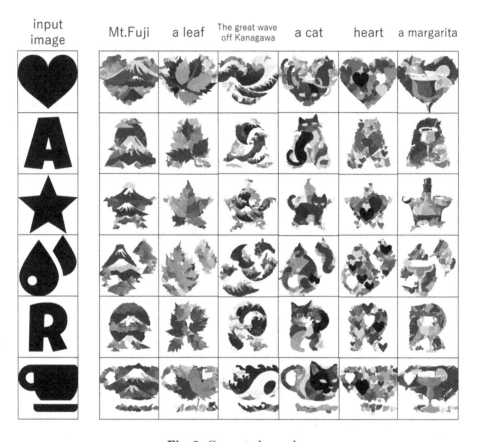

Fig. 9. Generated samples.

4.3 Ablation Studies

Inference of Tone Loss. The results of applying different weights to the Tone Loss are shown in Fig. 10. It can be seen that the larger the weight, the closer the shape of the output image is to the input image. Therefore, it is considered possible to control the shape of the output image by adjusting this loss.

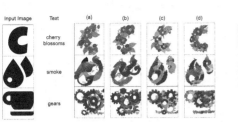 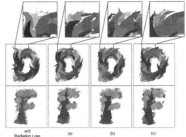

Fig. 10. An example of changing the Tone Loss weight. λ_{tone}. (a)λ_{tone} = 1.0, (b)λ_{tone} = 1.0 × 10, (c)λ_{tone} = 3.0 × 10²

Fig. 11. Example of varying the Radiation Loss weight λ_{rad}. (a)λ_{rad} = 1.0, (b)λ_{rad} = 1.0 × 10, (c)λ_{rad} = 1.0 × 10²

Inference of Radiation Loss. The results of applying different weights to Radiation Loss are shown in Fig. 11. The paths with Radiation Loss are completely convex, even when the weights are small. However, even when Radiation Loss is added, spine-like artifacts are still generated due to the increase in the size of the two points other than the starting control point.

A comparison of Xing Loss and Radiation Loss is also shown in Fig. 12, showing that path self-intersections are reduced in Radiation Loss compared to the Xing Loss case, with each path forming a larger structure.

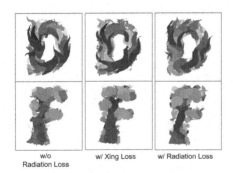

Fig. 12. Compare of Xing Loss and Radiation Loss

4.4 Other Experiments

In order to examine the effect of manipulating text prompts on output results, experiments were conducted by adding colors to adjectives and by changing the prefix and suffix of the prompts. Figure 13 shows the results of generating text prompts as "{color} {concept}". It can be seen that the output examples with colored input reflect the color, while for achromatic colors such as "black" and "white", the color is either ignored or only partially applied.

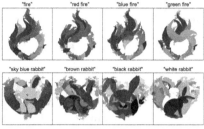

Fig. 13. Example output when color is added to adjectives

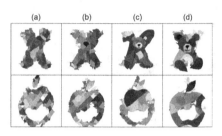

Fig. 14. Example output when i number of parameter updates is changed. (a) $i = 500$, (b) $i = 1000$, (c) $i = 1500$, (d) $i = 2000$.

Figure 14 shows an example of output when the number of parameter updates is changed. It can be seen that at $i = 1000$ most of the shape is formed, and at $i = 1500$ an almost completely stable shape is output. Therefore, it is advisable to choose an appropriate value between $1000 < i < 1500$.

An example of the generation when the number of paths N is changed is shown in Fig. 15. It can be seen that by deleting paths, unnecessary paths are deleted when an excessive number of paths are specified.

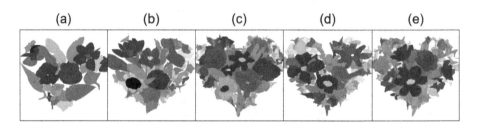

Fig. 15. Example output when the number of paths N is varied. The number of paths changes as follows by applying path deletion. (a) 32 to 32, (b) 64 to 52, (c) 128 to 116, (d) 256 to 141, and (e) 512 to 166.

5 Discussions

5.1 About the Output Results

Observing the sample of output results (Fig. 9), we can see that in some parts, structures like those shown in the text appear, while in other parts, paths of little significance appear to satisfy the shape. This decrease is thought to be caused by the Tone Loss limiting the divergence between the sample results obtained with SDS Loss and the shape image. In addition, the color scheme of some of the samples deviates from the hue imagined from the text, and it is thought that the shape constraint imposed by Tone Loss distorts the shapes that could be generated. Since it is impossible to change the shape of the sampling result by SDS Loss with the shape constraint by Tone Loss, it is thought that these effects are similar to cropping to the shape of the input image. Therefore, it is necessary to extend the sampling results to maintain the shape of the input image to some extent by using the input image for conditioning the diffusion model.

5.2 Influence of Radiation Loss

As shown in Fig. 11, Radiation Loss effectively suppressed the self-intersection problem and contributed to stabilizing the path shape. However, it remains inadequate for generating perfectly smooth paths, and artifacts often emerge where certain segments are elongated, resembling spines. This issue is likely caused by the optimization process attempting to minimize the effect of unnecessary segments on the drawing area within the constraints imposed by Radiation Loss.

Furthermore, because Radiation Loss incorporates information about the starting control point of each segment, it introduces unnecessary restrictions on its positioning, leading to impossible path shapes. Since this method employs SDS Loss to capture content based on the raster image post-rendering, the use of multiple Bézier curves to handle these impossible shapes results in suboptimal vector paths. To address this, an appropriate reference position corresponding to the center point c in Radiation Loss must be established.

6 Conclusions

In this paper, we propose a method for generating shape-preserving vector graphics by integrating the capabilities of a differentiable renderer [6] with SDS Loss [9] and Tone Loss [4]. Additionally, we introduce Radiation Loss, an extension of Xing Loss [8], to address the self-intersection problem in vector graphics generation. Our experimental results demonstrate that the proposed method can effectively incorporate textual content into vector graphics while preserving the shape of the input image. However, the output is influenced by the input conditions, indicating room for further improvement.

For future work, we consider employing image-to-image conditioning in Stable Diffusion, rather than directly controlling the shape, to produce outputs that more closely match the input image. Furthermore, we plan to refine Radiation Loss by adopting central-axis transformation [1] instead of the current central coordinate c, which will help to alleviate unnecessary restrictions on the starting control point.

Acknowledgments. This work was supported by JSPS KAKENHI Grant Numbers, 22H00540 and 22H00548.

References

1. Blum, H.: A Transformation for Extracting New Descriptors of Shape, pp. 362–380. MIT Press (1967)
2. Frans, K., Soros, L., Witkowski, O.: CLIPDraw: exploring text-to-drawing synthesis through language-image encoders. In: Advances in Neural Information Processing Systems, vol. 35, pp. 5207–5218 (2022)
3. Hertzmann, A.: Painterly rendering with curved brush strokes of multiple sizes. In: Proceedings of the 25th Annual Conference on Computer Graphics and Interactive Techniques - SIGGRAPH '98, pp. 453–460 (1998)
4. Iluz, S., Vinker, Y., Hertz, A., Berio, D., Cohen-Or, D., Shamir, A.: Word-as-image for semantic typography. ACM Trans. Graph. **42**(4), 1–11 (2023)
5. Jain, A., Xie, A., Abbeel, P.: VectorFusion: Text-to-SVG by abstracting pixel-based diffusion models. In: 2023 IEEE/CVF Conference on Computer Vision and Pattern Recognition (CVPR), pp. 1911–1920 (2023)
6. Li, T., Lukáč, M., M, G., Ragan-Kelley, J.: Differentiable vector graphics rasterization for editing and learning. ACM Trans. Graph. (Proc. SIGGRAPH Asia) **39**(6), 193:1–193:15 (2020)
7. Li, T.M., Aittala, M., Durand, F., Lehtinen, J.: Differentiable Monte Carlo ray tracing through edge sampling. ACM Trans. Graph. (Proc. SIGGRAPH Asia) **37**(6), 222:1–222:11 (2018)
8. Ma, X., et al.: Towards layer-wise image vectorization. In: Proceedings of CVF/IEEE International Conference on Computer Vision and Pattern Recognition, pp. 16314–16323 (2022)
9. Poole, B., Jain, A., Barron, J.T., Mildenhall, B.: DreamFusion: Text-to-3D using 2D diffusion. In: The Eleventh International Conference on Learning Representations (2022)
10. Radford, A., et al.: Learning transferable visual models from natural language supervision. In: Proceedings of the International Conference on Machine Learning, vol. 139, pp. 8748–8763 (2021)
11. Rombach, R., Blattmann, A., Lorenz, D., Esser, P., Ommer, B.: High-resolution image synthesis with latent diffusion models. In: 2022 IEEE/CVF Conference on Computer Vision and Pattern Recognition (CVPR), pp. 10674–10685 (2022)
12. Ronneberger, O., Fischer, P., Brox, T.: U-Net: convolutional networks for biomedical image segmentation. In: Medical Image Computing and Computer-Assisted Intervention–MICCAI 2015: 18th International Conference, pp. 234–241 (2015)
13. Saharia, C., et al.: Photorealistic text-to-image diffusion models with deep language understanding. In: Advances in Neural Information Processing Systems (2022)
14. Schaldenbrand, P., Liu, Z., Oh, J.: StyleCLIPDraw: coupling content and style in text-to-drawing translation. In: Proceedings of the Thirty-First International Joint Conference on Artificial Intelligence, pp. 4966–4972 (2022)
15. Selinger, P.: Potrace : a polygon-based tracing algorithm (2003). http://potrace.sourceforge.net/potrace.pdf

HYPNOS: Highly Precise Foreground-Focused Diffusion Finetuning for Inanimate Objects

Oliverio Theophilus Nathanael[1], Jonathan Samuel Lumentut[1,2](✉), Nicholas Hans Muliawan[1], Edbert Valencio Angky[1], Felix Indra Kurniadi[4], Alfi Yusrotis Zakiyyah[3], and Jeklin Harefa[1]

[1] Computer Science Department, School of Computer Science, Bina Nusantara University, Jakarta 11480, Indonesia
jlumentut@gmail.com
[2] Department of Computer Science, Edge Hill University, Ormskirk L39 4QP, UK
[3] Mathematics Department, School of Computer Science, Bina Nusantara University, Jakarta 11480, Indonesia
[4] Università di Pisa, 56126 Pisa, Italy

Abstract. In recent years, personalized diffusion-based text-to-image generative tasks have been a hot topic in computer vision studies. A robust diffusion model is determined by its ability to perform near-perfect reconstruction of certain product outcomes given few related input samples. Unfortunately, the current prominent diffusion-based finetuning technique falls short in maintaining the foreground object consistency while being constrained to produce diverse backgrounds in the image outcome. In the worst scenario, the overfitting issue may occur, meaning that the foreground object is less controllable due to the condition above, for example, the input prompt information is transferred ambiguously to both foreground and background regions, instead of the supposed background region only. To tackle the issues above, we proposed Hypnos, a highly precise foreground-focused diffusion finetuning technique. On the image level, this strategy works best for inanimate object generation tasks, and to do so, Hypnos implements two main approaches, namely: (i) a content-centric prompting strategy and (ii) the utilization of our additional foreground-focused discriminative module. The utilized module is connected with the diffusion model and finetuned with our proposed set of supervision mechanism. Combining the strategies above yielded to the foreground-background disentanglement capability of the diffusion model. Our experimental results showed that the proposed strategy gave a more robust performance and visually pleasing results compared to the former technique. For better elaborations, we also provided extensive studies to assess the fruitful outcomes above, which reveal how personalization behaves in regard to several training conditions.

Supplementary Information The online version contains supplementary material available at https://doi.org/10.1007/978-981-96-2644-1_15.

© The Author(s), under exclusive license to Springer Nature Singapore Pte Ltd. 2025
M. Cho et al. (Eds.): ACCV 2024 Workshops, LNCS 15483, pp. 211–227, 2025.
https://doi.org/10.1007/978-981-96-2644-1_15

Keywords: Generative model · Stable Diffusion · Dreambooth

1 Introduction

Personalization for text-to-image (T2I) Diffusion Model has been a hot field of study and rapidly popularized among researchers in computer vision studies as well as practitioners and even hobbyist. Personalization is the key for generative models to be widely used as a tool for various use cases. This personalization in the Diffusion Model has the same spirit as grounding Large Language Models (LLMs), where it opens the possibility of being reliably incorporated in diverse real-world environments. With reliable personalization in the Diffusion Model, it opens several potentials for recreational purposes and even in the business world, such as creating product advertisements and campaign photos, which mostly highlight the foreground objects (*foreground-focused objective*).

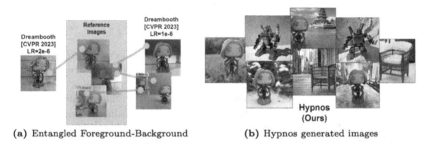

(a) Entangled Foreground-Background (b) Hypnos generated images

Fig. 1. The current (a) widely used method [21] is prone to subject-scene entanglement and overfitting as shown by the yellow lines. Our proposed method visually (b) shows a realistic disentanglement effect between subject and scene at the foreground and background, respectively.

One particular distinctive family of these techniques, namely Dreambooth [21], enables the ability to finetune Diffusion models on a specific image without losing prior knowledge. At the moment Dreambooth is one of the most popular personalization techniques because of its efficiency and minimum requirements of input samples (3–5 reference images) to be able to synthesize great quality, with high similarities to the reference input.

Despite that, there is still room for improvement, as the current method is still prone to structure and color distortions, which are mostly caused by the coarse attempt of the model to blend the object in the foreground and the scene in the background [21]. Visually, there is an entanglement of background and foreground semantic information as shown in Fig. 1a (content relations are highlighted with yellow lines). To overcome this issue, one might increase the learning rate of [21] to enforce more consistency to the reference images. Unfortunately, such approach causes another problem to arise [21], which is the overfitting issue.

To our observation, the work of [21] tends to produce the entangled version of both foreground and background outputs from the given foreground reference input (Fig. 1a). Adjusting the learning rate to the higher (or lower) values in [21] finetuning affects the variance of both foreground and background information. This can be perceived as an entangled foreground-background issue in the T2I task, which is improper for the foreground-focused objective mentioned above.

In our pursuit of creating a reliable T2I finetuning technique, we introduce HYPNOS, a highly precise foreground-focused diffusion finetuning approach for inanimate objects. We propose several strategies that work synergically to disentangle foreground-background information. These additions are proven to produce visually consistent and pleasing images as seen on Fig. 1b. In detail, our strategies comprises of: straightforward image dataset augmentation, explicit foreground-background prompts procedures, and sets of new supervision mechanisms to quantify the foreground deviation from the reference images. Importantly, those method are implemented while still maintaining fast finetuning time and low latency architecture. We also conducted an extensive evaluation of both qualitative and quantitative analysis to further support our studies. These evaluations are also used to explain some distinct behaviors of Hypnos compared to Dreambooth and Textual Inversion techniques. Our main contributions can be summarized as follows:

- We propose a novel technique with competitive result by leveraging content-centric augmentation and new sets of supervision mechanisms. This improvement reliably ensures foreground-background disentanglement, significantly lowers noise, and enables semantic level tuning for the T2I task.
- We introduce prompt-invariant and prompt-varying metrics as a new approach to quantitatively assess the Dreambooth family technique and how to interpret the results. We show that these new metrics can be used to gain a better insight of the finetuned model.
- We conduct analysis and experiments on the introduced hyperparameters to elaborate further on their effects and how to tune them properly.

2 Related Works

2.1 Vision-Based Generative Model

Diffusion Model is a groundbreaking image generation method that leverages thermodynamics based process [27]. It was then further popularized by the introduction of the Denoising Diffusion Probabilistic Model [8]. It is a probabilistic model that is capable of modeling a complex distribution using a denoising process. Denoising Diffusion Implicit Model later eliminated the markovian assumption which speeds up the denoising process [28]. Other popular improvement includes the adaptation of Cosine scheduling on the noise diffusion process to ensure the image is not too rapidly destroyed throughout the diffusion process [15]. Other physics-inspired generative models include Poisson Flow Generative

Models (PFGM), which is inspired by the physics of Electrodynamics [30]. Its improvement, PFGM++ [31], unified PFGM Model with Diffusion Model.

The current trend involves the strategy of text-prompt-based input to guide the denoising process, hence called as T2I diffusion model. CLIP, a transformer-based model that aligns text and image embedding [18], is introduced to satisfy such task. By utilizing CLIP, The work of GLIDE [14] showed to produce great quality results. Recent text transformer-based works, e.g. BERT [4] and T5 [19], also showed great performance in running the image generative task [23].

Diffusion process recently involved the denoising procedure on Variational Autoencoder's [10] latent space to make the model resource-effective. This method is commonly known as the Latent Diffusion Model [20]. The following are among the popular latent-based works: Stable Diffusion [20], Stable Diffusion XL [16], Dall-E 3 [26], and the recent video-based (temporal) diffusion generative task [1].

2.2 Personalization

As explained on Sect. 1 personalization aims to enable T2I models to be able to generate a specific object based on some reference images. Some of the methods include textual inversion [6] and controlnets [32]. Among many personalization techniques, the one that arguably stands out the most is Dreambooth [21]. Dreambooth finetunes a diffusion model using 3–5 reference images and some prior class images to prevent the loss of prior ability to produce diverse images on that particular class. Dreambooth also can be trained relatively fast.

In recent years, numerous works have been proposed to better fit Dreambooth in some use cases. LoRA [9] speeds up the finetuning process at the expense of image quality. Google's HyperDreamBooth [22] focuses on fast personalization for human faces using Hypernetworks to adjust main network weights [3]; further accelerated by LoRA. DreamCom [12] had shown remarkable results on the image inpainting task by reusing some good generated images to retrain the model. Other variations of Dreambooth include the procedure of learning other concepts besides the object appearance itself [13]. In this work, we opt to scope the usage of Dreambooth [21] (comprises of Stable Diffusion [20]) as the baseline for our foreground-focused diffusion finetuning task on inanimate objects.

3 Methodology

The proposed method is crafted to accommodate Dreambooth in terms of producing a more consistent foreground subject. Hence, we utilized the same backbone model, but with additional proposed functionalities to better disentangle the foreground-background information. As seen on Fig. 2, there are two main part of the process. The *first one* is the datasets refinement procedure that undergoes our specific image augmentation and prompt engineering functionalities (left side of Fig. 2). The *second part* is the Diffusion model itself that is trained on our diverse supervision mechanisms (right side of Fig. 2).

3.1 Dataset Refinement

Similar to Dreambooth, Hypnos is a finetuning technique that only need 3–5 image samples. We limit the scope of the generation task to only for inanimate objects. Our images that are used within this work are taken from personal images and Unsplash (https://unsplash.com/).

3.2 Text to Image Model

To generate the image we use a stable diffusion v1.5 model [20]. The model itself comprises of several networks that work synergically. These models usually utilize U-Net as the network to model the denoising process.

Stable Diffusion is a latent diffusion model which means that the diffusion and denoising process happens on a latent representation of an image. Furthermore, as a Text to Image model, it also incorporates CLIP embeddings into the U-Net model to guide the generated image based on a specific prompt. Following the common procedure, the images in our method are encoded by the Variational Autoencoders (VAE), while the text prompts are processed by the CLIP method.

In terms of finetuning, there are several approaches to run such task using the Dreambooth method. In this work, we opt to freeze the weight of the VAE, meaning that Hypnos only optimize both the denoising model and the text encoder. We also preserve the denoising procedure from the Stable Diffusion which maintain the lower latency characteristic [20].

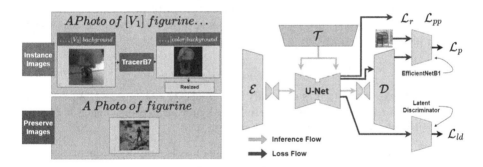

Fig. 2. We proposed to apply image augmentation and prompt engineering to the instance dataset, we also apply sets of losses as shown by the blue arrows (Color figure online)

3.3 Image Augmentation and Prompt Engineering Functionalities

Based on the original Dreambooth paper there are two sets of dataset to be extracted for supervision (Fig. 2 (left)). The *first one* (purple regionof Fig. 2 (left)) is the self generated class images paired with standard prompt such as "a photo of [C]" where [C] is the class name of the subject. The *second set* (blue regionof Fig. 2 (left)) is the specific provided subject images and traditionally labeled as "a photo of [V] [C]" where [V] is a rare word that acts as a name to tell our object apart from others. Both extracted datasets are utilized in supervision to prevent the lost of the prior general understanding of that particular class.

Unfortunately, doing the above's strategy in Dreambooth [21] invokes the entanglement effect in both foreground and background, which is unfavorable for foreground-focused objective. To avoid this, we propose a content-centric augmentation that explicitly introduces background-foreground disentanglement. We then utilized TracerB7 [11,17] to randomly replace the background with monotone colors. We empirically set the augmentation proportion to 0.66 to avoid losing the ability to blend the subject with the background. To further introduce the foreground diversity, we also resized a small portion of the sampled dataset spatially. Moreover, we provided *prompting adjustment strategy* to allow the model to recognize the foreground and background via first and second clauses, respectively, as seen on Fig. 2 To describe the original image's background, it is impractical to use a particular color name as used in the background changed images; hence, we opted to utilize another rare word as a placeholder.

3.4 Supervision Mechanisms

One of our proposed method includes specialized losses to make sure that the model will preserve the foreground structure and color while still letting the background to exhibit a high variance. Our proposed method is a combination of 4 different losses which are reconstruction losses, prior preservation loss, perceptual loss, and latent discriminator loss as shown with the blue arrows on Fig. 2, which is written as Eq. (1):

$$\mathcal{L} = \lambda_r \mathcal{L}_r + \lambda_{pp} \mathcal{L}_{pp} + \lambda_p \mathcal{L}_p + \lambda_{ld} \mathcal{L}_{ld}. \tag{1}$$

Based on experiments it is generally best to set $\lambda_r = 1$, $\lambda_{pp} = 1$, $\lambda_p = 0.003$, $\lambda_{ld} = 0.5$. Note that these weights itself are hyperparameters, hence it can be adjusted based on the intended model output.

Reconstruction Loss ($\mathcal{L}_r$) Traditionally this loss is simply the mean squared error of the predicted noise of the instance latent images which denotes by z_i at timestep t where $t \sim \mathcal{U}(1, 1000)$. This loss ensures the generated image is as similar as possible to the instance image. Though mean squared error is a widely known loss for reliably quantifying deviation from a target value, our observation suggests that exponential-growth-base loss function is preferable than quadratic-growth-base one. This is done as measuring the deviation from a noise on a latent space is focused on the data with normal distribution. One intuitive alternative is to use the inverse of gaussian distribution as shown on Eq. (2) as it is based of the VAE overall distribution itself.

$$\mathcal{L}_r = \sigma\sqrt{2\pi}(\mathbb{E}_{z_i,t,c,\epsilon} e^{\|\epsilon_\theta(z_i,t,c)-\epsilon\|_2^2/2\sigma^2} - 1) \tag{2}$$

Based on our experiment, this is very effective for getting rid of the noise artefact that is often seen in a higher learning rate of Dreambooth method (details on *Supplementary Material*). By adjusting the σ it is possible to tune on how steep the loss is. In detail, the L2 loss can be re-approximated by tweaking σ, particularly by minimizing the least squared regression of both function for bound $\pm a$ as shown on Eq. (3).

$$\min_{\sigma} \int_0^a (x^2 - \sigma\sqrt{2\pi}(e^{x^2/2\sigma^2} - 1))^2 dx \quad (3)$$

By solving Eq. (3) for a equals to 1, where reconstruction loss are very unlikely to exceed 1, it is empirically found that $\sigma = 1.382$ is the closest to L2 loss, with standard deviation larger than that, meaning that the loss is flatter than L2 loss.

Prior Preservation Loss ($\mathcal{L}_{pp}$) While Reconstruction Loss computes the deviation between the generated image with respect to the subject latent images, Prior Perceptual loss works with respect to the class latent images which denotes by z_p as written in Eq. (4):

$$\mathcal{L}_{pp} = \mathbb{E}_{z_p,t,c,\epsilon} e\|\epsilon_\theta(z_p,t,c) - \epsilon\|_2^2. \quad (4)$$

In this work, we opt to preserve the original MSE loss on this prior preservation function, as this allows the model to focus its learning process on the instance images rather than the class images.

Perceptual Loss ($\mathcal{L}_p$) Perceptual Loss is widely used for style transfer task. The advantage of using this loss is that it can semantically quantify the deviation of the generated image. We utilized EfficientNetB1 [29] (a classifier network) as the encoder and take the L2 loss across several of its activation. To preserve low-level semantic information, we gave larger weight to the activations of its shallow layers Furthermore, using classifier network to guide a diffusion model is a common practice. One justification includes the use of classifier output to aid an ambiguous text to produce an image that is better represent the intended picture [25].

Since classifier networks process decoded denoised images, thus, to be able to apply perceptual loss, it is required to denoise the images straight to z_0 and then decode the image beforehand. The resulting equation is described on Eq. (5):

$$\mathcal{L}_p = \mathbb{E}_{z_i,t,c,x}\|C_\theta(x) - C_\theta(D_\theta(z_0))\|_2^2, \quad (5)$$

where C_θ is a classifier model, D denotes the VAE Decoder, x denotes the reference image, and z_0 as the denoised latent.

Based on our observation, to avoid overfitting, there are two approaches that can be done, *the first one* is to apply Latent Discriminator that balances the loss and *second one* is by limiting its influence on the loss, rather than decreasing the loss weight. The latter can be implemented by limiting the number of steps that are influenced by Perceptual Loss. Intuitively, this is done to avoid local minima of the original Dreambooth loss that exhibit a high perceptual dissimilarity as described on Fig. 1a. It is also important to avoid a perfect perceptual match. Therefore, Perceptual Loss should guide optimization for only a certain number of steps, with the remaining steps optimized without it. To do so, we introduced a new hyperparameter s_p that directly control the perceptual similarity of the generated image by limiting the maximum step influenced by Perceptual loss. We set $s_p = 500$ out of 800 steps of total training cycle on the re-defined Perceptual loss $\mathcal{L}_p$ in Eq. (6).

$$\mathcal{L}_p = \begin{cases} \mathbb{E}_{z_i,t,c,x}\|C_\theta(x) - C_\theta(D_\theta(z_0))\|_2^2, & \text{if } s \leq s_p \ . \\ 0, & \text{otherwise} \end{cases} \quad (6)$$

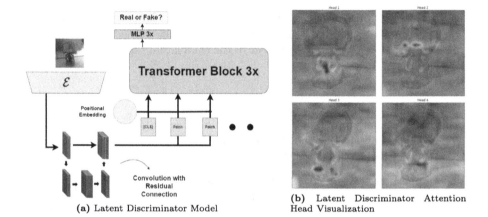

(a) Latent Discriminator Model

(b) Latent Discriminator Attention Head Visualization

Fig. 3. Latent Discriminator is designed to be lightweight and has low latency. Despite the small size, the discriminator successfully identify important feature as shown by the attention head visualization

Latent Discriminator Loss ($\mathcal{L}_{ld}$) Similar to perceptual loss, this loss aims to quantify the foreground deviation semantically to preserve the foreground information. Rather than computing the loss of the decoded image, this loss computes the difference straight from the denoised latent space. To achieve this, we make a discriminator trained on the latent representation of the instance image and the monotone colored background version as the real images while fake images comprises of class images and altered instance images such as removed foreground and negative colored foreground (details on *supplementary material*). The discriminator itself is a 3 layer vision transformer [5] with additional convolutional layers to ensure efficient semantic information extraction (visualized in Fig. 3a).

Similar to Adversarial Diffusion Distillation [24] that applied adversarial learning on diffusion model, our latent discriminator is also provided with the flexibility to compete with main model. It is written in Eq. (7) with $LD_\theta(.)$ and

z_0 denote the Latent Discriminator model and denoised latent image, respectively:
$$\mathcal{L}_{ld} = \mathbb{E}_{z_i,t} \|1 - LD_\theta(z_0)\|_2^2. \tag{7}$$

The initial intention of this loss ($\mathcal{L}_{ld}$) is to provide the diffusion model with an explicit semantic similarity loss referenced to all instance images. By this we can utilize all image references rather than just one sample as in the other Losses ($\mathcal{L}_r$, $\mathcal{L}_{pp}$, $\mathcal{L}_p$). Thus, adversarial learning can be viewed as an effort for the discriminator model only to adapt with the everchanging image quality of the generated image. We design our discriminator with light architecture and train it initially with 600 steps, and then followed by the combined training.

The implementated architecture for Latent Discriminator is shown on Fig. 3a. As seen on Fig. 3b, this simple architecture is enough to detect important features of the object.

3.5 Evaluation Metrics

Quantifying Dreambooth method family is one of the most tricky task to handle there are at least two main arguments to support this claim. The first one is that Dreambooth family falls into few-shot learning category which by definition has a scarce amount of ground truth, hence it is not suitable to represent the approximate intended data distribution, rather it is expected to be able extrapolate from that given data distribution. This imply that the measure of how good the extrapolation result vary a lot across preferences and use cases. Second, ground truth incorporating diverse prompt are non-existent, unlike what seen in classic Diffusion model task or even controlnet model [32]. This leads to some evaluation blind spot on how well model can adapt to diverse prompts and how does the model adapt to the prompt and what are the impact to the object.

Prompt Invariant and Prompt Varying Evaluation. In this work, we proposed a modification to the existing metric evaluation to open new perspective and used it alongside the existing method. This is reasonable since Dreambooth family techniques usually use standard prompt to generate all the images which we called as *Prompt Invariant Evaluation*. To better accommodate the current evaluation approach, we propose that it is also important to assess the adaptibility of the resulting model on diverse prompt, which we coined as *Prompt Varying Evaluation*.

To minimize bias that emerges while handcrafting every single prompt, we proposed a solution that samples parts of a generic image generation prompt structure as seen on Fig. 4. There are three main considerations, the first one is the ***image type*** such as photo or painting, the second part is the ***background information***. In this context, background information is not only limited on the background scene but also the image composition other than the main object itself. The third one is the ***styles***; style descriptions are crucial to generate high quality image in smaller diffusion models as they require longer prompt compared to larger model such as SDXL [16]. Based on our understanding, style

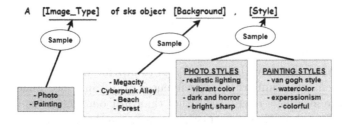

Fig. 4. Prompts for Varying Prompt Evaluation are sampled from the combination of predefined list for image type, background, and style

descriptions may vary across image types, hence its description is acquired from the previously sampled image type.

Dreambooth Metrics. As perceived from the Dreambooth work, their fine-tuning technique is a unique task that is inadequate to be measured using classic image generation metrics because of its nature that have limited reference image as explained earlier. Following their scheme, we use DINO [2], CLIP-I, and CLIP-T to perform the baseline quantitative measurement.

Image Quality and Perceptual Metrics. Apart from the metrics above, image generation tasks are also evaluated using metrics suvh as FID [7], LPIPS [33], SSIM, and PSNR These metrics are rarely used to evaluate the Dreambooth method family as it is prone to biased score. However, we still opt to include them as they still can provide additional insight on the generated image quality.

4 Result and Discussion

4.1 Evaluation

As mentioned on Sect. 3.5, the evaluation of Dreambooth method family can vary across preference and use cases. Therefore, we conduct both qualitative and quantitative analysis of the results. As shown later on this section, solely analyzing quantitative metrics on this particular task has a potential to be misleading, hence it is advised to interpret the quantitative assessment with a grain of salt. We resolve the issues above by employing the procedure of always combining quantitative and qualitative analysis throughout the evaluation process.

Prompt Invariant Evaluation. As seen on Table 1, we evaluated each metrics on three subject samples which are a Funko figurine (●), a rattan chair (●), and a Lego robot (●). This approach can be observed on Table 1. We opt to experiment on these samples to minimize the bias effect while assessing the metrics. The prior method of Dreambooth utilized a small number of instance images that

Table 1. *Prompt Invariant* quantitavie metrics evaluated on 3 datasets, Funko figurine (●), Rattan chair (◐), and Lego Robot (●).

Method		DINO	CLIP-I	CLIP-T	FID	SSIM	PSNR	LPIPS
Hypnos (Ours)	●	**0.7851**	**0.8635**	0.0094	3,6032	0.5974	11.8504	0.3850
	◐	**0.6502**	**0.8015**	0.0067	**2.3840**	**0.2225**	**9.4634**	**0.4166**
	●	0.6589	0.8369	0.0183	**5.6330**	0.3876	**10.5387**	0.4624
Dreambooth	●	0.6422	0.7935	0.0183	**2.9873**	**0.6056**	**12.2883**	**0.3663**
(LR=1e-6)	◐	0.5012	0.7404	**0.0549**	13.1933	0.1645	9.0604	0.4583
	●	**0.7130**	**0.8753**	**0.0458**	5.7367	0.3429	9.3005	0.4679
Dreambooth	●	0.5311	0.7468	0.0153	14.7671	0.4781	11.4756	0.4513
(LR=2e-6)	◐	0.2647	0.4742	0.0224	42.7634	0.1433	9.3789	0.5128
	●	0.5704	0.8323	0.0175	14.2261	0.3060	9.3813	**0.4622**
Textual	●	0.4934	0.6469	**0.0417**	12.2159	0.4565	9.9478	0.4917
Inversion	◐	0.4397	0.7134	0.0308	4.6512	0.2125	8.8142	0.4785
	●	0.3904	0.6118	0.0312	6.4942	**0.3929**	9.6875	0.5160

were insufficient to be referred to as the perfect target distribution. This yields the metrics results that vary vastly across datasets, which we desire to avoid. By this reasoning, we performed a comparison if and only if those methods were trained with the same instance images.

As shown in Table 1, Hypnos is consistently superior in DINO and CLIP-I metrics. In contrast, Dreambooth mainly dominates CLIP-T. However, CLIP-T's metric is the least aligned with qualitative outcomes (refer to CLIP-T scores in Table 1 vs. results in Fig. 5). We suspect that this is caused by the large domain gap between post-training CLIP text embedding and pre-trained CLIP image embedding, which, most of the time, each embedding deviates and becomes unreliable. DINO and CLIP-I, on the other hand, are rather in line with what was observed qualitatively. Visually, Hypnos exhibits minimum noise effect in the generated images while preserving most foreground information similar to the dataset. Dreambooth, even with a lower learning rate case, produces high-noise results (*first* and *second column* in Fig. 5) and tends to overfit on the instance image. This is reasonable because both the image and prompt are what exactly the model learned in the training phase. In contrast, Hypnos introduced higher variance in the training background with its respective prompt, resulting in arbitrary background results when no background information is provided in the prompt input. At the same time, Hypnos is still able to maintain structurally similar foreground outputs (*first* and *second column* in Fig. 5).

Based on our observation, Dreambooth tended to yield high noise when the object or scene comprises of mostly homogeneous regions (e.g. figurine case). On the Lego robot case, Dreambooth gave perfect generation results although lower learning rate is utilized and thus it produced higher scores in the metrics. Nonetheless, the noise level becomes aggravated in the Lego case if a higher learning rate is employed. Hypnos tackles the above problem as shown in the first row of Fig. 5. In the case of longer prompts input (yellowish prompt in Fig. 5),

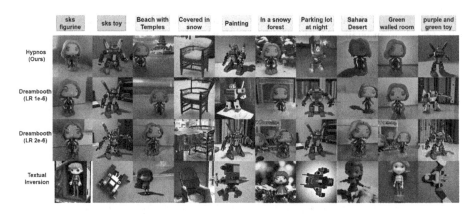

Fig. 5. Image generation comparison, red prompt denotes *prompt invariant*, yellow prompt denotes *prompt varying*, green prompt denotes specific prompting to analyze foreground-background disentanglement ability and highlight semantic leaking

unlike Dreambooth, Hypnos avoids the overfit and underfit effects, which yields the production of highly similar and high-quality foreground results.

Furthermore, as seen on Fig. 5 (*fourth row*), Textual Inversion produced a poor output in terms of similarity even with the standard prompt usage. From this point onward, Textual Inversion acted as a baseline and control to our metrics analysis as it is already widely known to underperform compared to the classic Dreambooth.

Table 1 also shows standard image generation metrics with the prompt invariant case. The metrics provided useful insights, yet they are insufficient to determine model superiority. Nevertheless, Hypnos still dominates most of the metrics. In terms of FID, Hypnos consistently generated a perceptually similar image to the instance images. This is further supported by higher SSIM and PSNR scores (highlighted with **bolds** in Table 1; **bolds** indicate best scores).

Prompt Varying Evaluation. Table 2 shows the metrics score across varying prompts. Varying prompts is expected to force the models to produce diverse images. This implies that the similarity index should decrease significantly. Those performances are observed in Hypnos when comparing the scores in Tables 2 and 1. A noticeable change is that the higher learning rate Dreambooth dominates most metrics, suggesting high similarity. This phenomenon can be explained by observing the resulting images as seen on the *third row* in Fig. 5. Qualitatively, it is apparent that Dreambooth (LR=2e-6) is overfitted, and the generated images fail to adapt to the prompt. Thus, Dreambooth's high scores are obtained from overfitting, which is inadequate.

On the other hand, Dreambooth, with a lower learning rate, seems to be able to better adapt to the prompt. However, as mentioned earlier, they came with a downside effect on the foreground quality. As seen on the *fifth* and *eighth column* of Fig. 5, it is capable of adapting to the prompt input, but its foreground object

Table 2. *Prompt Varying* quantitavie metrics evaluated on 3 datasets, Funko figurine (●), Rattan chair (○), and Lego Robot (●).

Method	DINO	CLIP-I	CLIP-T	FID	SSIM	PSNR	LPIPS
Hypnos (Ours)	● **0.7851**	**0.8635**	0.0094	3,6032	0.5974	11.8504	0.3850
	○ **0.6502**	**0.8015**	0.0067	**2.3840**	**0.2225**	**9.4634**	**0.4166**
	● 0.6589	0.8369	0.0183	**5.6330**	0.3876	**10.5387**	0.4624
Dreambooth	● 0.6422	0.7935	0.0183	**2.9873**	**0.6056**	**12.2883**	**0.3663**
(LR=1e-6)	○ 0.5012	0.7404	**0.0549**	13.1933	0.1645	9.0604	0.4583
	● **0.7130**	**0.8753**	0.0458	5.7367	0.3429	9.3005	0.4679
Dreambooth	● 0.5311	0.7468	0.0153	14.7671	0.4781	11.4756	0.4513
(LR=2e-6)	○ 0.2647	0.4742	0.0224	42.7634	0.1433	9.3789	0.5128
	● 0.5704	0.8323	0.0175	14.2261	0.3060	9.3813	**0.4622**
Textual	● 0.4934	0.6469	**0.0417**	12.2159	0.4565	9.9478	0.4917
Inversion	○ 0.4397	0.7134	0.0308	4.6512	0.2125	8.8142	0.4785
	● 0.3904	0.6118	0.0312	6.4942	**0.3929**	9.6875	0.5160

is highly altered. Hypnos, in contrast, exhibited a more consistent and robust foreground similarity.

Another downside of Dreambooth is that it is often overfit on the background. As seen on the *third column* of Dreambooth case of Fig. 5, the beach looks like a room, and on the *sixth column*, Dreambooth fails to produce a forest scene. It instead overfits the dataset background (*second row* of Fig. 5). Another example is shown on the *fifth* and *seventh column* where the toy clearly always stands in front of books and fails to follow the prompt; this is because some of the training images are taken in front of the books. These issues are effectively resolved on Hypnos-generated images as they are trained to have content-centric knowledge.

Hypnos also displayed a clear distinction between foreground and background. This statement is justified by how it preserves the image color and adapts to the scene naturally. On the other hand, Dreambooth showed the foreground-background entanglement effect, in which its phenomenon is observed on the *sixth* and *eighth column* of Fig. 5, where the base of the figurine's colors are affected by the scenes' color.

Furthermore, Hypnos' robust foreground structure preservation allows it to produce more imaginative images, such as snow on a rattan chair (*first row fourth column* of Fig. 5). In that case, both components are rarely seen together in the real world. Dreambooth, however, generated a very different chair and struggled with high-frequency features like a rattan chair (*fourth column* of Fig. 5). This explains why Hypnos dominates the metrics for Rattan Chair as seen on both tables (Tables 1 and 2).

Foreground-Background Disentanglement Analysis. Hypnos also solved another limitation of Dreambooth [21], mentioned in their work, which is the color and theme of the scene that leaks information from the foreground object or vice versa (entangled). As seen on Fig. 5 with green prompts, Dreambooth produced a highly overfitted image with noises when prompted with green walled room. Rather than coloring the room wall green, it painted the base green instead, and this phenomenon is frequently observed on Dreambooth. This phenomenon is also clearly observed in textual inversion, where the clothes are green instead of the wall. By leveraging content-centric strategies, Hypnos, in contrast, can correctly make a green walled room background while still having accurate foreground color (refer to the second column from right in Fig. 5).

Besides excelling in background modification, Hypnos also excels in foreground modification. This is observed to be challenging for Dreambooth, especially for the more complex objects where they often end up overfitting even with a lower learning rate (first column from the right of Fig. 5). Dreambooth structural consistency dropped with longer prompt and having difficulty adapting to the prompt whereas Hypnos successfully altered the object color to purple and green without altering much of the structure. This again proves the robustness of Hypnos in generating inanimate objects in various scenes and prompts.

4.2 Ablation Studies

Perceptual Loss. We evaluate Hypnos on two configurations to analyze the effect of Perceptual loss. The First one is when we exclude any limit to the maximum step influenced by Perceptual Loss. In this case, it tends to highly overfit, such as the same lighting and pose information directly pasted from the original image. We also observed the emergence of background overfitting because of this uninformed subject-scene loss. Secondly, we excluded the Perceptual Loss. Without perceptual loss, structural preservation is downgraded and some noise started to showed up although still far less than Dreambooth since Hypnos also utilized inverse gaussian reconstruction loss. Refer to suppl. material for details.

Latent Discriminator Loss. By removing the Latent Discriminator, the color consistency and background generation are severely impacted. This is reasonable as two-third of the training image scenes are just monotone colors, hence making the generated image have bad subject-scene blending.

4.3 Limitations

As shown on Fig. 5, Hypnos does not guarantee perfect image reconstruction nor 3D consistency. Hypnos also never intended to be used on humans nor animals hence it may not be optimized for those tasks. There also seems to be a tradeoff between high foreground reconstruction and subject-scene blending capability.

4.4 Societal Impact

Our method enables small businesses, up to the well-established businesses, to easily gain access to a more reliable and flexible high-quality product image generation for marketing purposes compared to background image inpainting as the most common approach at the time of writing of this work. On the other end of the spectrum, Hypnos can also potentially produce misleading images without any reality backing, which may lead to unintentional or even deliberate fraud or other malicious activities.

5 Conclusion

We propose Hypnos, a content-centric personalization diffusion finetuning technique that addresses high noise and foreground-background entanglement in T2I tasks, which are prevalent problems of its predecessors. Given 3–5 image input samples, Hypnos consistently preserves the foreground's structure and color while enabling diverse background modification. We also show that it is also possible to adjust the diversity of the generated image semantically, with inanimate object constrain, by utilizing our proposed supervision mechanisms. Furthermore, with the newly introduced hyperparameters, Hypnos offers a wider space of adjustment on the target distribution to support diverse use cases. We believe that the development of Hypnos finetuning strategy paves the way for various foreground-focused downstream computer vision-based generative tasks.

Acknowledgment. This research project is supported by the grant of Penelitian Pemula Binus No: 069A/VRRTT/III/2024 (Universitas Bina Nusantara).

References

1. Blattmann, A., et al.: Stable video diffusion: scaling latent video diffusion models to large datasets. arXiv preprint arXiv:2311.15127 (2023)
2. Caron, M., et al.: Emerging properties in self-supervised vision transformers. In: Proceedings of the IEEE/CVF International Conference on Computer Vision, pp. 9650–9660 (2021)
3. Chauhan, V.K., Zhou, J., Lu, P., Molaei, S., Clifton, D.A.: A brief review of hypernetworks in deep learning. arXiv preprint arXiv:2306.06955 (2023)
4. Devlin, J., Chang, M.W., Lee, K., Toutanova, K.: BERT: pre-training of deep bidirectional transformers for language understanding. arXiv preprint arXiv:1810.04805 (2018)
5. Dosovitskiy, A., et al.: An image is worth 16×16 words: transformers for image recognition at scale (2021)
6. Gal, R., et al.: An image is worth one word: personalizing text-to-image generation using textual inversion. arXiv preprint arXiv:2208.01618 (2022)
7. Heusel, M., Ramsauer, H., Unterthiner, T., Nessler, B., Hochreiter, S.: GANs trained by a two time-scale update rule converge to a local Nash equilibrium. In: Advances in Neural Information Processing Systems, vol. 30 (2017)
8. Ho, J., Jain, A., Abbeel, P.: Denoising diffusion probabilistic models. Adv. Neural. Inf. Process. Syst. **33**, 6840–6851 (2020)
9. Hu, E.J., et al.: LoRa: low-rank adaptation of large language models. arXiv preprint arXiv:2106.09685 (2021)

10. Kingma, D.P., Welling, M., et al.: An introduction to variational autoencoders. Found. Trends® Mach. Learn. **12**(4), 307–392 (2019)
11. Lee, M.S., Shin, W., Han, S.W.: Tracer: extreme attention guided salient object tracing network (student abstract). In: Proceedings of the AAAI Conference on Artificial Intelligence. vol. 36, pp. 12993–12994 (2022)
12. Lu, L., Zhang, B., Niu, L.: DreamCom: finetuning text-guided inpainting model for image composition. arXiv preprint arXiv:2309.15508 (2023)
13. Motamed, S., Paudel, D.P., Van Gool, L.: Lego: learning to disentangle and invert concepts beyond object appearance in text-to-image diffusion models. arXiv preprint arXiv:2311.13833 (2023)
14. Nichol, A., et al.: Glide: towards photorealistic image generation and editing with text-guided diffusion models. arXiv preprint arXiv:2112.10741 (2021)
15. Nichol, A.Q., Dhariwal, P.: Improved denoising diffusion probabilistic models. In: International Conference on Machine Learning, pp. 8162–8171. PMLR (2021)
16. Podell, D., et al.: SDXL: improving latent diffusion models for high-resolution image synthesis. arXiv preprint arXiv:2307.01952 (2023)
17. Qin, X., Zhang, Z., Huang, C., Dehghan, M., Zaiane, O.R., Jagersand, M.: U2-Net: going deeper with nested u-structure for salient object detection. Pattern Recogn. **106**, 107404 (2020)
18. Radford, A., et al.: Learning transferable visual models from natural language supervision. In: International Conference on Machine Learning, pp. 8748–8763. PMLR (2021)
19. Raffel, C., et al.: Exploring the limits of transfer learning with a unified text-to-text transformer. J. Mach. Learn. Res. **21**(140), 1–67 (2020)
20. Rombach, R., Blattmann, A., Lorenz, D., Esser, P., Ommer, B.: High-resolution image synthesis with latent diffusion models. In: Proceedings of the IEEE/CVF Conference on Computer Vision and Pattern Recognition (CVPR), pp. 10684–10695 (2022)
21. Ruiz, N., Li, Y., Jampani, V., Pritch, Y., Rubinstein, M., Aberman, K.: DreamBooth: fine tuning text-to-image diffusion models for subject-driven generation. In: Proceedings of the IEEE/CVF Conference on Computer Vision and Pattern Recognition, pp. 22500–22510 (2023)
22. Ruiz, N., et al.: HyperDreamBooth: hypernetworks for fast personalization of text-to-image models. In: Proceedings of the IEEE/CVF Conference on Computer Vision and Pattern Recognition, pp. 6527–6536 (2024)
23. Saharia, C., et al.: Photorealistic text-to-image diffusion models with deep language understanding. Adv. Neural. Inf. Process. Syst. **35**, 36479–36494 (2022)
24. Sauer, A., Lorenz, D., Blattmann, A., Rombach, R.: Adversarial diffusion distillation. arXiv preprint arXiv:2311.17042 (2023)
25. Schwartz, I., Snæbjarnarson, V., Chefer, H., Belongie, S., Wolf, L., Benaim, S.: Discriminative class tokens for text-to-image diffusion models. In: Proceedings of the IEEE/CVF International Conference on Computer Vision, pp. 22725–22735 (2023)
26. Shi, Z., Zhou, X., Qiu, X., Zhu, X.: Improving image captioning with better use of captions. arXiv preprint arXiv:2006.11807 (2020)
27. Sohl-Dickstein, J., Weiss, E., Maheswaranathan, N., Ganguli, S.: Deep unsupervised learning using nonequilibrium thermodynamics. In: International Conference on Machine Learning, pp. 2256–2265. PMLR (2015)
28. Song, J., Meng, C., Ermon, S.: Denoising diffusion implicit models. arXiv preprint arXiv:2010.02502 (2020)

29. Tan, M., Le, Q.: EfficientNet: rethinking model scaling for convolutional neural networks. In: International Conference on Machine Learning, pp. 6105–6114. PMLR (2019)
30. Xu, Y., Liu, Z., Tegmark, M., Jaakkola, T.: Poisson flow generative models. Adv. Neural. Inf. Process. Syst. **35**, 16782–16795 (2022)
31. Xu, Y., Liu, Z., Tian, Y., Tong, S., Tegmark, M., Jaakkola, T.: PFGM++: unlocking the potential of physics-inspired generative models. In: International Conference on Machine Learning, pp. 38566–38591. PMLR (2023)
32. Zhang, L., Rao, A., Agrawala, M.: Adding conditional control to text-to-image diffusion models. In: Proceedings of the IEEE/CVF International Conference on Computer Vision, pp. 3836–3847 (2023)
33. Zhang, R., Isola, P., Efros, A.A., Shechtman, E., Wang, O.: The unreasonable effectiveness of deep features as a perceptual metric. In: Proceedings of the IEEE Conference on Computer Vision and Pattern Recognition, pp. 586–595 (2018)

GraVITON: Graph Based Garment Warping with Attention Guided Inversion for Virtual-Tryon

Sanhita Pathak[1](✉), Vinay Kaushik[2], and Brejesh Lall[1]

[1] Indian Institute of Technology, Delhi, New Delhi, India
Sanhita.Pathak@dbst.iitd.ac.in, brejesh@ee.iitd.ac.in
[2] Indian Institute of Information Technology Sonepat, Khewra, Haryana, India
vkaushik@iiitsonepat.ac.in

Abstract. Virtual try-on, a rapidly evolving field in computer vision, is transforming e-commerce by improving customer experiences through precise garment warping and seamless integration onto the human body. Existing methods such as TPS and flow address the garment warping, but overlook the finer contextual details. In this paper, we introduce a novel graph based warping technique which emphasizes the value of context in garment flow. Our graph based warping module generates warped garment as well as a coarse person image, which is utilised by a simple refinement network to give a coarse virtual tryon image. We then exploit a latent diffusion model to generate the final tryon, treating garment transfer as an inpainting task. The diffusion model incorporates a Decoupled Garment Attention Adaptor(DGAA) for attention based diffusion inversion of visual and textual information. Our method, validated on VITON-HD and Dresscode datasets, showcases substantial state-of-the-art qualitative and quantitative results showing considerable improvement in garment warping, texture preservation, and overall realism.

Keywords: Virtual tryon · Optical Flow · Graph · Latent Diffusion models

1 Introduction

With the evolving shopping trends, ecommerce platforms have started catering to the customer needs keeping in sync with the emerging requirements. In the apparel industry, this has come into view as virtual tryon, which can provide a real inshop experience to the customers. The image based tryon methods [3,12] have proven to be more practical when compared to the 3D [15] models which require modelling of the person for a realistic tryon synthesis which is quite labor-some.

To produce a perfect tryon result, the person and garment variability has to be prioritised while formulating the tryon pipeline. Although various studies have synthesized compelling results on the benchmarks [6,11,21], there still exist some paucity in terms of realism.

The tryon technique was first introduced by VITON [11], which used TPS warping for solving the problem of warping garments in virtual tryon. CPV-TON [25] preserved texture, but lacked perfect alignment, while the flow based approaches [3,16,33] learnt robust structural alignment but lacked texture consistency. Other methods [20,26] focused more on improving the generation by using various synthesis models such as GANs and recently diffusion [14,20,28]. Amid all the advancements in various stages of virtual tryon, there are still considerable gaps such as learning better garment warp, handling occlusion, pose transformations, generating consistent texture, etc. present, that leave a great scope of improvement.

The current methods [3,33] typically model the flow as result of correlations (utilising either a simple convolution network or feature correlation) between features across garment and reference images(pose,agnostic). These approaches mainly encode the point wise correspondence between an image feature pair(s) while neglecting the intra-relations among pixels within regions [19]. There's a need to capture discriminative features for region and shape representations. Thus, decoupling the garment context from the warping procedure, and simultaneously transferring the region and shape prior of garment context to warping network can aid in learning an optimal garment warp.

Motivated from AGFLOW [19], which introduces iterative graph based flow estimation, we propose a solution to the aforementioned problem on warping by building a novel graph based garment warping module, which embeds context into learning garment warp onto the warping pipeline. The proposed Graph based flow warping module (GFW) learns to match features conditioned on garment context. This allows object's spatial neighbourhood to be well aggregated and thus largely decreases the uncertainty of ambiguous warping of garment.

Diffusion models [7] currently stand as the top-performing models; when compared to the flow and TPS based counterparts [2,25,33]. However, maintaining texture consistency during warping poses a challenge. Recent diffusion-based approaches, exemplified by LaDI-VTON [20], StableViton [14], dci-vton [8], CAT-DM [32] address this challenge by leveraging textual and visual context for virtual try-on generation, treating it as a conditional image inpainting task. To achieve this, LaDI-VTON [20] proposes an inversion module, where image features are extracted from an image encoder and mapped to new word embeddings by a trainable network and then concatenated with text embeddings. StableV-TON [14] utilises a ControlNet model that is directly conditioned on straight garment, incorporating a zero-conv cross attention block. CAT-DM [32] initiates a reverse denoising process, utilising an LDM, with an implicit distribution generated by a pre-trained GAN-based model, thereby reducing the sampling steps without compromising generation quality. In the cross-attention module of LaDI-VTON [20], merging straight cloth features and text features into the cross-attention layer only accomplishes the alignment of image features to text features, and potentially misses some image-specific information and eventually leads to only coarse-grained controllable generation with the reference image. This leads to texture transfer artefacts in some scenarios. For a better tryon inversion, we propose Decoupled Garment Attention Adaptor(DGAA), which adds an additional cross-attention layer only for image features [31].

The contributions of our proposed work are as follows:

- We introduce a Graph based flow warping module(GFW), that guides the appearance flow by providing garment pixel neighbourhood context into flow prediction.
- We propose a Decoupled Garment Attention Adaptor (DGAA), enriching latent space diffusion inversion for a realistic tryon.
- Extensive experimentation and rigorous validation demonstrates that our method achieves state-of-the-art performance compared to existing prominent methods.

2 Related Works

2.1 Virtual Tryon

Given a set of straight cloth and a person image, the goal of virtual tryon is to seamlessly warp the garment and overlay it onto the target person image. The initial work that introduced the garment warping and a generated complete person tryon was VITON [11]. Other methods [10,12,25,30,33] followed a similar two stage warping and generation pipeline, which learnt TPS or affine transformation parameters for computing garment warp. Although TPS preserves the texture of warped garment better than to that of it's flow based counterparts, incorporating flow achieves optimal garment alignment with the changing human pose. In order to achieve the realism in the final tryon, it is crucial to formulate a robust garment deformation module. This is usually achieved by the deformation of control points with an energy function (radial basis function) in TPS based pipelines (Thin Plate Spline) [25], and by computing per pixel appearance flow followed by target view synthesis in flow based pipelines. The flow based warping learns dense per-pixel correspondence [3,12,16,33], when compared to the TPS based methods which are unable to capture such local warp details.

2.2 Graph Neural Networks in Flow

Optical flow is the task of estimating dense per-pixel correspondence between images. GMFlow [27] introduced vision transformers for computing optical flow, but its heavy computational dependencies made it less diversely applicable. AGFlow [19] exploited the scene/context information, utilising graph convolutional networks, and incorporated it in the matching procedure to robustly compute optical flow. Virtual tryon entails computation of appearance flow [10,33], to warp the source cloth based on the reference person features (pose, densepose, etc.) GPVTON [33] tried to address the local deformations by applying a part wise flow based deformation, where the garment is disintegrated and deformed separately into three regions, one for each upper body part. GPVTON is not able to jointly optimise the local and global deformations. Another work KGI [17] utilised graph to predict the garment pose points guided by human pose which inpainted the predicted region using human segmentation. The method failed to achieve the precision in tryon alignment due to sparse guiding points to guide the dense pixel warping for garment texture unlike in flow methods. Hence, motivated by AGFlow [19], in this work we have shown that GCNs can help the garment warping by focusing on the pixel level deformations establishing a dense correlation that helps in preserving the local details post deformation, which is ideally faced by all the flow based garment warping methods.

2.3 Diffusion Models

Diffusion models marked research has become a foundational area in the field of image synthesis [7] because of its high quality image generation. Tasks such as image-to-image translation [24], image editing [1], text-to-image synthesis [9], and inpainting [18,22] have seen significant progress due to their realistic generation results. [13] concentrated on creating full-body images by sampling from a trained texture-aware codebook, given human position and textual descriptions of clothing shapes and textures. Furthermore, in order to address the problem of pose-guided human prediction, [5] created a texture diffusion block that was conditioned by multi-scale texture patterns from the encoded source image. Adding to the tryon generation features, [4] introduced using the model pose, the garment sketch, and a textual description of the garment to condition the tryon generation process. Building on these methods and to improve the texture generation in person tryon, LaDI-VTON [20] utilised a textual inversion component, enabling mapping of garment visual features to the CLIP [23] token embedding space. This process generates a set of pseudo-word token embeddings, effectively conditioning the generation process. DCI-VTON [8] leverages a warping module to combine the warped clothes with clothes-agnostic person image and adds noise to guide the diffusion model's generation. Other methods on diffusion such as StableVITON [14] and CAT-DM [32] utilises a ControlNet based approach conditioned on straight garment for tryon.

3 Proposed Approach

Our model uses a two-stage pipeline. The first stage involves warping, with a graph-based warping module followed by a refinement module. It takes the source garment (I_g), a reference input (reference pose I_{pose} and agnostic image $I_{agnostic}$) as input. This stage computes dense flow f_o using graph correlation, producing a warped garment (I_{warp_g}) and a coarse try-on (I_{tryon_c}). The second stage generates the final try-on result using a diffusion model with an inpainting approach. Inputs include the person segmentation mask (I_{coarse_b}) from the coarse try-on, warped output (I_{warp_g}), human pose keypoints (I_{pose}), agnostic image $(I_{agnostic})$, and noise (I_z). The diffusion process is conditioned on the

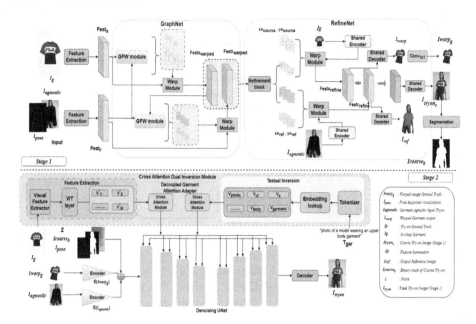

Fig. 1. Architecture Diagram of GraVITON. The top module utilizes GCNs for generating warped cloth and coarse tryon image. These outputs are processed to condition the Stable Diffusion model. The inversion model efficiently computes Cross-Modal attention to improve texture and structural consistency, generating the final tryon image.

source cloth texture (I_g) and produces final try-on image(I_{tryon}). The diffusion process is conditioned with the attention based inversion between textual data (T_{gar}) and source cloth for texture (I_g). The calculated decoupled attention conditions the latent space to generate final tryon(I_{tryon}).

3.1 Graph Based Coarse Tryon

The coarse tryon stage caters to the generation of warped garment I_{warp_g}) along with coarse tryon (I_{tryon_c}) that is further used in final tryon generation in stage 2. The input to the first stage, is source garment(I_g), reference pose (I_{pose}) and agnostic image ($I_{agnostic}$). The network employs a feature extraction module in form of convolution layers with N=3, N being the number of conv layers and a stride 2. The features extracted for both source($Feat_s$) and reference($Feat_r$) input are fed to the GraphNet module, that returns the warped source $Feat_{s_{warped}}$ and reference features $Feat_{r_{warped}}$ that are fed to the RefineNet for predicting final offsets ($x_{o_{source}}, y_{o_{source}}$) and ($x_{o_{ref}}, y_{o_{ref}}$) as shown in Fig. 1.

GraphNet. The overall working of GraphNet is similar to SDAFN [3], with the major difference is appearance flow estimation. The convolutional deformable flow warping stage in SDAFN is replaced by our novel Graph based Flow Warping (GFW) module as shown in Fig. 2. The features extracted for both source ($Feat_s$) and reference ($Feat_r$) further act as an input to GFW module. The dense flow offsets ($x_{o_{source}}, y_{o_{source}}$) along with the computed attention maps are utilised by the warping module to warp $Feat_s$ feature to compute source warped feature $Feat_{s_{warped}}$. Similarly, the source warped feature $Feat_{s_{warped}}$ and reference feature $Feat_r$ are fed to the GFW module to compute reference warped feature $Feat_{r_{warped}}$ from offsets ($x_{o_{ref}}, y_{o_{ref}}$).

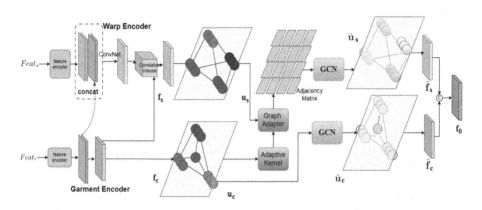

Fig. 2. Our Graph based Flow Warping (GFW) Module.

RefineNet. The RefineNet module computes the refined offsets for predicting the warped garment I_{warp_g} and coarse tryon I_{tryon_c}.

The concatenated source and reference warped features $Feat_{s_{warped}}$ and $Feat_{r_{warped}}$ are fed as the input to refinement block to compute final offsets $x_{o_{source}}, y_{o_{source}}$ and $x_{o_{ref}}, y_{o_{ref}}$ which are the final warping directives for source garment (I_g) and reference input ($I_{agnostic}$) respectively. The refinement block is a simple four layer convolutional network based on [3]. The cloth I_g is sent to the shared encoder to compute garment features which are sent to the warp module along with the final source offsets $x_{o_{source}}, y_{o_{source}}$ to compute final garment warped feature $Feat_{s_{refine}}$. Similarly, the image agnostic $I_{agnostic}$ is sent to the shared encoder to compute agnostic features which are sent to the warp module along with the final reference offsets $x_{o_{ref}}, y_{o_{ref}}$ to compute warped reference feature $Feat_{r_{refine}}$. Both source and reference refined features are summed and sent to a shared decoder to compute the warped output tryon image(I_{tryon_c}). Similarly, the source refined feature $Feat_{s_{refine}}$ is fed to the shared decoder to compute generated warped garment image (I_{warp}), which is further refined by being passed through a 1 × 1 convolution layer to compute I_{warp_g}.

Graph Based Flow Warping Module(GFW). The graph network provides a highly connected space utilising the dense pixel context for appearance flow estimation. The source and reference features $Feat_s$, $Feat_r$ are sent to a shared feature encoder, whose corresponding output is then utilized to construct a 4D correlation volume capturing the statistical similarity between the two as shown in Fig. 2. The resulting value is sent to four convolutions to capture source feature $\mathbf{f}_s$. The reference feature ($Feat_r$) is fed to the garment encoder network to compute context feature $\mathbf{f}_c$ as shown in Fig. 2. Both features are utilised to perform a holistic warp reasoning by computing offsets $\mathbf{f}_o = (x_o, y_o)$.

The graph based module in stage 1 consists of nodes(N) and edges(E) formulated in a directed graph as $G=(N,E)$. The node embeddings are mapped to the graph space using a simple projection function, $\mathbf{u} = \mathcal{P}_{f \to u}(\mathbf{f})$, where $\mathbf{u}$ denotes the nodes in graph space, $\mathcal{P}$ is the projection function and $\mathbf{f}$ depicts the feature space. We define the nodes mapped into context (garment) feature $\mathbf{f}_c$ and warp feature $\mathbf{f}_s$ encoded as, $\mathbf{u}_c = (u_c^1, u_c^2, ..., u_c^n)$ and $\mathbf{u}_s = (u_s^1, u_s^2, ..., u_s^n)$, where $\mathbf{u}_c$ is context nodes for garment warping while $\mathbf{u}_s$ is the warp nodes computed from the normalized feature correlation between source and reference features $Feat_s$ and $Feat_r$ in the graph space.

The process of node creation for both the source and context entails the computation of the adjacency matrix, which measures the similarity between all nodes denoted as $\mathbf{u}_c$ and $\mathbf{u}_s$. To facilitate adaptive graph learning, we employ $\mathcal{L}()$ as a graph learner, comprising of a two-layer convolutional network with ReLU activation. The first layer focuses on channel-wise learning for $\mathbf{u}_s$, while the second layer introduces node-wise interaction learning, resulting in a refined node representation for the source denoted as $\hat{\mathbf{u}}_s^{(t)}$.

$$\breve{\mathbf{A}}_s = \mathcal{L}(\mathbf{u}_s; \Theta(\mathbf{u}_c)); \hat{\mathbf{u}}_s^= \mathcal{F}_{\text{AG}}(\mathbf{u}_s, \breve{\mathbf{A}}) \qquad (1)$$

$$\hat{\mathbf{u}}_c = \mathcal{F}_{\text{Graph}}(\mathbf{u}_c, \mathbf{A}), \text{where } \mathbf{A} = \mathbf{u}_c^\top \mathbf{u}_c, \qquad (2)$$

The final adjacency matrix for context and warp nodes is formulated in Eq. 1 giving the modified nodes for the source, with $\Theta()$ signifying a parameter learner and $\mathcal{F}_{\text{AG}}$ is adaptive graph learning function for warping. The context nodes are computed as in Eq. 2 where $\mathcal{F}_{\text{Graph}}$, is graph learner function in the Graph Adapter block defining the warping context.

The projection function $\mathcal{P}$ preserves the spatial details during the first(initial) conversion to the graph space, and utilising this, the modified nodes are projected back from graph to feature space using the projection function $\mathcal{P}$ as shown in Eq. 3 and Eq. 4, giving $\hat{\mathbf{f}}_c$ garment (context) and source warp feature $\hat{\mathbf{f}}_s$.

$$\hat{\mathbf{f}}_c = \mathbf{f}_c + h\mathcal{P}_{\text{v} \to \text{f}}(\hat{\mathbf{u}}_c), \qquad (3)$$

where, h denotes a learnable parameter that is initialized as 0 and gradually performs a weighted sum. Similarly, the source warp feature $\hat{\mathbf{f}}_s$ is produced by

$$\hat{\mathbf{f}}_s = \mathbf{f}_s + l\mathcal{P}_{\text{s} \to \text{f}}(\hat{\mathbf{u}}_s). \qquad (4)$$

where, l denotes a learnable parameter The resultant features are then concatenated to give the resulting offsets on the original grid from source image.

$$\mathbf{f_o} = (x_o^g, y_o^g) = (1 + F_{ch}(\hat{\mathbf{f}}_s)) * concat(\hat{\mathbf{f}}_c, \hat{\mathbf{f}}_s) \tag{5}$$

where, F_{ch} signifies the channel attention.

The overall loss for training stage 1 is defined below, where $L_{style}, L_{prec}, L_{L1}$ are style, perceptual and L1 losses.

$$\mathcal{L} = (\lambda_{L1}\mathcal{L}_{L1} + \lambda_{perc}\mathcal{L}_{perc} + \lambda_{style}\mathcal{L}_{style}) \tag{6}$$

3.2 Cross Modal Attention for Inversion

We utilise the coarse tryon output I_{tryon_c} from stage 1 to compute all the preprocessing inputs at stage 2 including person agnostic $I_{agnostic}$, binary person segmentation mask I_{coarse_b}, as well as pose keypoints I_{pose}. The preprocessed inputs go into the diffusion model for training.

Diffusion Model: The model consists of a latent encoder **E** and latent decoder **D** block, from a pretrained VAE. A time conditioned U-net is used with a denoising parameter ϵ. The diffusion encoder takes in the warped garment I_{Warp_g} and person agnostic processed by a VAE encoder E giving the warped encoded garment $E(I_{Warp_g})$ and encoded person agnostic $E(I_{agnostic})$. The additional inputs: pose I_{pose}, mask I_{coarse_b} and noise z are resized to the latent size and concatenated.

The resulting inputs to the network are combined as:
$\beta = [Z; I_{coarse_b}; I_{pose}; E(I_{Warp_g}); E(I_{agnostic})]$ and used for latent learning. As virtual tryon aims to transfer the given warped garment to the person, it is treated as an inpainting task, inspired by [20]. The stable diffusion model is used as an in-painting approach where the latent space is conditioned with our DGAA adaptor. Our proposed framework focuses to inpaint the masked area, but instead of being guided by a TPS based warped garment, our diffusion model is guided by the warped garment computed from stage 1.

A CLIP encoder is employed for textual inversion which takes textual data T_{gar} as an input. Similarly, input straight cloth I_g is fed to a pretrained variational encoder, and the features are fed to a ViT layer to compute texture feature for the same. The texture features from image are represented in CLIP token embedding space, similar to [20]. The token embeddings from the textual data acts as a textual prompt that guides the garment texture positioning. To enhance this, we introduce a Decoupled garment attention adaptor to condition the Denoising UNet giving realistic tryon results.

Decoupled Garment Attention Adaptor (DGAA). Although LaDI-VTON [20] enhances diffusion with inversion, it generates try-ons with erroneous texture details due to ineffective embedding of image features as they simply feed

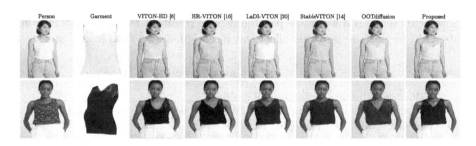

Fig. 3. Qualitative results generated by proposed method in comparison with recent state-of-the-art approaches.

the concatenated features to the cross-attention layers. To address this, we propose the Decoupled Garment Attention Adaptor. Similar to [31], which improves conditioning in text-to-image generation, we utilise DGAA for conditioning the inpainting task of virtual tryon.

The textual features obtained from the CLIP embedding x_t are fed into the cross attention layer along with the query features z, given by latent. Hence, the cross-attention equation is given as,

$$\mathbf{z}' = \text{Attention}(\alpha, \beta, \gamma) = \text{Softmax}(\frac{\alpha \beta^\top}{\sqrt{d}})\gamma, \qquad (7)$$

where, $\alpha = zW_\alpha$, $\beta = x_iW_\beta$ and $\gamma = x_iW_\gamma$ are the query, key, and values matrices from the text features and W_β, W_γ are the corresponding weight matrices. In DGAA, the cross attention layers for text features and garment features are separate. We add a new cross attention layer, for each cross attention layer in the original UNet model to insert garment features. Given the garment features g_i, the output of new cross attention $\mathbf{z}''$ is computed as follows:

$$\mathbf{z}'' = \text{Attention}(\alpha, \beta', \gamma') = \text{Softmax}(\frac{\alpha(\beta')^\top}{\sqrt{d}})\gamma', \qquad (8)$$

where, $\alpha = zW_\alpha$, $\beta' = g_iW'_\beta$ and $\gamma' = g_iW'_\gamma$ are the query, key, and values matrices from the image features and W'_β, W'_γ are the corresponding weight matrices.

We use the same query for image cross-attention as for text cross-attention. Consequently, we only need to add two paramemters W'_β and W'_γ for each cross-attention layer. In order to speed up the convergence, W'_β and W'_γ are initialized from W_β and W_γ.

Combining both the equations, 7 and 8 we get the final cross attention equation as below,

$$\mathbf{z}^{new} = \text{Softmax}(\frac{\alpha\beta^\top}{\sqrt{d}})\gamma + \text{Softmax}(\frac{\alpha(\beta')^\top}{\sqrt{d}})\gamma' \quad (9)$$
$$\text{where } \alpha = \mathbf{z}\mathbf{W}_\alpha, \beta = \mathbf{x}_t\mathbf{W}_\beta, \gamma = \mathbf{x}_t\mathbf{W}_\gamma,$$
$$\beta' = \mathbf{x}_i\mathbf{W}'_\beta, \gamma' = \mathbf{x}_i\mathbf{W}'_\gamma$$

Here, W'_k and W'_v are the only trainable weights.

Loss: The diffusion model learns from the l1 loss function over noise as in [20].

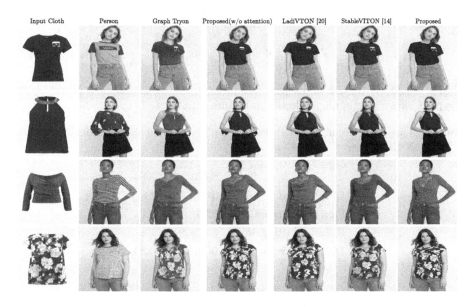

Fig. 4. Qualitative results showing the successive visual enhancement in results and analysis with LaDI-VTON and StableVITON.

3.3 Dataset

The experiments were conducted on VITON-HD and Dresscode datasets. VITON-HD is a high resolution dataset with resolution of 1024 × 768. The train set consists of 11,647 train pairs and 2,032 test pairs. DressCode is composed of 48,392/5,400 training/testing pairs of front-view full-body person and garment from different categories (i.e., upper, lower, dresses). The model is trained for both datasets in a paired setting on upper body garments and tested on both paired and unpaired setting. The same garment tryon is tested on the model as it is wearing in paired. While, a different garment tryon is tested on the model in an unpaired setting.

Fig. 5. Qualitative results of our proposed methodology on VITON-HD Dataset depicting pose, hair, sleeve length and texture variations.

Fig. 6. Qualitative results of our proposed methodology on Dresscode Dataset depicting pose, sleeve length, upper/dress and texture variations.

Table 1. Quantitative comparison between proposed method and incremental modules on VITON-HD dataset for paired setting

	SSIM	FID	KID
Graph based tryon	0.857	10.32	1.8
Flow based tryon	0.851	10.77	2.1
Diffusion with flow	0.873	9.21	1.2
Diffusion with graph	**0.881**	**8.37**	**0.81**

3.4 Implementation Details and Training

The model is trained in two stages successively. The graph based warping stage is trained first for 200 epochs, for a batch of 6 with a learning rate(LR) of 0.000035 on a V100 GPU. Weights for the loss functions are $\lambda_{L1} = 1, \lambda_{prec} = 1, \lambda_{style} = 100$. We used AdamW as training optimiser with $\beta1 = 0.9$, $\beta2 = 0.999$ and weight decay equal to 1e-2.

For training stage two, the inputs derived from stage 1 are utilised to construct the preprocessed inputs. This requires training our decoupled attention

Table 2. Flow method comparison on VITON-HD dataset

	SSIM	FID
Iterative flow	0.852	9.08
Single stage flow	0.874	8.91
Deformable flow	0.881	8.37

adaptor for 160 epochs with Adam optimizer, batch size 8 and $1e^{-5}$ LR. We employ SSIM, FID and KID metrics in both paired and unpaired settings for evaluation.

3.5 Qualitative Results

To qualitatively assess our findings, we present sample images generated by our model alongside those by competing methods in Fig. 3. While VITON-HD and HR-VITON have limitations in texture and warp accuracy, LaDI-VTON slightly improves texture details but looses colour consistency in garment. StableVITON doesnt preserve the garment shape accurately but improves colour and texture consistency. OOTDiffusion further improves texture details yet struggles to keep the garment person alignment intact as can be seen in row 2(right shoulder). Our approach produces highly realistic images, preserving the intricate textures and details of the original garments and garment warp using a decoupled attention-based inversion module and graph-based flow estimation.

In Fig. 4, we compare both stages of our approach. Graph Tryon output from stage one shows initial garment warp, providing a baseline for refinement. In Stage 2, the diffusion model generates the warped garment on the agnostic image. This stage uses graph-based flow warping for preprocessing, generating rich textures and ensuring correct global warp. Through visual inspection, we discern improvements in texture preservation, micro-texture retention in green top(last row), spatial coherence in black dress(second row), and consistent boundary warp(third row) by our proposed approach. Figure 4 also compares our stage 1 and stage 2 results with the existing state-of-the-art methods LaDI-VTON and StableVITON. We observe that even without attention and inversion, our approach performs slightly better than LaDI-VTON. This is due to incorporation of our graph based warping stage, which predicts much better warps than TPS utilised in LaDI-VTON. The proposed approach as can be seen in last column, retains better texture and aligns garment optimally according to person's pose, thereby giving the best results.

Figure 5 depicts the garment tryon in an unpaired setting for VITON-HD dataset with texture variations, sleeve lengths, pose and hair. The generated images provide visual effectiveness of our method to handle self occlusion due to complex arm positions as can be seen in four images from the left. The proposed method also generates realistic garment textures retaining the fine details of text and symbols in the images. Figure 6 shows realistic tryon generation for Dresscode dataset. Our work generates realistic tryon for garments in unpaired setting. The results preserve texture, sleeve length and are agnostic to pose variations.

Table 3. Effect of attention in inversion module

	SSIM	FID
Attention	0.881	8.37
W/O Attention	0.873	9.24

Table 4. Quantitative Ablation of our proposed modules on VITON-HD dataset

Graph	Diffusion	Inversion	DGAA	SSIM	FID
✗	✗	✗	✗	0.851	11.25
✓	✗	✗	✗	0.857	10.32
✓	✓	✗	✗	0.868	9.78
✓	✓	✓	✗	0.873	9.24
✓	✓	✓	✓	**0.881**	**8.37**

Table 5. Quantitative results on the VITON-HD dataset [6]. The best results are reported in **bold**.

Method	LPIPS ↓	SSIM ↑	FID ↓	KID ↓
VITON-HD [6]	0.116	0.863	12.13	3.22
HR-VITON [16]	0.097	0.878	12.30	3.82
LaDI-VTON [20]	0.091	0.875	9.31	1.53
GP-VTON [33]	0.083	**0.892**	9.17	0.93
StableVITON [14]	0.084	0.862	9.13	1.20
OOTDiffusion [28]	0.071	0.878	8.81	0.82
Proposed	**0.070**	0.881	**8.37**	**0.81**

3.6 Quantitative Results

We describe the robustness and correctness of our proposed approach by conducting extensive experiments and ablation on Dresscode and VITON-HD datasets. Table 1 demonstrates that the affect of introduction of graph for garment warping and coarse try-on prediction improves the accuracy of try-on module significantly when compared with the flow based traditional counterparts. It also describes the improvement in final try-on after utilising our diffusion model for target person generation. As we see, combination of Graph and Diffusion achieves the best result quantitatively.

Table 2 describes how various flow modules aid in warping input garment. The iterative flow which was motivated from RAFT [19] is unable to learn optimal warp, as the flow being learnt is an intermediate component of our network. While, RAFT [19] being a supervised framework introduced a flow consistency constraint which utilises ground truth flow that aids in learning of the iterative flow. We learn flow as an intermediate component in self-supervised manner.

We also see that introduction of deformable flow [3] to our graph based flow estimation framework drastically improves learning of warped garment. This enhancement can be attributed to the fusion of features warped using multiple flows, resulting in the creation of a single optimized try-on. Consequently, while individual warped features may exhibit slight discrepancies, the fusion process aggregates the most favorable attributes from all features to generate an optimal try-on output.

The introduction of decoupled cross attention between text embedding and garment texture feature embedding improves the consistency of texture learnt in final tryon. This can be seen as improvement in FID and SSIM scores in Table 3.

Table 6. Quantitative results on the Dress Code dataset [21]. The best results are reported in **bold**.

Method	LPIPS ↓	SSIM ↑	FID ↓	KID ↓
PSAD [21]	0.058	0.918	17.51	7.15
Paint-by-Example [29]	0.078	0.851	18.63	4.81
LaDI-VTON [20]	0.067	0.910	12.30	1.30
GP-VTON [33]	0.051	0.921	12.20	1.22
Proposed	**0.041**	**0.925**	**10.86**	**0.69**

We analyzed the impact of each component on the performance of the model. As shown in Table 4, incorporating graph-based flow estimation led to a notable improvement in SSIM scores, indicating enhanced spatial coherence and perceptual quality in the generated images. Similarly, the integration of diffusion mechanisms in the generation process resulted in significantly lower FID scores, demonstrating improved fidelity and realism in the synthesized outputs. The inclusion of attention mechanisms within the inversion module led to substantial gains in both SSIM and FID metrics, highlighting the importance of selective feature extraction and reconstruction in enhancing image quality and content preservation. Our comprehensive approach, combining all key components yielded the most impressive results.

As depicted in Table 5 and 6, we achieve highest SSIM and lowest FID scores among all prominent tryon methods on VITON-HD and Dresscode datasets, demonstrating the synergistic effects of our holistic technique.

4 Conclusion

Our paper introduces novel solutions to enhance virtual try-on, addressing critical challenges in garment warping and generation. By incorporating novel Graph-based Flow Warping module (GFW), we achieve accurate context reasoning, significantly reducing uncertainty in garment transfer. We introduce latent inversion for rich garment and text conditioning to a stable diffusion inpainting model. Our novel Decoupled Garment Cross-Attention Mechanism (DGAA) enriches latent space information of the diffusion model, leading to realistic try-on. Empirical validation on VITON-HD and DressCode datasets demonstrates substantial improvements in garment warping, texture preservation, and overall realism compared to existing methods.

References

1. Avrahami, O., Lischinski, D., Fried, O.: Blended diffusion for text-driven editing of natural images. In: Proceedings of the IEEE/CVF Conference on Computer Vision and Pattern Recognition, pp. 18208–18218 (2022)
2. Bai, M., Luo, W., Kundu, K., Urtasun, R.: Exploiting semantic information and deep matching for optical flow. In: Computer Vision–ECCV 2016: 14th European Conference, Amsterdam, The Netherlands, October 11-14, 2016, Proceedings, Part VI 14, pp. 154–170. Springer (2016)
3. Bai, S., Zhou, H., Li, Z., Zhou, C., Yang, H.: Single stage virtual try-on via deformable attention flows. In: European Conference on Computer Vision (2022). https://api.semanticscholar.org/CorpusID:250644446
4. Baldrati, A., Morelli, D., Cartella, G., Cornia, M., Bertini, M., Cucchiara, R.: Multimodal garment designer: human-centric latent diffusion models for fashion image editing. arXiv preprint arXiv:2304.02051 (2023)
5. Bhunia, A.K., et al.: Person image synthesis via denoising diffusion model. In: Proceedings of the IEEE/CVF Conference on Computer Vision and Pattern Recognition, pp. 5968–5976 (2023)
6. Choi, S., Park, S., Lee, M.G., Choo, J.: VITON-HD: high-resolution virtual try-on via misalignment-aware normalization. In: 2021 IEEE/CVF Conference on Computer Vision and Pattern Recognition (CVPR), pp. 14126–14135 (2021). https://api.semanticscholar.org/CorpusID:232427801
7. Dhariwal, P., Nichol, A.: Diffusion models beat GANs on image synthesis. ArXiv **abs/2105.05233** (2021). https://api.semanticscholar.org/CorpusID:234357997
8. Gou, J., Sun, S., Zhang, J., Si, J., Qian, C., Zhang, L.: Taming the power of diffusion models for high-quality virtual try-on with appearance flow. In: Proceedings of the 31st ACM International Conference on Multimedia (2023)
9. Gu, S., Chen, D., Bao, J., Wen, F., Zhang, B., Chen, D., Yuan, L., Guo, B.: Vector quantized diffusion model for text-to-image synthesis. In: Proceedings of the IEEE/CVF Conference on Computer Vision and Pattern Recognition, pp. 10696–10706 (2022)
10. Han, X., Huang, W., Hu, X., Scott, M.R.: ClothFlow: a flow-based model for clothed person generation. In: 2019 IEEE/CVF International Conference on Computer Vision (ICCV), pp. 10470–10479 (2019). https://api.semanticscholar.org/CorpusID:204959889
11. Han, X., Wu, Z., Wu, Z., Yu, R., Davis, L.S.: Viton: an image-based virtual try-on network. In: 2018 IEEE/CVF Conference on Computer Vision and Pattern Recognition, pp. 7543–7552 (2017). https://api.semanticscholar.org/CorpusID:4532827
12. He, S., Song, Y.Z., Xiang, T.: Style-based global appearance flow for virtual try-on. In: 2022 IEEE/CVF Conference on Computer Vision and Pattern Recognition (CVPR), pp. 3460–3469 (2022). https://api.semanticscholar.org/CorpusID:247939336

13. Jiang, Y., Yang, S., Qiu, H., Wu, W., Loy, C.C., Liu, Z.: Text2Human: text-driven controllable human image generation. ACM Trans. Graph. (TOG) **41**(4), 1–11 (2022)
14. Kim, J., Gu, G., Park, M., Park, S., Choo, J.: StableVITON: learning semantic correspondence with latent diffusion model for virtual try-on. In: Proceedings of the IEEE/CVF Conference on Computer Vision and Pattern Recognition, pp. 8176–8185 (2024)
15. Lal Bhatnagar, B., Tiwari, G., Theobalt, C., Pons-Moll, G.: Multi-garment Net: learning to dress 3D people from images. arXiv e-prints arXiv–1908 (2019)
16. Lee, S., Gu, G., Park, S., Choi, S., Choo, J.: High-resolution virtual try-on with misalignment and occlusion-handled conditions. arXiv preprint arXiv:2206.14180 (2022)
17. Li, Z., Wei, P., Yin, X., Ma, Z., Kot, A.C.: Virtual try-on with pose-garment keypoints guided inpainting. In: Proceedings of the IEEE/CVF International Conference on Computer Vision (ICCV), pp. 22788–22797 (2023)
18. Lugmayr, A., Danelljan, M., Romero, A., Yu, F., Timofte, R., Van Gool, L.: Repaint: inpainting using denoising diffusion probabilistic models. In: Proceedings of the IEEE/CVF Conference on Computer Vision and Pattern Recognition, pp. 11461–11471 (2022)
19. Luo, A., Yang, F., Luo, K., Li, X., Fan, H., Liu, S.: Learning optical flow with adaptive graph reasoning. In: Proceedings of the AAAI Conference on Artificial Intelligence, vol. 36, pp. 1890–1898 (2022)
20. Morelli, D., Baldrati, A., Cartella, G., Cornia, M., Bertini, M., Cucchiara, R.: LaDI-VTON: latent diffusion textual-inversion enhanced virtual try-on. arXiv preprint arXiv:2305.13501 (2023)
21. Morelli, D., Fincato, M., Cornia, M., Landi, F., Cesari, F., Cucchiara, R.: Dress code: high-resolution multi-category virtual try-on. In: 2022 IEEE/CVF Conference on Computer Vision and Pattern Recognition Workshops (CVPRW), pp. 2230–2234 (2022). https://api.semanticscholar.org/CorpusID:248240016
22. Nichol, A., et al.: GLIDE: towards photorealistic image generation and editing with text-guided diffusion models. arXiv preprint arXiv:2112.10741 (2021)
23. Radford, A., et al.: Learning transferable visual models from natural language supervision (2021)
24. Saharia, C., et al.: Palette: image-to-image diffusion models. In: ACM SIGGRAPH 2022 Conference Proceedings, pp. 1–10 (2022)
25. Wang, B., Zheng, H., Liang, X., Chen, Y., Lin, L., Yang, M.: Toward characteristic-preserving image-based virtual try-on network. In: Proceedings of the European Conference on Computer Vision (ECCV), pp. 589–604 (2018)
26. Xie, Z., Huang, Z., Zhao, F., Dong, H., Kampffmeyer, M.C., Liang, X.: Towards scalable unpaired virtual try-on via patch-routed spatially-adaptive GAN. In: Neural Information Processing Systems (2021). https://api.semanticscholar.org/CorpusID:244478414
27. Xu, H., Zhang, J., Cai, J., Rezatofighi, H., Tao, D.: GMFlow: learning optical flow via global matching. In: Proceedings of the IEEE/CVF Conference on Computer Vision and Pattern Recognition, pp. 8121–8130 (2022)
28. Xu, Y., Gu, T., Chen, W., Chen, C.: OOTDiffusion: outfitting fusion based latent diffusion for controllable virtual try-on. arXiv preprint arXiv:2403.01779 (2024)
29. Yang, B., et al.: Paint by example: exemplar-based image editing with diffusion models. In: Proceedings of the IEEE/CVF Conference on Computer Vision and Pattern Recognition, pp. 18381–18391 (2023)

30. Yang, H., Zhang, R., Guo, X., Liu, W., Zuo, W., Luo, P.: Towards photo-realistic virtual try-on by adaptively generating-preserving image content. In: Proceedings of the IEEE/CVF Conference on Computer Vision and Pattern Recognition, pp. 7850–7859 (2020)
31. Ye, H., Zhang, J., Liu, S., Han, X., Yang, W.: IP-Adapter: text compatible image prompt adapter for text-to-image diffusion models. arXiv preprint arXiv:2308.06721 (2023)
32. Zeng, J., Song, D., Nie, W., Tian, H., Wang, T., Liu, A.: CAT-DM: controllable accelerated virtual try-on with diffusion model. arXiv preprint arXiv:2311.18405 (2023)
33. Zhenyu, X., et al.: GP-VTON: towards general purpose virtual try-on via collaborative local-flow global-parsing learning. In: Proceedings of the IEEE/CVF Conference on Computer Vision and Pattern Recognition (CVPR) (2023)

Image and Video Compression Using Generative Sparse Representation with Fidelity Controls

Lebin Zhou, Wei Wang, and Wei Jiang[✉]

Futurewei Technologies Inc., San Jose, CA 95131, USA
{lzhou,rickweiwang,wjiang}@futurewei.com

Abstract. We propose a framework for learned image and video compression using the generative sparse visual representation (SVR) guided by fidelity-preserving controls. By embedding inputs into a discrete latent space spanned by learned visual codebooks, SVR-based compression transmits integer codeword indices, which is efficient and cross-platform robust. However, high-quality (HQ) reconstruction in the decoder relies on intermediate feature inputs from the encoder via direct connections. Due to the prohibitively high transmission costs, previous SVR-based compression methods remove such feature links, resulting in largely degraded reconstruction quality. In this work, we treat the intermediate features as fidelity-preserving control signals that guide the conditioned generative reconstruction in the decoder. Instead of discarding or directly transferring such signals, we draw them from a low-quality (LQ) fidelity-preserving alternative input that is sent to the decoder with very low bitrate. These control signals provide complementary fidelity cues to improve reconstruction, and their quality is determined by the compression rate of the LQ alternative, which can be tuned to trade off bitrate, fidelity and perceptual quality. Our framework can be conveniently used for both learned image compression (LIC) and learned video compression (LVC). Since SVR is robust against input perturbations, a large portion of codeword indices between adjacent frames can be the same. By only transferring different indices, SVR-based LIC and LVC can share a similar processing pipeline. Experiments over standard image and video compression benchmarks demonstrate the effectiveness of our approach.

Keywords: learned image compression · learned video compression · sparse visual representation · generative controls

1 Introduction

Image and video compression has been a decades-long research topic, and great success has been achieved recently by using neural networks (NN) for both learned image compression (LIC) [9,22] and learned video compression (LVC) [14,33]. Most existing LIC methods [9,18,22,30,40] use a hyperprior framework

[3], which combines classical entropy coding with NN-based representation learning in a Variational AutoEncoder (VAE) structure. An entropy model is used to encode the quantized latent feature for easy transmission. Most existing LVC methods follow the pipeline of traditional video coding [16,19], while replacing processing modules like motion estimation, motion compensation, residue coding *etc.* by learned NNs.

In this paper, we explore a different compression pipeline for both LIC and LVC, based on controlled generative modeling using the Sparse Visual Representation (SVR) (as shown in Fig. 1 and Fig. 2). We learn discrete generative priors as visual codebooks, and embed images into a discrete latent space spanned by the codebooks. By sharing the learned codebooks between the encoder and decoder, images can be mapped to integer codeword indices in the encoder, and the decoder can use these indices to retrieve the corresponding codewords' latent features for reconstruction.

The SVR-based compression has several benefits. (1) Transferring integer indices is very robust to heterogeneous platforms. One caveat of the hyperprior framework is the extreme sensitivity to small differences between the encoder and decoder in calculating the hyperpriors [4]. Even perturbations caused by floating round-off error can lead to catastrophic error propagation in the decoded latent feature. By encoding codeword indices instead of latent features, SVR-based compression does not suffer from such sensitivity. (2) Transferring indices gives the freedom of expanding latent feature dimension (often associated with better representation power for better reconstruction) without increasing bitrate, in comparison to transferring latent features or residues. (3) Generative SVR increases robustness to input degradations. Realistic and rich textures can be generated using hiqh-quality (HQ) codebooks even for low-quality (LQ) inputs.

However, SVR-based HQ restoration [7,8,25] relies on the dense connection of multi-scale features between the embedding network (encoder) and reconstruction network (decoder). Such intermediate features are too large to transfer, defeating the purpose of the compression task. As a result, previous SVR-based compression methods remove such direct feature links. By applying to specific content like human faces [20,42], a code transformer is used to recover an aligned structured code sequence for HQ face restoration without intermediate features. For general images, M-AdaCode [21] compensates the performance loss of removed feature links by using data-adaptive weights to combine multiple semantic-class-dependent codebooks and uses weight masking to reduce transmitted weight parameters. However, although the restored images may look okay perceptually, important fidelity details are usually lost. As shown in Fig. 4, without direct feature links images generated by M-AdaCode often lack rich details.

In our opinion, the generative SVR-based reconstruction aims at high perceptual quality, and the multi-scale intermediate features provide complementary fidelity details to the reconstruction. Such details should NOT be ignored for applications like compression. Therefore, we focus on how to obtain effective and transmission-friendly fidelity information to balance bitrate and quality.

Our work is inspired by the success of ControlNet [38] where conditioning controls are used to guide image generation. We view the multi-scale intermediate features as fidelity-preserving control signals that guide the conditioned reconstruction in the decoder. As control conditions, such signals do NOT have to come from the original input. Instead, we draw these control signals from an LQ alternative of the original input in decoder. This LQ alternative is computed in decoder based on highly compressed easy-to-transmit fidelity-preserving information, which is generated by fidelity-preserving methods like the previous NN-based or traditional image and video compression methods.

Based on this idea, we propose a framework (Fig. 2) that combines generative SVR-based restoration with fidelity-preserving compression. A highly compressed LQ alternative is transmitted with efficient bits, from which LQ control conditions are extracted to guide the reconstruction process. A conditioned generation network with weighted feature modulation is used to combine the SVR-based latent features with the LQ control features. The quality of the LQ control features is determined by the bitrate of the LQ alternative. The strength of the LQ control features in the conditioned generation process balances the importance between the HQ codebook and LQ fidelity details, which can be tuned based on the current reconstruction target to pursue high perceptual quality or high fidelity. As shown in Fig. 4, with our LQ control features, the restored results have largely improved fidelity with rich details.

In addition, we extend the SVR-based LIC framework into an effective LVC framework. Since SVR is robust against input degradation and small perturbations, a substantial amount of codeword indices between adjacent frames can be the same (47% in our experiments). We only need to transfer different indices for most frames. No motion estimation or motion compensation is involved and there is no error propagation. Comparing to previous LIC and LVC, our SVR-based LIC and LVC share a similar processing pipeline, which makes it possible to simplify industrial productive optimization.

We evaluate our approach using benchmark datasets for image and video standardization. Specifically, SVR-based LIC is tested over the JPEG-AI dataset [2]. SVR-base LVC is tested over a combined dataset comprising of video sequences from AOM [1], MPEG [17], JVET [34], and AVS [15]. Also, we evaluate the performance of different SVR-based restoration methods, based on a single codebook [8] or multiple codebooks [25]. Experimental results demonstrate the effectiveness of our method.

2 Related Works

2.1 Sparse Visual Representation Learning

Discrete generative priors have shown impressive performance in image restoration tasks like super-resolution [8], denoising [11] *etc.* By embedding images into a discrete latent space spanned by learned visual codebooks, SVR improves robustness to various degradations. For instance, VQ-VAE [28] learns a highly compressed codebook by a vector-quantized VAE. VQGAN [11] further improves restoration quality by using GAN with adversarial and perceptual loss. In general, natural images have very complicated content, and it is difficult to learn a single class-agnostic codebook for all image categories. Therefore, most methods focus on specific categories. In particular, great success has been achieved in face generation due to the highly structured characteristics of human faces [37,42].

For general images, the recent AdaCode [25] uses image-adaptive codebook learning. Instead of learning a single codebook for all categories of images, a set of basis codebooks are learned, each corresponding to a semantic partition of the latent space. A weight map to combine such basis codebooks is adaptively determined for each input image. By learning the semantic-class-guided codebooks, the semantic-class-agnostic restoration performance can be largely improved.

2.2 Learned Image Compression

There are two main research topics for LIC: how to learn a latent representation, and how to quantize and encode the latent representation. One most popular framework is based on hyperpriors [3], where the image is transformed into a dense latent feature, and an entropy model encodes/decodes the quantized latent feature for efficient transmission. Many improvements have been made to improve the transformation for computing the latent [9,27,43], the entropy model [13,27,29], or the quantization strategy [30,40].

One vital issue of the hyperprior framework is the extreme sensitivity to small differences between the encoder and decoder in calculating the hyperpriors [4]. Most works simply assume homogeneous platforms and deterministic CPU calculation. Some work uses integer NN to prevent non-deterministic GPU computation [4] or designs special NN module that is computational friendly to CPU [41]. However, such solutions cannot be easily generalized to arbitrary network architectures. Also, it is well known that there are complex relations among bitrate, distortion, and perceptual quality [5,6], and it is difficult to pursue high perceptual quality and high pixel-level fidelity at the same time.

2.3 Learned Video Compression

Existing LVC methods [14,26,31,33] follow the traditional video coding pipeline by replacing processing modules like motion estimation, motion compensation, post-enhancement by NNs. Generally the independent (I) frames in a GoP (group of pictures) are compressed as images, and the predictive (P) frames and the bidirectional predictive (B) frames are compressed based on motion estimation and residue coding. This pipeline is not designed for LVC, resulting in error accumulation from different modules. Also the computation cost is generally very high due to the complicated framework.

2.4 SVR-Based Compression

SVR is intuitively suitable for compression, since the integer codeword indices are easy to transfer and are robust to small computation differences in heterogeneous hardware and software platforms. However, HQ SVR-based restoration relies on direct links of multi-scale features between the encoder and decoder. Such features are too expensive to transfer, which often cost more bits than the original input. To make SVR-based compression feasible, previous approaches remove such feature connections. For example, when applied to specific categories like aligned human faces [20,36,42], it is possible to predict a cohesive code sequence for HQ restoration without direct feature links. However, for general images the reconstruction quality is severely impacted without such features. As a result, most methods focus on very low-bitrate scenarios [10], where reconstruction with low fidelity yet good perceptual quality is tolerated. The recent M-AdaCode [25] compensates the performance loss of the removed feature links by using data-adaptive weights to combine multiple semantic-class-dependent codebooks and trades off bitrate and distortion by weight masking to reduce transmitted weight parameters. Unfortunately, for general image content, without the feature connections the restoration quality is overall unsatisfactory.

3 SVR-Based Compression with Conditional Controls

The general framework of SVR-based restoration can be summarized in Fig. 1. An input image $X \in \mathbb{R}^{w \times h \times c}$ is embedded into a latent space as latent feature $Y \in \mathbb{R}^{u \times v \times d}$ by an embedding network E^{emb}. Using a learned codebook $\mathcal{C} = \{c_l \in \mathbb{R}^d\}$, the latent Y is further mapped into a discrete quantized latent feature $Y^q \in \mathbb{R}^{u \times v \times d}$. Each super-pixel $y^q(l)$ ($l = 1,\ldots,u \times v$) in Y^q corresponds to a codeword $c_l \in \mathcal{C}$ that is closest to the corresponding latent feature $y(l)$ in Y:

$$y^q(l) = c_l = argmin_{c_i \in \mathcal{C}} D(c_i, y(l))).$$

$y^q(l)$ can be represented by the index z_l of codeword c_l, and the entire Y^q can be mapped to an n-dim vector Z of integers, $n = u \times v$. Based on indices Z, the quantized feature Y^q can be retrieved from the codebook $\mathcal{C}$. Also, multi-scale features F are computed from several downsampling blocks in E^{emb}, which are fed to the corresponding upsampling blocks in a reconstruction network E^{rec} as residual inputs. E^{rec} then reconstructs the output image $\hat{x}$ based on the quantized latent Y^q and features F.

To improve the performance for general image restoration, instead of using one codebook as in [24,42], AdaCode [25] learns a set of basis codebooks $\mathcal{C}_1,\ldots,\mathcal{C}_K$, each corresponding to a semantic partition of the latent space. A weight map $W \in \mathbb{R}^{u \times v \times K}$ is computed to combine the basis codebooks for adaptive restoration. The quantized latent Y^q is a weight combination of individual quantized latents $Y_1^q,\ldots,Y_K^q$ using each of the basis codebooks:

$$y^q(l) = \sum_{j=1}^{K} w_j(l) y_j^q(l), \qquad (1)$$

and $w_j(l)$ is the weight of the j-th codebook for the l-th super-pixel in W.

For the purpose of compression, previous methods [20,21] remove the direct skip connections of the multi-scale features F that are too heavy to transfer. Only the indices $Z_1, \ldots, Z_K$ are sent to the decoder to retrieve the quantized latent Y^q for reconstruction. However, the multi-scale features F provide important fidelity details of the input, and without F the reconstructed result may lack details and may not be consistent with the original input.

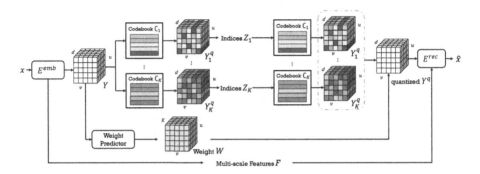

Fig. 1. The general workflow of SVR-based restoration. Discarding the multi-scale features F will sacrifice restoration quality significantly.

3.1 SVR-Based Compression Using LQ Control Conditions

In the realm of image generation, ControlNet [38] has been developed to enable different levels of control over generated results. The key idea is to provide data-specific conditions to a pre-trained generative model to control the generation process. This is analogous to SVR-based compression, where the reconstruction network E^{rec} generates the output based on quantized feature Y^q, and the multi-scale features F provide additional control conditions drawn from the current input. This perspective motivates our compression framework in Fig. 2. When used as control conditions, the multi-scale features do not have to come from the original input, and therefore we can avoid transmitting the heavy F. Instead, they can be drawn from an LQ substitute of the input x^{LQ} in the decoder, and the LQ substitute can be computed in decoder based on fidelity-preserving information calculated by existing compression methods like [19,22] with high compression rates and low bitrate. With the help of additional controls, this framework not only improves the restoration fidelity and quality, but also enables flexible quality control. By tuning the bitrate of the LQ substitute, we can tune the quality of the LQ substitute and change the quality of the control condition.

SVR-Based LIC. As shown in Fig. 2(a), in the encoder, the input image x is embedded into the latent space as latent feature Y, which is further quantized into $Y_1^q, \ldots, Y_K^q$ with associated codeword indices $Z_1, \ldots, Z_K$, by using codebooks $\mathcal{C}_1, \ldots, \mathcal{C}_K$, respectively. At the same time, x is encoded into a highly

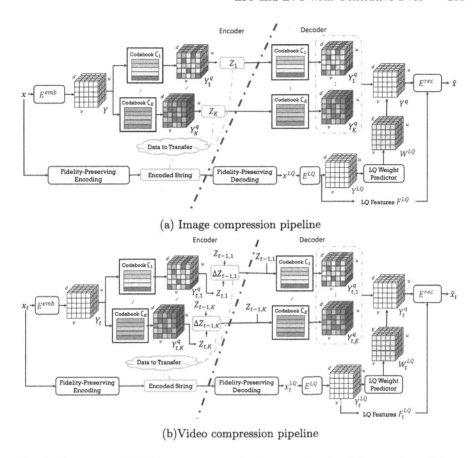

Fig. 2. The proposed SVR-based compression framework using LQ control conditions.

compressed string with low bitrate using a fidelity-preserving image compression method (*e.g.*, an existing LIC method [10]), which is transferred to the decoder together with indices $Z_1, \ldots, Z_K$. Then in the decoder, the quantized latents $Y_1^q, \ldots, Y_K^q$ are retrieved from the corresponding codebooks using the codeword indices, and the LQ substitute x^{LQ} is decoded from the corresponding image compression method. Albeit low quality, x^{LQ} carries important fidelity information about the original x to guide reconstruction.

In the decoder, x^{LQ} is fed into an LQ embedding network E^{LQ} to compute the multi-scale LQ features F^{LQ} from the multiple downsampling blocks in E^{LQ}, where F^{LQ} has the same size as the original multi-scale features F (if computed from the original E^{emb}). When $K > 1$, an LQ weight map $W^{LQ} \in \mathbb{R}^{u \times v \times K}$ is also computed by an LQ weight predictor. Both F^{LQ} and W^{LQ} are control conditions drawn from the fidelity-preserving LQ substitute x^{LQ}, which are used by the reconstruction network E^{rec} to guide the reconstruction process. Since F^{LQ} and W^{LQ} are calculated in decoder using x^{LQ}, they do not increase bitrate.

In detail, to reconstruct the output, when $K>1$, the final quantized latent Y^q is a weighted combination of $Y_1^q, \ldots, Y_K^q$ similar to Eq. (1), using LQ weight W^{LQ}. When $K=1$, $Y_1^q = Y^q$. Then the multi-scale LQ feature F^{LQ} is used as modulating conditions to the multiple upsampling blocks in E^{rec} to guide the reconstruction from Y^q. In this work, we use the Controllable Feature Transformation (CFT) module from [42] to apply modulating conditions. Let Θ_c denote the parameters of the CFT module CFT_{code} to combine the codebook-based quantized feature Y^q and the LQ control feature F^{LQ}. F^{LQ} tunes Y^q into a modulated $Y^{mod} = Y^q + \alpha_c * (\beta_c * Y^q + \gamma_c)$, where β_c, γ_c are affine parameters $\beta_c, \gamma_c = \Theta_c(concat(Y^q, F^{LQ}))$, and $concat(\cdot)$ is the concatenation operation. α_c determines the strength of the control feature F^{LQ} in conditioning the codebook-based feature Y^q, which can be flexibly set according to the actual compression needs, i.e., to pursue high perceptual quality or high fidelity.

One advantage of our SVR-based LIC method is the flexibility in accommodating different scenarios. For homogeneous computing platforms, our method combines the strength of the fidelity cue from existing fidelity-preserving image compression methods (classic or learning-based) and the perceptual cue from SVR-based restoration, enables bitrate control by tuning the bitrate of the LQ substitute, and allows tradeoff between perceptual quality and fidelity. For heterogeneous computing platforms where previous LIC methods may have difficulty to apply, our method can still give a decent low-bitrate baseline reconstruction with good perceptual quality using SVR-based restoration alone, or can pair with classic compression methods for improved reconstruction.

SVR-Based LVC. The above SVR-based LIC method can be easily extended to an effective SVR-based LVC method, whose workflow is shown in Fig. 2(b). Since SVR is robust to input degradation and perturbations, for most frames in a video, a large portion of the codeword indices can be the same between adjacent frames. Therefore, for a video frame x_t at time stamp $t>1$, only the different indices $\Delta Z_{t,1}, \ldots, \Delta Z_{t,K}$ from the previous frame need to be transmitted for the decoder to restore the quantized latent Y_t^q for SVR-based reconstruction. Actually our experiments show that 47% of the codeword indices remain unchanged on average, leading to effective bit reduction for LVC. The remaining processing modules are similar to the LIC method, with the difference that the LQ substitute x_t^{LQ} of frame x_t comes from a fidelity-preserving video compression method (e.g., classic VVC [19] or learning based DVC [26]) or a fidelity-preserving image compression method.

Similar to SVR-based LIC, our SVR-based LVC method provides flexibility to accommodate different scenarios, where we can choose different methods to generate the LQ substitutes by considering different factors like computation and transmission requirements, reconstruction targets, etc. In addition, there is no error propagation in recovering the codeword-based quantized feature for every frame, and a decent low-bitrate baseline with good perceptual quality can be mostly guaranteed. Furthermore, in comparison to previous LIC and LVC methods that usually have completely different processing pipelines, the SVR-based LIC and LVC have a similar workflow with similar processing modules, making it possible to simplify industrial productive optimization.

It is worth mentioning that in our implementation the reconstruction network E^{rec} has the same architecture for both LIC and LVC. A video-oriented network like C3D [35] can be used for LVC to better ensure temporal consistency. We found it unnecessary in experiments as the result is quite consistent temporally due to the deterministic generation and temporal-consistent fidelity control.

Complexity. Our approach is very efficient in computation. For SVR-based LIC, our encoding time includes inference through E^{emb} with time $\mathcal{T}(E^{emb})$ and compressing for X^{LQ} with time $\mathcal{T}(Enc(X^{LQ}))$. In comparison, previous LIC methods compress for the original X, and the encoding time $\mathcal{T}(Enc(X))$ includes both the embedding inference $\mathcal{T}(E^{emb})$ and the hyperprior coding time $\mathcal{T}(Enc(X_{hyper}))$. Usually $\mathcal{T}(Enc(X_{hyper}))$ is much larger than $\mathcal{T}(E^{emb})$ due to the expensive autoregressive process using CPU. Since our X^{LQ} can be highly compressed, our $\mathcal{T}(Enc(X^{LQ}))$ is much less than $\mathcal{T}(Enc(X_{hyper}))$ (*e.g.*, by using a much smaller embedding latent and hyperprior to encode to largely reduce bitrate and computation).

Similarly, our decoding time mainly includes inference through E^{rec} with $\mathcal{T}(E^{rec})$, decoding for X^{LQ} with $\mathcal{T}(Dec(X^{LQ}))$, and inference through E^{LQ} with $\mathcal{T}(E^{LQ})$. In comparison, previous LIC methods needs to decode for the original X where the decoding time includes $\mathcal{T}(E^{rec})$ and $\mathcal{T}(Dec(X))$. Again, due to the expensive hyperprior decoding process, $\mathcal{T}(Dec(X))$ is usually much larger than $\mathcal{T}(Dec(X^{LQ}))$ and $\mathcal{T}(E^{LQ})$ combined.

For SVR-based LVC, our computation is basically a linear extension of the computation for SVR-based LIC according to the number of frames. This is much more efficient than previous LVC methods, which not only require multiple inference processes through multiple networks, but also require the entire decoding computation in the encoder to obtain residues.

3.2 Training Strategy

Stage 1: Pretrain The embedding network E^{emb}, codebooks $\mathcal{C}_1,\ldots,\mathcal{C}_K$, reconstruction network E^{rec}, and weight predictor (for $K > 1$) are pretrained for single-codebook-based restoration [8] or multi-codebook-based restoration [25].
Stage 2: Train SVR-based LIC The embedding network E^{emb} and codebooks $\mathcal{C}_1,\ldots,\mathcal{C}_K$ are fixed, and we train the LQ embedding network E^{LQ}, the CFT module CFT_{code}, the LQ weight predictor (for $K>1$), the reconstruction network E^{rec}, and the GAN discriminator for the LIC pipeline. The training loss comprises of pixel-level L_1 loss and SSIM, the perceptual loss [23], LPIPS [39], and the GAN adversarial loss [12]. The straight-through gradient estimation is used for back-propagation through the non-differentiable vector quantization process during training. The strength of control is set as $\alpha_c = 1$ for all inputs.
Stage 3: Train SVR-based LVC The embedding network E^{emb}, the codebooks $\mathcal{C}_1,\ldots,\mathcal{C}_K$, and the reconstruction network E^{rec} are fixed, and we train

the LQ embedding network E^{LQ}, the CFT module CFT_{code}, the LQ weight predictor (when $K > 1$), and the GAN discriminator by finetuning from the corresponding LIC version in the previous stage. One benefit of funetuning from the LIC counterparts is to benefit from the large variety of image training content to avoid overfitting, due to the limited amount of training videos from the standardization community.

3.3 Bit Reduction for Integer Codeword Indices

To transfer codeword indices, naively we need $b(Z_k) = u \times v \times \text{floor}(\log_2 n_k)$ bits for each codebook $\mathcal{C}_k$ of size n_k. This number can be further reduced to save bit consumption of the whole system. We propose an effective arithmetic coding method that can losslessly compress the integer indices by 5× on average. For natural images, codewords normally show up with different frequencies. For instance, codewords of natural scenes may be used more frequently than those of human faces. We can assign less bits to more frequently used indices to reduce the total bitrate. Specifically, we first calculate the frequency of codewords' usage in training data and reorder the codewords in descending order. Then for each particular indices string of each datum, we convert each odd index ix to a negative integer as $ix^* = -(ix+1)/2$ and rescale even indices by $1/2$. Such operations transform the indices distribution to a Gaussian style bell shape, which can be efficiently encoded by Gaussian Mixture-based arithmetic coding [32].

4 Experiments

Datasets. We tested the proposed SVR-based LIC and LVC method, respectively, over the JPEG-AI dataset [2,18] and a mixed video dataset combining test video sequences from several video compression standards including AOM [1], MPEG [17], JVET [34], and AVS [15]. The JPEG-AI dataset had 5664 images with a large variety of visual content and resolutions up to 8K. The training, validation, and test set had 5264, 350, and 50 images, respectively. The mixed video set contained 150 videos, which were used as test sequences by the standardization community. We removed the duplicate sequences, e.g., the same sequences used by different standards, or the same sequences resized to different resolutions, where we kept videos with different resolutions ranging from 240 × 400 to 4K. 134 and 16 video sequences were used for training and test respectively.

The training patches was 256 × 256, randomly cropped from randomly resized training images or video frames, augmented by random flipping and rotation. For evaluation, the maximum inference tiles was 1080 × 1080. For training SVR-based LVC modules, video frames were randomly sampled from videos, and were used as images in the same way as training SVR-based LIC modules. For all tested methods, each training stage had 500K iterations with Adam optimizer and a batch size of 32, using 8 NVIDIA Tesla V100 GPUs. The learning rate for the generator and discriminator were fixed as 1e−4 and 4e−4, respectively.

Evaluation Metrics. For reconstruction distortion, we measured PSNR and SSIM, as well as the perceptual LPIPS [39]. The bitrate was measured by bpp (bit-per-pixel): $bpp = B/(h \times w)$, and the overall bits $B = b_c + b_{LQ}$ consisted of b_c for sending codebook indices and b_{LQ} for sending the encoded string to compute the LQ substitute x^{LQ} using previous image/video compression methods. In detail, for LIC $b_c = \sum_{k=1}^{K} b(Z_k)$, and for LVC $b_c = \sum_{k=1}^{K} [b(Z_{1,k}) + \sum_{t=2}^{T} b(\Delta Z_{t,k})]/T$. $b(Z_k)$ is computed based on the indices reduction method described in Sect. 3.3. In terms of b_{LQ}, it determined the quality of the LQ substitute x^{LQ}. We chose a low b_{LQ} (< 0.1 bpp) to roughly match b_c.

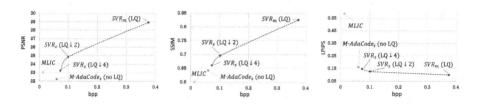

Fig. 3. Rate-distortion performance for image compression.

Evaluated Methods. We evaluated two configurations for our approach using SOTA SVR-based restoration algorithms: the single-codebook-based FeMaSR [8] and the multi-codebook-based AdaCode [25]. With only a single codebook, FeMaSR gave very low bitrate. Using multiple codebooks, AdaCode gave improved restoration quality but consumed more bits.

To generate fidelity-preserving LQ substitute x^{LQ}, for SVR-based LIC we used previous SOTA LIC method MLIC [22]. The pre-trained MLIC model with the lowest available bitrate setting was used, which corresponded to the quality-1 model in [22]. In order to get lower bitrate for b_{LQ} to match b_c, we first downsampled the input x by 2× or 4×, and then used MLIC to encode the downsampled input and then upsampled the decoded x^{LQ} back to the original size. The bicubic filter was used for downsampling/upsampling. For SVR-based LVC we used the SOTA VVC video compression method [19] with $qp = 42$ to generate x^{LQ}, which gave reasonable low-bitrate reconstruction in general.

4.1 LIC Results

Figure 3 gives the rate-distortion performance for image compression. For our SVR-based LIC, we tested 3 different settings: single-codebook SVR with 2× and 4× downsampled-upsampled x^{LQ} as "SVR$_s$(LQ↓2×)" and "SVR$_s$(LQ↓4×)", and multi-codebook SVR with x^{LQ} without downsampling-upsampling SVR$_m(LQ)$.

We also compared with M-AdaCode without x^{LQ} with the 1-codebook setting [21] ("M-AdaCode$_s$(no LQ)") and compared with MLIC [22] generated x^{LQ}. From the figures, "SVR$_s$(LQ↓4×)" outperformed M-AdaCode with 1 dB, 2.3% and 8.9% improvements over PSNR, SSIM and LPIPS, respectively, using only a 0.013bpp increase. Compared to MLIC, "SVR$_s$(LQ↓4×)" improved LPIPS and SSIM by 74.8%, and 10.1% respectively. Among methods having $< 0.1bpp$, our SVR-based LIC gave balanced results with good fidelity and perceptual quality.

Figure 4 gives some restoration examples, which clearly show the strength of our method. With the help of x^{LQ}, our SVR-based LIC largely improved the reconstruction fidelity and perceptual quality with rich visually pleasing details, compared to "M-AdaCode$_s$(no LQ)". Also, our framework can be flexibly configured to different settings to tradeoff bitrate and reconstruction quality.

4.2 LVC Results

For video compression, we tested the single-codebook SVR$_s$. The LQ substitute x^{LQ} was generated by the VVC standard [19] using $qp = 42$ over the original resolution. This is basically the lowest bpp configuration of VVC ($bpp = 0.06$) with a reasonable reconstruction quality for x^{LQ}. Table 1 gives the rate-distortion performance, and Fig. 5 gives some restoration examples. Our "SVR$_s$(LQ)" achieved much better perceptual quality with a 64.8% improvement over LPIPS given only 0.035bpp increase. As expected, improvements over PSNR and SSIM are less significant as VVC is tailored to optimize such pixel-level distortions. With overall $bpp < 0.1$, the SVR-based LVC can generate rich visually pleasing details compared to the overly smoothed results from VVC. On average, 47% of codeword indices remain unchanged, verifying the effectiveness of using the similar pipeline for both SVR-based LIC and LVC.

Table 1. SVR-based LVC performance

	PSNR	SSIM	LPIPS	bpp
SVR$_s$(LQ)	28.15	0.812	0.109	0.095
VVC (x^{LQ})	28.09	0.806	0.310	0.06

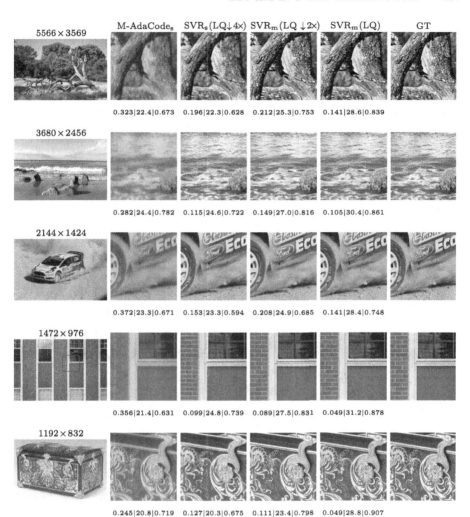

Fig. 4. "LPIPS|PSNR|SSIM" under each result. "SVR$_m$"/ "SVR$_s$": multi-codebook-based/single-codebook-based SVR. "LQ"/"LQ↓ 2×"/"LQ↓ 4×": x^{LQ} of original size/x^{LQ} with 2× downsampling-upsampling /x^{LQ} with 4× downsampling-upsampling.

Fig. 5. "LPIPS|PSNR|SSIM" under each result. Single-codebook SVR was used. x^{LQ} was computed by VVC with $qp = 42$. The average $b_{LQ} = 0.06$ and $b_c = 0.035$. Our SVR-based LVC can largely improve reconstruction fidelity and perceptual quality.

5 Conclusions

We proposed a general SVR-based compression framework for both LIC and LVC. Based on the idea of guided image generation with conditional controls, our method drew fidelity cues as control signals from a low-bitrate LQ version of the original input to guide reconstruction. Compared with previous approaches that relied on SVR-based generation alone, the fidelity cues largely improved the reconstruction quality. By tuning the bitrate of the LQ input, we could trade off bitrate, reconstruction fidelity and perceptual quality. By transferring the difference of codeword indices between adjacent frames, a similar processing pipeline was used for both SVR-based LIC and LVC. Experimental results showed improved performance over SOTA image and video compression methods.

References

1. Alliance for open media, press Release. http://aomedia.org/press-release/. Git repositories on aomedia: https://aomedia.googlesource.com/
2. Ascenso, J., Akyazi, P., Pereira, F., Ebrahimi, T.: Learning-based image coding: early solutions reviewing and subjective quality evaluation. SPIE Photonics Europe - Optics, Photonics and Digital Technologies for Imaging Applications VI (2020)
3. Balle, J., Minnen, D., Singh, S., Hwang, S., Johnston, N.: Variational image compression with a scale hyperprior. In: ICLR (2018)
4. Ballé, J., Johnston, N., Minne, D.: Integer networks FRO data compression with latent-variable models. In: ICLR (2019)
5. Blau, Y., Michaeli, T.: The perception-distortion tradeoff. In: CVPR, pp. 6228–6237 (2018)
6. Blau, Y., Michaeli, T.: Rethinking lossy compression: the rate-distortion-perception tradeoff. In: ICML, pp. 675–685 (2019)
7. Chan, K., Wang, X., Xu, X., Gu, J., Loy, C.: GLEAN: generative latent bank for large-factor image super-resolution. In: CVPR (2021)
8. Chen, C., et al.: Real-world blind super-resolution via feature matching with implicit high-resolution priors. ACM Multimedia (2022)
9. Cheng, Z., Sun, H., Takeuchi, M., Katto, J.: Learned image compression with discretized gaussian mixture likelihoods and attention modules. In: CVPR (2020)
10. El-Nouby, A., Muckle, M., Ullrich, K., Laptev, I., Verbeek, J., Jegou, H.: Image compression with product quantized masked image modeling. arXiv preprint: arXiv:2212.07372 (2022)
11. Esser, P., Rombach, R., Ommer, B.: Taming transformers for high-resolution image synthesis. In: CVPR (2021)
12. Goodfellow, I., et al.: Generative adversarial nets. In: NeurIPS (2014)
13. He, D., Zheng, Y., Sun, B., Wang, Y., Qin, H.: Checkerboard context model for efficient learned image compression. In: CVPR (2021)
14. Hu, Z., Lu, G., Xu, D.: FVC: a new framework towards deep video compression in feature space. In: CVPR (2021)
15. IEEE: Audio video coding standard. IEEE Standard for Second-Generation IEEE 1857 Video Coding (2019). ISBN 978-1-5044-5461-2
16. Int. Telecommun. Union-Telecommun. (ITU-T) and Int. Standards Org./Int/Electrotech. Commun. (ISO/IEC JTC 1): High efficiency video coding, rec. ITU-T H.265 and ISO/IEC 23008-2 (2019)
17. ISO/IEC JTC 1/SC 29: Moving picture experts group (MPEG)
18. ISO/IEC JTC 1/SC29/WG1: Report on the JPEG AI call for proposals results. ISO/IEC JTC1/SC29 WG1, N100250, July 2022
19. ITU-T and ISO: Versatile video coding, rec. ITU-T H.266 and ISO/IEC 23090-3 (2020)
20. Jiang, W., Choi, H., Racapé, F.: Adaptive human-centric video compression for humans and machines. In: CVPRW on NTIRE, June 2023

21. Jiang, W., Wang, W., Chen, Y.: Neural image compression using masked sparse visual representation. In: WACV (2024)
22. Jiang, W., Yang, J., Zhai, Y., Ning, P., Gao, F., Wang, R.: MLIC: multi-reference entropy model for learned image compression. In: ACM Multimedia (2023)
23. Johnson, J., Alahi, A., Fei-Fei, L.: Perceptual losses for real-time style transfer and super-resolution. In: Leibe, B., Matas, J., Sebe, N., Welling, M. (eds.) ECCV 2016. LNCS, vol. 9906, pp. 694–711. Springer, Cham (2016). https://doi.org/10.1007/978-3-319-46475-6_43
24. Li, T., Chang, H., Mishra, S., Zhang, H., Katabi, D., Krishna, D.: MAGE: masked generative encoder to unify representation learning and image synthesis. In: CVPR (2023)
25. Liu, K., Jiang, Y., Choi, I., Gu, J.: Learning image-adaptive codebooks for class-agnostic image restoration. In: ICCV (2023)
26. Lu, G., Ouyang, W., Xu, D., Zhang, X., Cai, C., Gao, Z.: DVC: an end-to-end deep video compression framework. In: CVPR, pp. 11006–11015 (2019)
27. Mentzer, F., et al.: VCT: a video compression transformer. arXiv preprint: arXiv:2206.07307 (2022)
28. Oord, A., Vinyals, O., Kavukcuoglu, K.: Neural discrete representation learning. In: NeurIPS (2017)
29. Qian, Y., Lin, M., Sun, X., Jin, Z.T.R.: Entroformer: a transformer-based entropy model for learned image compression. arXiv preprint: arXiv:2202.05492 (2022)
30. Feng, R., Guo, Z., Li, W., Chen, Z.: NVTC: nonlinear vector transform coding. In: CVPR (2023)
31. Rippel, O., Nair, S., Lew, C., Branson, S., Anderson, A., Bourdev, L.: Learned video compression. In: ICCV (2019)
32. Rissanen, J., Langdon, G.: Arithmetic coding. IBM J. Res. Dev. **23**(2) (1979). https://doi.org/10.1147/rd.232.0149
33. Shi, Y., Ge, Y., Wang, J., Mao, J.: AlphaVC: high-performance and efficient learned video compression. In: Avidan, S., Brostow, G., Cissé, M., Farinella, G.M., Hassner, T. (eds.) ECCV 2022. LNCS, vol. 13679, pp. 616–631. Springer, Cham (2022). https://doi.org/10.1007/978-3-031-19800-7_36
34. Suehring, K., Li, X.: JVET common test conditions and software reference configurations. document JVET-B1010 of JVET (2016)
35. Tran, D., Bourdev, L., Fergus, R., Torresani, L., Paluri, M.: Learning spatiotemporal features with 3d convolutional networks. In: ICCV (2015)
36. Wang, T., Mallya, A., Liu, M.: One-shot free-view neural talking-head synthesis for video conferencing. In: CVPR (2021)
37. Wang, Z., Zhang, J., Chen, R., Wang, W., Luo, P.: RestoreFormer: high-quality blind face restoration from undegraded key-value pairs. In: CVPR (2022)
38. Zhang, L., Rao, A., Agrawalau, M.: Adding conditional control to text-to-image diffusion models. ArXiv, arXiv:2302.05543 (2023)
39. Zhang, R., Isola, P., Efros, A., Shechtman, E., Wang, O.: The unreasonable effectiveness of deep features as a perceptual metric. In: CVPR (2018)
40. Zhang, X., Wu, X.: LVQAC: lattice vector quantization coupled with spatially adaptive companding for efficient learned image compression. In: CVPR (2023)
41. Zheng, Z., Wang, X., Lin, X., Lv, S.: Get the best of the three worlds: real-time neural image compression in a non-GPU environment. In: ACM Multimedia (2021)
42. Zhou, S., Chan, K., Li, C., Loy, C.: Towards robust blind face restoration with codebook lookup transformer. In: NeurIPS (2022)
43. Zou, R., Song, C., Zhang, Z.: The devil is in the details: window-based attention for image compression. In: CVPR (2022)

Enhancing Continuous Skeleton-Based Human Gesture Recognition by Incorporating Text Descriptions

Thi-Lan Le[1](✉)[iD], Viet-Duc Le[1][iD], and Thuy-Binh Nguyen[2][iD]

[1] SigM Lab, School of Electrical and Electronics Engineering Hanoi University of Science and Technology, Hanoi, Vietnam
`lan.lethi1@hust.edu.vn`
[2] University of Transport and Communications, Hanoi, Vietnam

Abstract. Continuous gesture recognition is a crucial task in human-computer interaction. Unlike isolated gesture recognition, where individual gestures are analyzed independently, continuous recognition involves detecting and classifying multiple gestures seamlessly from continuous video streams. In this paper, we propose a method for continuous gesture recognition. Our proposed model operates in two stages: isolated gesture recognition and a sliding window-based approach for continuous gesture recognition. For isolated gesture recognition, we propose a dual encoder method named TDDNet, stand for Text-Enhanced DDNet, that integrates a skeleton encoder based on the DDNet model [6] with a text encoder based on CLIP. We evaluate our model on a self-collected dataset comprising 19 gestures relevant to human-COBOT interaction, collected from 50 subjects. Experimental results demonstrate that our model improves isolated gesture recognition accuracy from 84.2% to 85.5%, while for continuous gesture recognition, the model achieves a performance of 66.60%, compared to 66.00% of the baseline model. The source code is publicly available at https://github.com/duclvQ/improved_DDNet.

Keywords: continuous action recognition · text description · skeleton-based action recognition

1 Introduction

Continuous gesture recognition is a crucial task in human-computer interaction (HCI) [19]. Unlike isolated gesture recognition, which analyzes individual gestures independently, continuous recognition involves detecting and classifying multiple gestures seamlessly from continuous video streams [8]. Although significant progress has been made, most existing approaches rely on deep models trained on RGB sequences. These models, while accurate, often contain a large number of parameters, making them computationally heavy and challenging to deploy in real-world applications.

Recently, the increasing availability of low-cost 3D cameras and advancements in human pose estimation have led to the rise of 3D skeleton-based action recognition as an active area of research. This approach offers a lightweight and informative representation for continuous gesture recognition [5,16]. However, despite considerable progress, skeleton-based action recognition remains challenging, particularly when dealing with highly similar gesture classes. Additionally, they mainly focus on modeling the relation of human joints in a unimodal training scheme [6]. Therefore, they do not fully exploit the semantic relationships between actions. In some cases, the primary distinction between different gestures lies solely in the shape or movement of the hands. This information may be ignored by visual models due to the low resolution of the hands. However, in such cases, action descriptions could reveal these details. Therefore, recent researches have attempted to incorporate action descriptions in isolated human action recognition from trimmed videos [24].

Recently, Contrastive Language-Image Pre-training (CLIP) has emerged as an efficient method for image representation learning using natural language supervision. CLIP jointly trains an image encoder and a text encoder to predict correct pairings between an image and its corresponding text. In this paper, we explore the use of CLIP for continuous skeleton-based gesture recognition. Our proposed model operates in two stages: isolated gesture recognition and a sliding window-based approach for continuous gesture recognition. For isolated gesture recognition, we propose a dual encoder method named TDDNet that integrates a skeleton encoder based on the DDNet model [6] with a text encoder based on CLIP. We evaluate our model on a self-collected dataset comprising 19 gestures relevant to human-COBOT interaction, collected from 50 subjects. Experimental results demonstrate that our model improves isolated gesture recognition accuracy from 84.2% to 85.5%, while for continuous gesture recognition, the model achieves a performance of 66.60%, compared to 66.00% of the baseline model.

The rest of this paper is structured as follows. In Sec. II, some notable work related to human gesture recognition (HGR) are briefly presented. The proposed model is described in Sec. 3. Experiments and results on our self-built dataset are provided in Sec. 4.

2 Related Works

In the literature, existing Hand Gesture Recognition (HGR) methods are generally categorized into two main approaches: isolated and continuous hand gesture recognition. While isolated recognition approach focuses on identifying gestures from segmented videos, each containing a single hand gesture. Continuous recognition approach involves recognizing multiple hand gestures from a continuous video stream, requiring the system to detect and differentiate gestures as they occur in real-time. Compared to isolated recognition, continuous recognition task is considered more challenging due to the added complexity of temporal action localization (TAL), identify when each gesture starts and ends within the video stream.

For isolated recognition, early studies primarily employed CNN-based models to extract spatial features from each frame of a segmented video and RNN-based architectures to capture the temporal consistency between consecutive frames. This combination aimed to model both spatial and temporal aspects of hand gestures [9,20,27]. More recent works have paid attention on building more effective spatio-temporal descriptors by using 3D-CNN networks to simultaneously capture spatial information. As a result, 3D-CNN networks have significantly enhanced the performance of isolated hand gesture recognition systems, providing more robust and accurate representations of hand gestures in a variety of contexts [1,13].

Although 3D-CNN networks have achieved important milestones by leveraging both spatial and temporal information, Yu et al. [27] declared that these networks are often too complex, leading to lower efficiency. Consequently, some other works focused on dealing with skeleton data to reduce computation complexity while ensuring the recognition rate [10,15,26,28]. In these work, skeleton data can be captured by using either a depth camera Microsoft Kinect or a deep-learning model that can estimate skeleton information from RGB images, such as OpenPose [4], MediaPipe [11], AlphaPose [7], etc. To build a hand gesture representation, spatial features can be extracted using Convolutional Neural Networks (CNNs), while the temporal correlation of skeleton sequences can be captured by deep learning models designed to handle sequential data. Among these models, Graph Convolutional Network (GCN) based approach including the Spatial Temporal Graph Convolutional Network (STGCN) [25] and its variants are regarded as one of the most effective approaches. The main objective of GCN-based approach is to learn simultaneously both spatial and temporal information from skeleton data by directly modeling the skeleton as a graph structure, with nodes representing joints and edges representing the connections between them. However GCN-based approach suffers from its heavy computation. Therefore, DDNet - a light weight model has been introduced for skeleton-based activity and gesture recognition [6].

For continuous recognition, besides the task of classifying which hand gesture is being performed, this task must determine the starting and ending times of each gesture. Noted that isolated hand gesture recognition is the first step in the framework handling continuous hand gesture recognition. For this, sliding window is one of the popular technique for video segmentation. Isolated recognition methods are applied into each video segmentation. After that, a postprocessing mechanism is utilized to filter and aggregate the obtained results of the isolated recognition to accurately locate the time duration of each hand gesture [5,12,16,23]. Recently, large language models (LLM) such as ChatGPT [3], LLaMA [22] have become relatively popular and have been widely used for dealing with numerous human language. These models are pre-trained over various human language to learn common language-related characteristics. As a consequence, large language models can be effectively and efficiently applied to even new human languages. Inspired by this, Qu et al. [17] proposed a novel framework for action recognition in which skeleton sequences treated as the description sentences and LLM is used to capture some useful characteristics for action recognition. This is a novel approach for action recognition, the relationship between consecutive frames is regarded as the connection between words in a sentence.

Based on the analysis above, this study introduces a novel framework for continuous gesture recognition that integrates text descriptions with skeleton information. This approach aims to overcome the limitations of isolated gesture recognition by managing multiple hand gestures in real time and leveraging natural language descriptions for enhanced interpretation and classification. The subsequent section will provide a comprehensive explanation of this framework and its components.

3 Proposed Method for Hand Gesture Recognition

The proposed method for continuous gesture recognition, illustrated in Fig. 1, consists of two phases: training phase and inference phase. In the training phase, a dual encoder model TDDNet is trained from pairs of segmented samples and their corresponding text description. In the inference phase, a sliding window method is used to allow the trained model to stride through a video and make predictions on each step. A threshold-based mechanism is used to aggregate predictions from all steps to a sequence of gesture.

3.1 Skeleton Data Preprocessing

Since the input data consists of RGB frames, we need to extract skeleton joints from these frames. It is important to note that any human pose estimation model can be applied. In our study, we utilize an off-the-shelf model, Google's Mediapipe, for this purpose.

Given an image of arbitrary resolution, the Mediapipe returns a $J \times D$ matrix, with $J = 21$ being the number of extracted joints for each hand and $D = 3$ being the spatial dimension of the joint. Because some gestures in our self-collected dataset requires the involvement of two hands simultaneously, resulting in two matrices of shape 21×3. Because some gestures involve arm movements over a long trajectory, we select six additional joints from the arm and shoulder, represented by a 6×3 matrix, to enhance the recognition accuracy. This matrix captures critical motion data, improving the system's ability to interpret and classify complex gestures effectively. Noted that, face and foot landmarks are neglected since they do not contribute to the meaning of the gestures. For every input frame, we perform pose estimation and receive two 21×3 matrices and a 6×3 matrix. These matrices are concatenated to form a 48×3 matrix. For frames where the pose estimators fail to extract any skeleton joints, zeros are replaced to guarantee the shape of the final matrix.

3.2 TDDNet for Isolated Hand Gesture Recognition

As analyzed in the related works, current approaches proposed for skeleton-based gesture recognition do not fully exploit the semantic relationships between actions. For instance, in some gestures, the primary distinction lies solely in the shape or movement of the hands. This information may be ignored by visual models due to the low resolution of the hands. However, in such cases, action descriptions could reveal these details. Therefore, recent research has attempted to incorporate action descriptions in human action recognition [24].

In our study, to leverage the information from text description, a dual encoder method TDDNet (illustrated in Fig. 2) that comprises of two encoders: one for

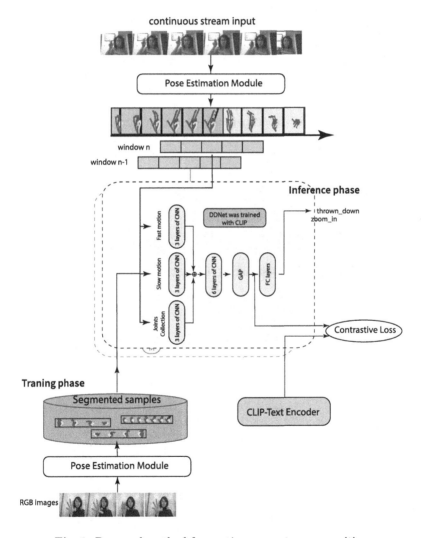

Fig. 1. Proposed method for continuous gesture recognition

skeleton and one for text is proposed. The skeleton encoder extracts features from human skeletons, and the text encoder provides auxiliary information from text descriptions. The text encoder helps the skeleton encoder learn effectively during the training stage, without affecting the inference time of the skeleton encoder. For skeleton encoder, we employ DD-Net [6], a lightweight 1D-CNN network for action recognition thanks to its superior performance obtained for action recognition. The text encoder relies on the model of CLIP [18] that takes a gesture description as input and results a text embedding feature.

We formulate the loss function in our proposed model as follows:

$$\mathcal{L}_{\text{total}} = \mathcal{L}_{\text{cls}}\ (E_s(\mathcal{S})) + \lambda \mathcal{L}_{\text{con}}\ (E_s(\mathcal{S}), E_t(\mathcal{T})) \tag{1}$$

where $\mathcal{L}_{\text{cls}}$ is the cross-entropy loss, $\mathcal{L}_{\text{con}}$ is the contrastive loss between the features generated from the text encoder and the skeleton encoder. The contrastive loss is similar to the one used in the original paper by Radford et al. [18]. The weights λ is chosen through experimentation. $E_s(\mathcal{S})$ and $E_t(\mathcal{T})$ refer to the skeleton encoder model and the natural language encoder model whereas $\mathcal{S}$ and $\mathcal{T}$ are skeleton sequence and text description, respectively.

In the training process, for each gesture, a corresponding description is given. As the dataset used in our experiment is collected in the context of human-robot interaction, a description that explain how to perform the gesture is already available. For example, the text descriptions for two gestures Pickpart and Report illustrated in Fig. 4 are *"A person places one arm in front of their chest, vertical to the body, with palms facing out in front of the body, fingers clustered in a claw shape"* and *"A person raises one arm in front of their chest, vertical to the body axis, with fingers spread out and coming together at one point, the tips of the hands facing the front of the body."*, respectively.

3.3 Continuous Hand Gesture Recognition

After having finished training on the isolated task, the best model will be saved and used in the continuous task. Similar to previous works [5,14], we use a sliding

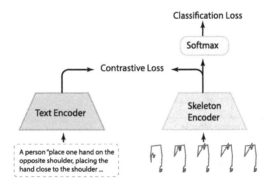

Fig. 2. TDDNet for isolated gesture recognition with dual encoder

window mechanism as the data loader to the model. For each step, the size of the sliding window is fixed and chosen via the isolated evaluation. This sequence of frames is fed to the trained model to provide a prediction, just like a single forward pass in the isolated recognition. After each step, this window strides forward by one frame to get another sequence of frames. The sliding window mechanism is described as in Fig. 3

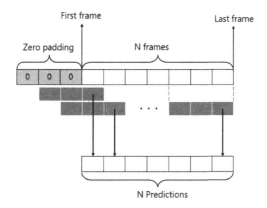

Fig. 3. Sliding window mechanism for continuous recognition

Given a sliding window with size T and an input video of N frame, the data fed to the model at step n will be the sequence of $(n - T + 1)^{th}$ to n^{th} frames. Using this approach, the first $T - 1$ steps will not have enough frames to fit in the sliding window. Therefore, frames that have the indices being smaller than one will be replaced by zeros, or in other words they are zero-padded.

The sliding window shifts forward frame by frame. At each step, the model produces a prediction based on the current position of window. The predicted label of the current window is assigned to the middle frame of the window.

4 Experiments and Results

4.1 Dataset

To evaluate the proposed method, we use our self-collected dataset. The dataset was gathered as part of our project, which aims to develop gesture-based human-COBOT interaction. It consists of 19 gestures, designed according to [2], and was collected from 50 subjects, resulting in a total of 300 videos. Each subject performed 19 different gestures continuously, with random breaks between gestures lasting two to five seconds. In addition to these gestures, "no gesture' sequences (labeled with ID 0) were also collected. Table 1 provides statistical information about the gesture sequences in our dataset. We observe that the duration of gesture classes varies significantly. Moreover, some gestures are quite similar, as illustrated in Fig. 4, which poses challenges for gesture recognition methods. For evaluation, we define a subject-independent protocol, where sequences from subjects with odd IDs are used for training, while sequences from the remaining subjects are reserved for testing.

For our method, we use the skeleton data instead of the original RGB frames. An off-the-shelf model, Mediapipe, is employed to extract skeleton sequences from the RGB video sequences. Then, we process skeleton sequence as explained in Sec. 3.1. The skeleton file is a tensor of shape $T \times J \times D$, where T is the

Table 1. Statistic information of 19 gesture classes in our dataset.

ID	Label	No.Samples	Ave duration(in frames)	ID	Label	No.Samples	Ave duration(in frames)
1	Start	188	94.86	11	PickPart	162	57.15
2	Stop	154	94.63	12	DepositPart	158	51.08
3	Slower	154	91.15	13	Report	138	55.41
4	Faster	154	82.64	14	Ok	153	46.57
5	Done	154	89.22	15	Again	154	47.96
6	FollowMe	120	92.67	16	Help	154	45.77
7	Lift	154	88.2	17	Joystick	154	50.41
8	Home	154	57.24	18	Identification	156	53.51
9	Interaction	151	49.16	19	Change	152	49.23
10	Look	157	48.18				

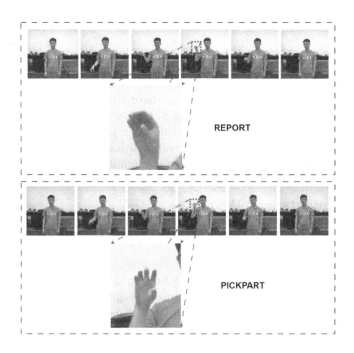

Fig. 4. Examples sequences of Report and PickPart gestures in our dataset.

number of frames, $J = 48$ is the number of skeleton joints, and $D = 3$ indicates the 3D joints.

4.2 Evaluation Metrics

Continuous hand gesture involves two individual tasks, different sets of metrics are needed for each task to evaluate the efficiency of the model.

Isolated recognition refers to the evaluation of individual instances of gestures, making it be similar to a classification task. Therefore, traditional metrics that is Accuracy is employed for evaluation.

While isolated recognition is performed at the instance level, continuous recognition is evaluated at the video level, where the input consists of multiple gesture instances. To assess the performance of continuous recognition, we use frame-wise accuracy.

Frame-wise accuracy is the average accuracy of every frame of a video. Specifically, given a video with T frames, the prediction for T frames being $P = \{P_1, P_2, \ldots, P_T\}$ and the corresponding ground-truth being $G = \{G_1, G_2, \ldots, G_T\}$ the frame-wise accuracy can be formulated as:

$$\text{Frame-wise accuracy} = \frac{1}{T} \sum_{n=1}^{T} S(P_n, G_n) \tag{2}$$

with

$$S(P_n, G_n) = \begin{cases} 1 & \text{if } P_n = G_n \\ 0 & \text{if } P_n \neq G_n \end{cases} \tag{3}$$

The accuracy for a set consists of N videos is calculated by the average Frame-wise accuracy of all videos.

4.3 Experimental Results

Isolated Recognition Results. Table 2 summarizes the results of our ablation study using different models from CLIP [18]. In the original CLIP implementation, the image encoder leverages two distinct architectures: one based on a ResNet and the other on a Vision Transformer (ViT). The results in Table 2 evaluate the performance of these architectures and their impact on isolated gesture recognition. The best performance, 85.50%, is achieved when using the ResNet architecture in CLIP, while employing the ViT model results in only a slight variation in performance. It is worth noting that RN50x4 and RN50x16 are variants of ResNet-50, scaled up by factors of 4 and 16, respectively, following the principles of EfficientNet [21] scaling rule. This scaling strategy allow the network to handle more complex tasks by improving its capacity to learn more useful feature representations. Therefore, in the remaining experiments, we utilizes ResNet architecture in CLIP model.

The performance of the proposed TDDNet model, in comparison to the baseline model, is shown in Table 3. Notably, the proposed method, leveraging a

Table 2. Ablation study of different models used for text encoder.

Model name	Accuracy (%)
DDNet+RN50	85.20
DDNet+RN101	84.60
DDNet+RN50 × 4	**85.50**
DDNet+RN50 × 16	84.80
DDNet+ViT-B/16	82.80
DDNet+ViT-L/14	84.40

dual encoder architecture, outperforms the baseline DDNet model by 1.3%. This improvement is significant, especially considering the challenging nature of the dataset, which follows a subject-independent testing protocol. The challenge arises from the high variability in how different subjects perform the same gesture and the substantial inter-similarity between different gestures.

Table 3. Comparison with the baseline method.

Model name	Accuracy (%)
DDNet (baseline) [6]	84.20
Proposed method TDDNet	**85.50**

Figure 5 and 6 present the confusion matrices for both the baseline model (DDNet) and the proposed model. Both models demonstrate good classification performance across most gesture classes. The proposed method outperforms the baseline method in majority classes except "Look" and "Pickpart". Although the results are promising, we can observe that for certain gestures, such as "Look" "PickPart" "Report" "Identification" and "Change", the performance of both methods is still low due to their similar execution patterns. Figure 7 illustrates two gestures with almost identical movements, where the key distinguishing feature is the positioning of the fingers, which makes the models more susceptible to misclassification.

Figure 8 shows the t-SNE visualization of the gesture class distribution for both the baseline and proposed models. The proposed model exhibits better feature clustering, with gestures of the same class forming tighter clusters, indicating improved feature representation.

Continuous Gesture Recognition Results. The results for continuous gesture recognition are shown in Table 4, while Fig. 9 illustrates an example of recognition for continuous videos from subject ID 20. It can be observed that the proposed method outperforms also the baseline model for continuous recognition. Although the improvement with 0.6% in term of frame wise accuracy is

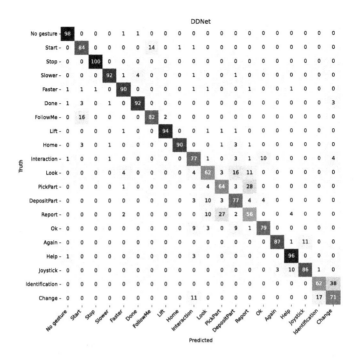

Fig. 5. Confusion matrix obtained by the baseline model DDNet

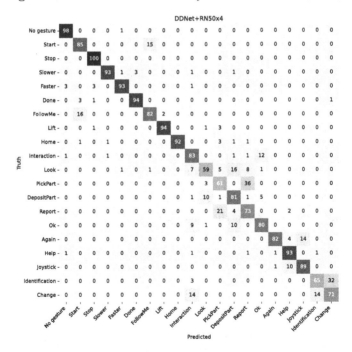

Fig. 6. Confusion matrix obtained by the proposed method TDDNet.

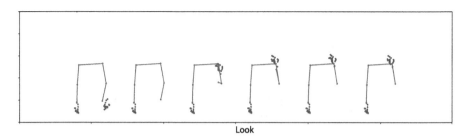

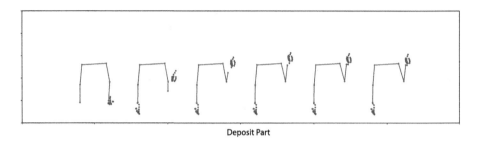

Fig. 7. Skeleton sequence of "Deposit Part" and "Look" gesture

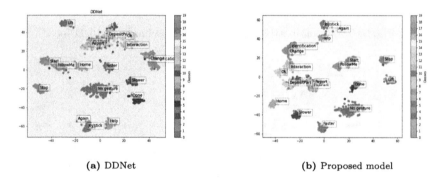

Fig. 8. The t-SNE visualization of gesture classes distribution when using the baseline model DDNet (a) and the proposed method (b).

not significant, however, from the visualization in Fig. 9, the proposed method does not results many fragments.

Table 4. Results obtained by the proposed method for continuous gesture recognition.

Model name	Frame wide accuracy (%)
DDNet (baseline)	66.00
Proposed method	66.60

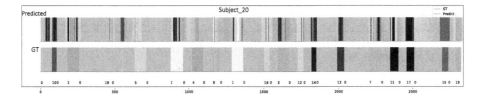

Fig. 9. Continuous recognition results for subject ID 20.

5 Conclusions and Future Works

In this paper, a lightweight deep learning model for continuous gesture recognition based on skeleton information was proposed. The model operates in two stages: isolated gesture recognition and a sliding window-based approach for continuous recognition. To improve performance in the isolated gesture recognition step, we incorporate a text encoder model based on CLIP into the DDNet architecture. Experimental results on the self-collected dataset of 19 gestures demonstrate that our model improves isolated gesture recognition accuracy from 84.2% to 85.5%. For continuous gesture recognition, the model achieves a performance of 66.60%, compared to 66.00% achieved by the baseline model. Although the results for isolated gesture recognition are promising, the performance in continuous gesture recognition still requires improvement. Future work should explore action detection frameworks that allow for simultaneous action localization and recognition from a single representation to enhance continuous gesture recognition.

Acknowledgements. This research was funded by the Vietnam Ministry of Education and Training under grant number B2024-GHA-11.

References

1. Abavisani, M., Joze, H.R.V., Patel, V.M.: Improving the performance of unimodal dynamic hand-gesture recognition with multimodal training. In: Proceedings of the IEEE/CVF Conference on Computer Vision and Pattern Recognition, pp. 1165–1174 (2019)
2. Barattini, P., Morand, C., Robertson, N.M.: A proposed gesture set for the control of industrial collaborative robots. In: 2012 IEEE RO-MAN: The 21st IEEE International Symposium on Robot and Human Interactive Communication, pp. 132–137. IEEE (2012)

3. Brown, T.B.: Language models are few-shot learners. arXiv preprint arXiv:2005.14165 (2020)
4. Cao, Z., Simon, T., Wei, S.E., Sheikh, Y.: Realtime multi-person 2D pose estimation using part affinity fields. In: Proceedings of the IEEE Conference on Computer Vision and Pattern Recognition, pp. 7291–7299 (2017)
5. Dallel, M., Havard, V., Dupuis, Y., Baudry, D.: A sliding window based approach with majority voting for online human action recognition using spatial temporal graph convolutional neural networks. In: Proceedings of the 2022 7th International Conference on Machine Learning Technologies, pp. 155–163. ICMLT 2022, Association for Computing Machinery, New York, NY, USA (2022). https://doi.org/10.1145/3529399.3529425
6. Fan Yang, Sakriani Sakti, Y.W., Nakamura, S.: Make skeleton-based action recognition model smaller, faster and better. In: ACM International Conference on Multimedia in Asia (2019)
7. Fang, H.S., et al.: AlphaPose: whole-body regional multi-person pose estimation and tracking in real-time. IEEE Trans. Pattern Anal. Mach. Intell. **45**(6), 7157–7173 (2022)
8. Gammulle, H., Ahmedt-Aristizabal, D., Denman, S., Tychsen-Smith, L., Petersson, L., Fookes, C.: Continuous human action recognition for human-machine interaction: a review. ACM Comput. Surv. **55**(13s) (2023). https://doi.org/10.1145/3587931
9. Gao, Q., Liu, J., Ju, Z.: Hand gesture recognition using multimodal data fusion and multiscale parallel convolutional neural network for human-robot interaction. Expert. Syst. **38**(5), e12490 (2021). https://doi.org/10.1007/978-3-031-13844-7_3
10. Hu, Q., Gao, Q., Gao, H., Ju, Z.: Skeleton-based hand gesture recognition by using multi-input fusion lightweight network. In: International Conference on Intelligent Robotics and Applications, pp. 24–34. Springer (2022)
11. Kim, J.W., Choi, J.Y., Ha, E.J., Choi, J.H.: Human pose estimation using MediaPipe pose and optimization method based on a humanoid model. Appl. Sci. **13**(4), 2700 (2023)
12. Kwolek, B.: Continuous hand gesture recognition for human-robot collaborative assembly. In: Proceedings of the IEEE/CVF International Conference on Computer Vision, pp. 2000–2007 (2023)
13. Li, Y., Miao, Q., Qi, X., Ma, Z., Ouyang, W.: A spatiotemporal attention-based Resc3D model for large-scale gesture recognition. Mach. Vis. Appl. **30**, 875–888 (2019)
14. Molchanov, P., Yang, X., Gupta, S., Kim, K., Tyree, S., Kautz, J.: Online detection and classification of dynamic hand gestures with recurrent 3D convolutional neural network. In: Proceedings of the IEEE Conference on Computer Vision and Pattern Recognition, pp. 4207–4215 (2016)
15. Narayan, S., Mazumdar, A.P., Vipparthi, S.K.: SBI-DHGR: skeleton-based intelligent dynamic hand gestures recognition. Expert Syst. Appl. **232**, 120735 (2023)
16. Nguyen, T.T., et al.: A continuous real-time hand gesture recognition method based on skeleton. In: 2022 11th International Conference on Control, Automation and Information Sciences (ICCAIS), pp. 273–278. IEEE (2022)
17. Qu, H., Cai, Y., Liu, J.: LLMs are good action recognizers. In: Proceedings of the IEEE/CVF Conference on Computer Vision and Pattern Recognition, pp. 18395–18406 (2024)
18. Radford, A., et al.: Learning transferable visual models from natural language supervision. In: International Conference on Machine Learning (2021). https://api.semanticscholar.org/CorpusID:231591445

19. Robinson, N., Tidd, B., Campbell, D., Kulić, D., Corke, P.: Robotic vision for human-robot interaction and collaboration: a survey and systematic review. ACM Trans. Hum. Robot Interact. **12**(1), 1–66 (2023)
20. Sincan, O.M., Keles, H.Y.: Using motion history images with 3D convolutional networks in isolated sign language recognition. IEEE Access **10**, 18608–18618 (2022)
21. Tan, M.: EfficientNet: rethinking model scaling for convolutional neural networks. arXiv preprint arXiv:1905.11946 (2019)
22. Touvron, H., et al.: LLaMA: open and efficient foundation language models. Preprint at arXiv:2302.13971 (2023)
23. Villani, V., Secchi, C., Lippi, M., Sabattini, L.: A general pipeline for online gesture recognition in human-robot interaction. IEEE Trans. Hum. Mach. Syst. **53**(2), 315–324 (2023)
24. Xiang, W., Li, C., Zhou, Y., Wang, B., Zhang, L.: Generative action description prompts for skeleton-based action recognition (2022).https://doi.org/10.48550/ARXIV.2208.05318, https://arxiv.org/abs/2208.05318
25. Yan, S., Xiong, Y., Lin, D.: Spatial temporal graph convolutional networks for skeleton-based action recognition. In: Proceedings of the AAAI Conference on Artificial Intelligence, vol. 32 (2018). https://doi.org/10.1609/aaai.v32i1.12328
26. Yang, C.L., Li, W.T., Hsu, S.C.: Skeleton-based hand gesture recognition for assembly line operation. In: 2020 International Conference on Advanced Robotics and Intelligent Systems (ARIS), pp. 1–6. IEEE (2020)
27. Yu, J., Qin, M., Zhou, S.: Dynamic gesture recognition based on 2D convolutional neural network and feature fusion. Sci. Rep. **12**(1), 4345 (2022)
28. Zhong, E., Del-Blanco, C.R., Berjón, D., Jaureguizar, F., García, N.: Real-time monocular skeleton-based hand gesture recognition using 3D-JointsFormer. Sensors **23**(16), 7066 (2023)

Towards Robust Video Frame Interpolation with Long-Term Propagation

Ziqi Huang[1], Kelvin C. K. Chan[1], Bihan Wen[2], and Ziwei Liu[1] (✉)

[1] S-Lab, Nanyang Technological University, Singapore, Singapore
{ziqi002,chan0899,ziwei.liu}@ntu.edu.sg
[2] School of EEE, Nanyang Technological University, Singapore, Singapore
bihan.wen@ntu.edu.sg

Abstract. Video frame interpolation aims to synthesize non-existent frames between two consecutive frames in a video, and its importance can be seen from its wide applications in computer vision. The key to video frame interpolation is predicting the intermediate motions between two given frames, so that the synthesized frames are coherent with the input video. However, the existing approaches of imposing assumptions on motions, such as linear and quadratic trajectories, are not generalizable to complex motions in real-world videos. In this work, we propose **long-term propagation** for **robust and general motion prediction** in video frame interpolation. To more thoroughly understand the motion trajectories, we propose to implicitly track the motion paths through a long sequence of video frames. The motion features are then used to refine the motion predicted from the primitive motion assumptions. Our proposed long-term propagation of motion can be *easily integrated into existing video frame interpolation approaches*. Quantitative and qualitative results demonstrate that long-term propagation effectively *improves interpolation performance* when incorporated into various state-of-the-art baselines. In addition, we present a series of analytical experiments to study the mechanism, advantages, and limitations of long-term propagation in video frame interpolation to inspire future works.

Keywords: Video frame interpolation · Motion modeling

1 Introduction

Given a low-frame-rate video, we can apply *video frame interpolation* as a post-processing step to synthesize non-existent frames between two consecutive video frames. Video frame interpolation is mainly applied to increase the frame

---

Supplementary Information The online version contains supplementary material available at https://doi.org/10.1007/978-981-96-2644-1_19.

rate and produce videos with smoother transitions. It is also useful in video compression [4] and super-slow-motion generation [9,12].

Video frame interpolation has attracted considerable attention in the computer vision community. There have been three major classes of solutions: flow-based [2,3,7,8,12,16,22,23,26,30,31], kernel-based [6,24,25], and phase-based [19,20] methods. In recent years, *flow-based methods* have seen notable advancement and achieved relatively impressive results. This class of methods usually consists of three steps: *1)* estimating the optical flows from the input frames to the non-existent frames, *2)* warping the input frames according to the reversed flow estimations, and *3)* fusing the warped frames. Since the hidden frame is unavailable at inference time, the intermediate flows (*i.e.*, the optical flows from the input frames to the non-existent hidden frame) are not directly accessible, and can only be estimated under certain assumptions.

Most flow-based methods [2,12,22,23,26] assume a linear motion model and approximate the intermediate flows using the two nearest given frames. A more recent work QVI [30] relaxes the linearity constraint and uses the four nearest input frames to model quadratic motions. However, *these motion models assuming constant velocities or accelerations severely simplified real-world situations*, and does not reflect the complexity of real-world motions, potentially resulting in misalignment and unnatural movements in the outputs. Recognizing the complexity of real-life object motions, which cannot be fully captured by simple polynomial models (*e.g.*, linear, quadratic, or even higher-order), we are interested in developing *a more robust and general motion estimation module*.

To this end, we propose *long-term propagation* for intermediate motion estimation in video frame interpolation, where we *understand the object motion through implicitly tracking it*. Specifically, we propagate motion features along a sequence of video frames. We adopt a recurrent neural network (RNN) to implicitly learn the motion trajectory in order to accurately estimate the intermediate optical flows. Our approach models motion without relying on prior assumptions about its characteristics, offering a more general method of motion modeling. By leveraging propagated motion information from the entire video sequence rather than just two input frames, we capture the true motion more accurately, and offer improved interpolation results.

Our work begins with a preliminary study of several existing motion models to better understand how motion models affect interpolation results, setting the stage for an in-depth exploration of long-term propagation. We then evaluate long-term propagation both qualitatively and quantitatively to show its effectiveness in improving video frame interpolation performance. We also investigate different factors that might influence the effectiveness of long-term propagation, and discuss its underlying mechanism, advantages, and limitations. Recommendations for future research are also provided.

Our contributions are summarized as follows:

- We propose *long-term propagation* for more robust video frame interpolation, moving beyond existing assumptions of linear or quadratic motions. This

approach allows for more accurate intermediate motion estimation, leading to improved interpolation outcomes.
- Through comprehensive experiments, we demonstrate the superiority of long-term propagation in accurate motion estimation and high-quality frame interpolation.
- Long-term propagation is a plug-and-play module that can be easily integrated into any flow-based video frame interpolation methods for more accurate motion estimation and better interpolation quality.

2 Related Work

2.1 Video Frame Interpolation

Video frame interpolation is a challenging problem in computer vision, whose objective is to generate frames that do not originally exist between two consecutive input frames. This field has seen diverse approaches, broadly categorized into flow-based, kernel-based, and phase-based methods.

Flow-Based Methods. Flow-based methods [2,3,7,8,12,15,16,22,23,26,30,31] first estimate the optical flows from the input frames to the non-existent frames, then synthesize preliminary frames via spatial warping, and finally fuse the preliminary frames guided by occlusion masks [3,12,31] or depth information [2]. Several techniques are shown effective in improving interpolation performance, including softmax splatting [23], contextual information [22], and cycle consistency [26]. Compared to backward warping [2,8,12,16,30], forward warping [1,22,23] is less adopted due to various challenges, such as multiple pixels being mapped to the same pixel in the non-existent frame. The performance of flow-based methods is limited by the accuracy of flow estimation to the non-existent frames. Most methods [2,12,22,23,26] assume uniform linear motion between two consecutive frames, which is far from ideal for real-world motions. Recent works relax this assumption by proposing quadratic [16,30] and cubic [7] motion models. In contrast, our *long-term propagation* does not pose polynomial assumptions on intermediate motions, and implicitly track the motion trajectories by propagating motion features through a long sequence of video frames, achieving more accurate motion estimation and interpolation quality.

Kernel-Based Methods. Kernel-based methods [6,24,25] use spatially adaptive filters to sample from neighboring pixels in input frames to synthesize a non-existent frame. Early attempts, such as DVF [17], first estimate a 3D voxel flow across space and time, followed by a trilinear sampling operation on the voxel volume. This approach implicitly assumes linear motion. More recent works like FLAVR [13] use 3D space-time convolutions to learn motion and occlusion patterns, facilitating direct interpolation.

Phase-Based Methods. Phase-based methods [19,20] consider each frame as a linear combination of wavelets, and interpolate the phase and magnitude of individual components. Since phase is modeled as a linear-time function [20], motions are implicitly assumed to be linear.

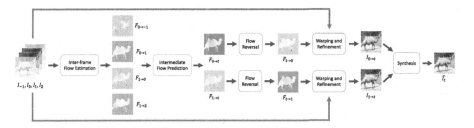

Fig. 1. Flow-Based Frame Interpolation Pipeline. Flow-based methods usually perform interpolation in four steps: (1) intermediate flow estimation, (2) flow reversal (optional), (3) warping, and (4) synthesis. Since most existing methods only use two input frames I_0 and I_1, we present I_{-1} and I_2 in gray color. In this work, to better estimate the intermediate flows, we work on: *(1) using more than two input frames to gather long-term motion information, and (2) better alternatives to the "Intermediate Flow Prediction" module.*(Images and optical flows in the figure are obtained from the original paper [30] for illustration purposes.)

2.2 Propagation

Propagation is an important aspect of video restoration because it allows leveraging information in other frames to enrich the understanding of video content. Existing video frame interpolation methods [2,3,6–8,12,16,17,19,20,22–26,30,31] typically reply on local information propagation, using only the nearest two to four frames. This approach results in a limited temporal receptive fields and confines the amount of motion information that can be utilized. In contrast, we propose to propagate information across a long temporal range to better solve the problem of video frame interpolation. Long-term propagation has been applied in other tasks like video super-resolution [10,11] and has been shown effective. In this work, we would like to initiate the study on long-term propagation in video frame interpolation.

3 Method

3.1 Preliminary Study on Motion Models

To deepen our understanding of how motion models influence video frame interpolation outcomes, we initiated a preliminary investigation into the impact of selecting various motion models on interpolation quality. Specifically, we introduce and study several ***polynomial motion models***, including *Linear, Quadratic, Quadratic Least Squares,* and *Cubic*. Detailed theoretical formulation of these model models, and their implementation details are provided in the supplementary file.

Problem Formulation. Given two consecutive input frames I_0 and I_1 in a video, the objective of video frame interpolation is to synthesize a non-existent frame I_t, where $t \in (0, 1)$. In this work, we focus on the flow-based approach. An example of flow-based approaches is shown in Fig. 1. In general, flow-based

Table 1. Interpretation of Motion Models. (1) **Order**: the polynomial degree that models the motion trajectory over time, (2) **Formula**: the mathematical representation for calculating an object's location $x(t)$ at a given time t, where a, b, c, and d are constant coefficients that define the motion's characteristics, (3) **Physical Meaning**: the nature of the motion that each model aims to capture, (4) **#Frames**: the number of timestamps t (*i.e.*, frames) needed to solve the constant coefficients in the motion formula.

Motion Model	Order	Formula	Physical Meaning	#Frames
Linear	1	$x(t) = at + b$	constant velocity	2
Quadratic	2	$x(t) = at^2 + bt + c$	constant acceleration	3
Quadratic Least Squares	2	$x(t) = at^2 + bt + c$	constant acceleration	4
Cubic	3	$x(t) = at^3 + bt^2 + ct + d$	linear-time acceleration	4
Long-Term Propagation (Ours)	N.A	N.A. (not applicable)	no constraint	any

methods [2,3,7,8,12,16,22,23,26,30,31] first estimate the optical flows, $F_{0 \rightarrow t}$ and $F_{1 \rightarrow t}$, from the input frames (*i.e.*, I_0 and I_1) to the non-existent frame (*i.e.*, the frame to be interpolated, I_t). The estimated flows are used to warp the given frames to produce two preliminary source frames $I_{0 \rightarrow t}$ and $I_{1 \rightarrow t}$, which are then fused to produce the final output $\hat{I}_t$.

In this work, we aim to explore better approaches (*i.e.*, long-term propagation) to estimate the intermediate optical flows $F_{0 \rightarrow t}$ and $F_{1 \rightarrow t}$. Particularly, we will explore different alternatives to the "Intermediate Flow Prediction" module in Fig. 1.

Polynomial Motion Models. We explore the effects of different polynomial motion models on the final interpolation results. We conduct experiments using four models: *Linear*, *Quadratic*, *Quadratic Least Squares*, and *Cubic*. Table 1 presents an overview of these four models, and Fig. 2 illustrates the process of computing the intermediate optical flows according to each motion model. The supplementary file offers detailed formulas for calculating intermediate optical flows for each model and their comprehensive analysis.

Analysis of Preliminary Study. We evaluate the performance of video frame interpolation when using four different polynomial motion models, on the Adobe240 dataset [28]. We report the PSNR and SSIM of the synthesized frames with respect to ground-truth frames in Table 2. For each data sample, given I_{-1}, I_0, I_1, I_2, we use the trained baselines to synthesize seven frames at $t = 0.125, 0.25, 0.375, 0.5, 0.625, 0.75, 0.875$ respectively, where "avg" denotes the mean of PSNR or SSIM at all seven possible t values, while "center" denotes PSNR and SSIM at $t = 0.5$ only. We provide more experimental implementations and detailed result analysis in the supplementary file. Through the preliminary study, we observe that *the choice of motion models can largely affect interpolation results*. For instance, the *Quadratic* intermediate flow estimation method (PSNR: 33.02) outperforms the *Linear* method (PSNR: 30.99) by 1.03 in terms of PSNR. This result encourages us to seek better motion models and intermediate flow estimation methods to further improve interpolation performance. In subsequent sections, we introduce our long-term propagation approach for estimating intermediate flows, and demonstrate its effectiveness to achieve more robust motion estimation and video frame interpolation.

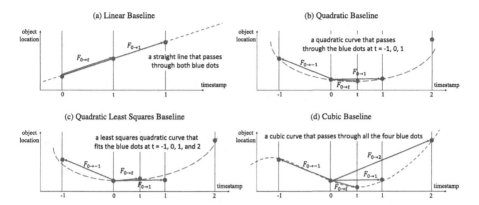

Fig. 2. Illustrations of Intermediate Flow Prediction. In each baseline, we first predict optical flows between input frames (*i.e.*, the blue arrows) using off-the-shelf optical flow estimators (*e.g.*, [27]). We then fit the object locations (*i.e.*, blue dots) using a motion trajectory curve (*i.e.*, the green dashed line) according to the motion model. After that, we compute the object location at time t (*i.e.*, the orange dot) on the motion trajectory curve. Finally, we take the object displacement from time 0 to t as the intermediate optical flow $F_{0 \to t}$ (*i.e.*, the orange arrow).

Table 2. Quantitative Comparisons of Polynomial Motion Models (Adobe240 Dataset [28]). We conduct a preliminary study on the video frame interpolation performance with different polynomial motion models. We find that the choice of motion models can largely affect interpolation results.

Baseline	PSNR(avg) ↑	PSNR(center) ↑	SSIM(avg) ↑	SSIM(center) ↑
Linear	30.99	29.46	0.9329	0.9113
Quadratic	**33.02**	**32.32**	**0.9558**	**0.9504**
Quadratic Least Squares	32.76	31.97	0.9537	0.9473
Cubic	32.70	31.84	0.9537	0.9467

3.2 Long-Term Propagation

Recurrent Network Overview. Our goal is to obtain an accurate estimation of the intermediate optical flows. We use a recurrent network, as illustrated in

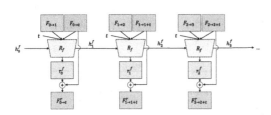
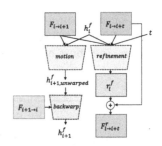

Fig. 3. Long-Term Propagation. We propose long-term propagation to achieve robust interpolation. Specifically, we propagate long-term motion information using a recurrent network. At each recurrent step, the network R_f takes the inter-frame flow (e.g., $F_{0\to 1}$), the preliminary intermediate flow (e.g., $F_{0\to t}$), the hidden state (e.g., h_0^f), and the timestamp t as inputs, and predicts a residue r that refines the preliminary intermediate flow. The hidden state propagates the motion information through the video sequence.

Fig. 4. Recurrent Step. Each recurrent step consists of three modules: *motion*, *refinement*, and *backwarp*. (1) The *motion* module takes motion information as input, and update the hidden state to propagate long-term motion information forward. (2) The *refinement* module predicts the residue that refines the intermediate flow. (3) The *backwarp* module aligns the updated hidden state to the next temporal interval for future information propagation.

Fig. 3, to propagate motion information through a sequence of frames, and use this knowledge to guide the intermediate flow estimation step shown in Fig. 1.

Given a sequence of input frames $I = [I_0, I_1, I_2, ..., I_n]$, we propagate the motion information forward, from I_0 to I_n. Specifically, the operation of the i-th recurrent step is as follows:

$$r_i^f, h_{i+1}^f = R_f(h_i^f, t, F_{i\to i+1}, F_{i\to i+t}), \tag{1}$$

where R_f is the forward propagation branch, h_i^f is the hidden state that carries the motion information forward, $t \in (0, 1)$ is the timestamp of the unknown frame I_{i+t} to be synthesized, $F_{i\to i+1}$ is the inter-frame optical flow between the input frames I_i and I_{i+1}, which is predicted by an off-the-shelf flow estimation model, and $F_{i\to i+t}$ is the preliminary intermediate flow to be refined by long-term propagation. Each recurrent step outputs the residue r_i^f, and the updated hidden state h_{i+1}^f. The residue r_i^f is used to refine the preliminary intermediate flow $F_{i\to i+t}$:

$$F_{i\to i+t}^r = F_{i\to i+t} + r_i^f, \tag{2}$$

where $F_{i\to i+t}^r$ is the refined intermediate flow. We choose to refine a preliminary intermediate flow instead of directly predicting the refined intermediate flow because of two reasons: (1) A coarse intermediate flow obtained from existing baselines such as *Linear* and *Quadratic* provides rough motion information of the object, serving as a good starting point, and (2) predicting only the residue to coarse motions eases training and accelerates convergence.

Bidirectional Propagation. To warp both source frames I_i and I_{i+1} to the target timestamp t, we need two intermediate optical flows $F^r_{i \to i+t}$ and $F^r_{i+1 \to i+t}$. Therefore, other than the forward propagation branch which estimates the forward intermediate flow $F^r_{i \to i+t}$, we also need a backward branch to estimate the backward intermediate flow $F^r_{i+1 \to i+t}$ similarly:

$$r_i^b, h_i^b = R_f(h_{i+1}^b, t, F_{i+1 \to i}, F_{i+1 \to i+t}), \tag{3}$$

$$F^r_{i+1 \to i+t} = F_{i+1 \to i+t} + r_i^b. \tag{4}$$

Both $F^r_{i \to i+t}$ and $F^r_{i+1 \to i+t}$ will be used in the subsequent stages of the flow-based framework (refer to Fig. 1). Both forward and backward branches use the same recurrent step R_f.

Recurrent Step. For clarity, we detail the recurrent steps using the forward branch as an example, and the backward branch operates in a similar fashion.

As illustrated in Fig. 4, the recurrent step consists of three key modules: *motion*, *refinement*, and *backwarp*:

$$h^f_{i+1,unwarped} = motion(h_i^f, F_{i \to i+1}), \tag{5}$$

$$r_i^f = refinement(h_i^f, t, F_{i \to i+1}, F_{i \to i+t}), \tag{6}$$

$$h^f_{i+1} = backwarp(h^f_{i+1,unwarped}, F_{i+1 \to i}). \tag{7}$$

The *Motion* Module. As shown in Eq. 5, the *motion* module aims to update the hidden state h_i^f to $h^f_{i+1,unwarped}$ in order to pass the motion information along. It does not take the timestamp t or the intermediate flow $F_{i \to i+t}$ as input, ensuring that the hidden state's update is not influenced by specific timestamps.
The *Refinement* Module. The *refinement* module (Eq. 6) predicts a residue r_i^f to refine the preliminary intermediate flow $F_{i \to i+t}$. Since the value of $F^r_{i \to i+t}$ varies with the timestamp t, the residue r_i^f should also depend on t. Therefore, the *refinement* module takes both the timestamp t and preliminary intermediate flow $F_{i \to i+t}$ as inputs. To guide intermediate flow refinement, the hidden state h_i^f provides long-term motion information, while the inter-frame flow $F_{i \to i+1}$ provides local motion information.
The *Backwarp* Module. Since the hidden state $h^f_{i+1,unwarped}$ is aligned to I_i rather than I_{i+1}, we need to warp it to align with I_{i+1} for the next recurrent step. As shown in Eq. 7, we use $F_{i+1 \to i}$ to *backwarp* (*i.e.*, the backward spatial warping operation) the hidden state $h^f_{i+1,unwarped}$ to obtain h^f_{i+1}, which is aligned to I_{i+1}.

3.3 Training Scheme

We train the network in three stages:

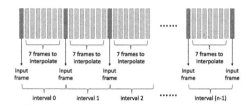

Fig. 5. Interval Definition. The blue frames are the given input frames. The yellow frames are the ones to be synthesized. Each segment has n intervals, where in each interval we synthesize seven in-between frames. When $n = 10$, the total number of frames per segment (including input frames) is 81.

1. **Pretrain**: We train a preliminary model (*e.g.*, *Quadratic+PWCNet*, which will be discussed in detail in Sec. 4), which estimates intermediate flows using one of the motion assumptions discussed in Sec. 3.1.
2. **Plug in**: We integrate the long-term propagation module (*i.e.*, the recurrent network in Sec. 3.2) into the preliminary model for intermediate flow refinement. We keep the weights of the preliminary model frozen, and only train the long-term propagation module.
3. **Finetune**: We perform finetuning on the entire network to optimize its performance.

4 Experiments

4.1 Implementation Details

Dataset. Due to the long training time[1] experienced during the preliminary study, we use a subset of the REDS dataset [21] for faster training and evaluation in our subsequent experiments. We use 100 video clips (*i.e.*, clip 000 to 099) as the training dataset, with spatial random crops for data augmentation. We use four video clips (*i.e.*, clip 240, 241, 246, and 257) as the evaluation dataset. We follow BasicVSR [5]'s choice on the four evaluation clips. These four clips have diverse video content, serving as a good benchmark for model performance.

Interpolation Intervals. For each interpolation interval (refer to Fig. 5 for its definition), we synthesize seven in-between frames (*i.e.*, $t \in \{0.125, 0.25, 0.375, 0.5, 0.625, 0.75, 0.875\}$) during both training and evaluation. We use ten intervals as one segment, resulting in 81 frames per segment. Information only propagates within a segment, and does not propagate across two different segments. During training, we use all possible 81-frame segments in each 500-frame video clip, and randomly flip the segments temporally for data augmentation. At test time, we only use the first 81-frame segment in each video clip for evaluation.

---

[1] For instance, training the *Quadratic Least Squares* model takes 17 GPU days on 32-GB V100 GPUs.

Hyperparameters. The hyperparameters for the three training stages are detailed below:

1. **Pretrain**: We adopt the Adam optimizer [14], initialize the learning rate as 10^{-4} and further decrease it by a factor of 0.1 at the end of the 50th and 75th epoch. We freeze the weights of the pretrained optical flow estimation model for 100 epochs before finetuning.
2. **Plug in**: We choose 5×10^{-5} as the initial learning rate, and use the Cosine Annealing scheme [18] for learning rate decay. For both the *motion* and *refinement* modules, we use five residual blocks.
3. **Finetune**: We choose 3×10^{-5} as the initial learning rate, and use the Cosine Annealing scheme [18] for learning rate decay. We apply a learning rate multiplier of 0.25 to the off-shelf optical flow estimation model.

The design choices (*e.g.*., hyperparameters and optimizer schemes) are carefully determined through extensive comparative experiments. We provide these experimental set-up and results in the supplementary file.

4.2 Comparison Methods

For the first stage (refer to Sec. 3.3) of our training scheme, we experiment on different preliminary models to assess (1) the ability of long-term propagation to enhance interpolation results across different preliminary models, and (2) the impact of these models' characteristics on the effectiveness of long-term propagation. We adopt (1) *Quadratic+PWCNet*, (2) *Quadratic+RAFT*, and (3) *Linear+PWCNet* as the preliminary models in the subsequent experiments. The naming convention used is *"Motion Model + Optical Flow Estimator"*.

Quadratic+PWCNet. We follow the QVI [30] baseline, employing the *Quadratic* motion assumption, and using the pretrained PWCNet [27] as the optical flow estimator.

Quadratic+RAFT. RAFT [29] is one of the state-of-the-art optical flow estimation methods. We use *Quadratic+RAFT* alongside *Quadratic+PWCNet* to evaluate the influence of different pretrained flow estimators on long-term propagation's performance.

Linear+PWCNet. Since *Linear* and *Quadratic* are two commonly adopted motion models, we use *Linear+PWCNet* alongside *Quadratic+PWCNet* to analyze how different preliminary motion models can affect the PSNR improvement introduced by long-term propagation.

4.3 Quantitative Comparisons

In Table 3, we present a comparison of PSNR for preliminary models before and after applying long-term propagation. "PSNR w/o" refers to results obtained using models from the first stage (refer to Sec. 3.3), whereas "PSNR w" represents results from the third stage, incorporating long-term propagation. We see that long-term propagation introduces PSNR improvement to all the three preliminary models. This indicates that *long-term is effective in improving interpolation accuracy*. Furthermore, a relatively less robust preliminary model tends to benefit more from long-term propagation. For instance, *Linear+PWCNet*, which adopts a simpler motion model compared to *Quadratic+PWCNet*, experiences a larger PSNR increase. Similarly, *Quadratic+PWCNet* sees a greater PSNR improvement over *Quadratic+RAFT*. This pattern will be explored more comprehensively in Sect. 4.5.

4.4 Qualitative Comparisons

The qualitative results on the REDS dataset [21] evaluation clips highlight the impact of long-term propagation on video frame interpolation.

(a) **Qualitative Comparisons (Alignment)**. The green lines are reference lines that show the discrepancy in object positioning relative to the ground truth when long-term propagation is absent. (1) "w/o": Before applying long-term propagation, objects are not well aligned with the ground-truth. (2) "w": The incorporation of long-term propagation leads to a noticeable improvement in object alignment, suggesting that our method enhances the accuracy of intermediate flow estimation, which is the source for maintaining alignment with the ground truth.

(b) **Qualitative Comparisons (Blur)**. The absence (*i.e.*, "w/o") of long-term propagation results in blurred synthesis outcomes, while applying it (*i.e.*, "w") yields clearer and sharper images. The cause of bluriness is that the estimated intermediate flows from two neighboring frames fail to point to the same pixel locations, and fusing the warpings results in ghosting residues which appear to be bluriness. Thus, the enhanced clarity and sharpness with long-term propagation indicate significant improvements in intermediate flow estimation accuracy.

Fig. 6. Qualitative Comparisons. We show the comparisons of before and after applying long-term propagation.

In Fig. 6a, the green lines are reference lines that show the discrepancy in object positions relative to the ground truth when long-term propagation is absent. We observe that before applying long-term propagation, objects are not well aligned with the ground truth. The incorporation of long-term propagation leads to a noticeable improvement in object alignment, suggesting that our method enhances the accuracy of intermediate flow estimation, which is the source for maintaining alignment with the ground truth.

Figure 6b shows that the absence of long-term propagation results in blurry outputs, while applying it yields clearer and sharper images. We conjecture that the two preliminary source frames (*i.e.*, $I_{0\to t}$ and $I_{1\to t}$) warped from the two input frames (*i.e.*, I_0 and I_1) respectively are not perfectly aligned, thus fusing them results in blurry outputs. The misalignment of preliminary source frames is due to inaccurate intermediate flow estimation. That is, $F_{i\to i+t}$ and $F_{i+1\to i+t}$ fail to accurately point to the same pixel locations.

*Please refer to the **demo video** in our supplementary materials for more qualitative results and comparisons.*

4.5 Improved Intermediate Flow Estimation

We conduct analytical experiments to understand the mechanism behind the improvements in interpolation results brought by long-term propagation. Specifically, we hope to answer these questions: (1) *How does long-term propagation improve PSNR? That is, what is the source of PSNR improvement?* (2) *Does long-term propagation offer a more accurate estimation of the motions from source frames to the intermediate frames?*

Table 3. PSNR with and without Long-Term Propagation. We report the PSNR of each preliminary model, without (denoted as "PSNR w/o") and with (denoted as "PSNR w") long-term propagation respectively.

Preliminary Model	PSNR w/o	PSNR w	PSNR Increase ↑
Linear+PWCNet	26.90	**28.03**	1.14
Quadratic+PWCNet	28.33	**28.80**	0.47
Quadratic+RAFT	28.97	**29.12**	0.16

To address these questions, we study the relationship between long-term propagation and the accuracy of intermediate flow estimation. We employ End Point Error (EPE) as the metric for flow estimation accuracy. EPE is obtained by first computing the Euclidean distance between the estimated flow and the ground-truth flow at each pixel location, and then taking the average of all Euclidean distances in an image. A lower EPE indicates a more accurate flow estimation. For generating the ground-truth optical flows, we use RAFT [29], one of the state-of-the-art optical flow estimators.

Table 4 shows the EPE of the intermediate flows predicted by the interpolation networks, before and after applying long-term propagation. Specifically, we compute the EPE for both $F^r_{i\to i+t}$ and $F^r_{i+1\to i+t}$, and then average these two values to report "avg EPE" for each model. Notably, across all models, incorporating long-term propagation results in a reduction of EPE, compared to the counterpart before adding long-term propagation. This reduction in EPE demonstrates the effectiveness of long-term propagation in producing more accurate intermediate flows. Consequently, *the observed PSNR improvements across all models further support that the enhanced performance can be attributed to* more accurate intermediate flow estimations facilitated by long-term propagation.

Table 5 illustrates how the accuracy of intermediate flow predictions—measured by End Point Error (EPE) without the use of long-term propagation—affects the degree of PSNR improvement when long-term propagation is applied. We observe that a higher initial EPE, reflecting less accurate flow predictions, correlates with a more substantial PSNR increase introduced by long-term propagation. This suggests that a larger initial EPE signifies greater potential for improvement in flow estimation accuracy. Consequently, this pattern supports the hypothesis that *long-term propagation enhances video frame interpolation results by enabling more accurate intermediate flow estimations*.

4.6 Further Analysis

We further study how different factors, namely sequence length (*i.e.*, temporal receptive field) and motion size, affect long-term propagation's effectiveness. Detailed analysis are provided in the supplementary file.

Table 4. Accuracy of Intermediate Flow Estimation. For each preliminary model before and after applying long-term propagation, we present the EPE of $F^r_{i \to i+t}$, $F^r_{i+1 \to i+t}$, and the average of these two. Lower "avg EPE" indicates more accurate intermediate flow estimation. We find that by adding long-term propagation to a preliminary model, the PSNR increases, and the "avg EPE" decreases, which implies that long-term propagation improves the accuracy of intermediate flow estimation.

Models	PSNR ↑	$F^r_{i \to i+t}$ EPE ↓	$F^r_{i+1 \to i+t}$ EPE ↓	avg EPE ↓
Linear+PWCNet w/o	26.90	1.52	1.51	1.51
Linear+PWCNet w	**28.03**	1.40	1.58	**1.49**
Quadratic+PWCNet w/o	28.33	1.39	1.53	1.46
Quadratic+PWCNet w	**28.80**	1.37	1.38	**1.37**
Quadratic+RAFT w/o	28.97	1.21	1.38	1.29
Quadratic+RAFT w	**29.12**	1.16	1.33	**1.25**

Sequence Length. We studied how sequence length, which defines the temporal receptive field, affects interpolation results using the *Quadratic+PWCNet* model with long-term propagation. Figure 8 shows that PSNR increases with sequence length up to 19 intervals, beyond which it plateaus, suggesting that longer sequences improve motion estimation up to a point. However, sequence lengths beyond 20 intervals show no further PSNR improvement, indicating that additional information becomes irrelevant and increases GPU memory demand and inference time. The optimal sequence length for our scenario is around 19 intervals.

Motion Size. We evaluated the impact of motion size on the effectiveness of long-term propagation by quantifying motion size using the optical flow length

Table 5. Initial EPE vs PSNR Increase. For each preliminary model, we show the "PSNR increase" brought by long-term propagation, and End Point Error (EPE) of intermediate optical flows as predicted by the network prior to the application of long-term propagation. We observe that a larger initial EPE is linked with a larger PSNR increase, indicating long-term propagation's ability to correct the initially inaccurate intermediate flow estimations, thereby directly contributing to improved interpolation quality.

Models	PSNR w/o	PSNR w	PSNR increase ↑	EPE w/o ↓
Linear+PWCNet	26.90	28.03	**1.14**	**1.51**
Quadratic+PWCNet	28.33	28.80	**0.47**	**1.46**
Quadratic+RAFT	28.97	29.12	**0.16**	**1.29**

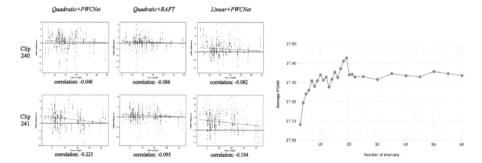

Fig. 7. Long-Term Propagation Effectiveness vs Motion Size. In each scatter plot, we show the PSNR increase brought by long-term propagation (y-axis) against motion size (x-axis). Different scatter plot represents a different preliminary model and evaluation video clip. We observe that long-term propagation is robust across different motion sizes.

Fig. 8. Long-Term Propagation vs Sequence Length. Interpolation performance (*i.e.*, PSNR) increases with sequence lengths from one to 19 intervals, beyond which the increase plateaus.

between consecutive frames. In Fig. 7, we plotted the PSNR increase due to long-term propagation against flow length for three models. The scatter plots show little correlation, indicating that long-term propagation remains robust across different motion sizes. Interestingly, the most significant PSNR improvements occur in the medium to small motion range, where long-term propagation is particularly effective.

5 Conclusion

In this work, we propose *long-term propagation* for robust video frame interpolation. Unlike existing methods that rely on strong assumptions on between-frame motions, we propagate the motion information through the video sequence for general and robust motion estimation. Our method consistently yields more accurate intermediate motion estimations, thereby improving video frame interpolation quality. As a plug-and-play module, long-term propagation can be seamlessly integrated into existing video frame interpolation methods for performance gain. Beyond the introduction of long-term propagation, we also conducted extensive experimental analysis on the influence of motion models on video frame interpolation, explored the underlying mechanisms of long-term propagation's effectiveness. We believe our study on long-term propagation is beneficial for future research in video frame interpolation, motion estimation, and the broader domain of video restoration.

Acknowledgments. This study is supported by the Ministry of Education, Singapore, under its MOE AcRF Tier 2 (MOET2EP20221-0012), NTU NAP, and under the RIE2020 Industry Alignment Fund - Industry Collaboration Projects (IAF-ICP) Funding Initiative, as well as cash and in-kind contribution from the industry partner(s).

References

1. Baker, S., Roth, S., Scharstein, D., Black, M.J., Lewis, J., Szeliski, R.: A database and evaluation methodology for optical flow. In: ICCV (2007)
2. Bao, W., et al.: Depth-aware video frame interpolation. In: CVPR (2019)
3. Bao, W., Lai, W.S., Zhang, X., Gao, Z., Yang, M.H.: MEMC-Net: motion estimation and motion compensation driven neural network for video interpolation and enhancement. IEEE TPAMI (2018)
4. Bégaint, J., Galpin, F., Guillotel, P., Guillemot, C.: Deep frame interpolation for video compression. In: 2019 Data Compression Conference (DCC) (2019)
5. Chan, K.C., Wang, X., Yu, K., Dong, C., Loy, C.C.: BasicVSR: the search for essential components in video super-resolution and beyond. In: Proceedings of the IEEE Conference on Computer Vision and Pattern Recognition (2021)
6. Cheng, X., Chen, Z.: Video frame interpolation via deformable separable convolution. In: AAAI (2020)
7. Chi, Z., Mohammadi Nasiri, R., Liu, Z., Lu, J., Tang, J., Plataniotis, K.N.: All at once: temporally adaptive multi-frame interpolation with advanced motion modeling. In: ECCV (2020)
8. Huang, Z., Zhang, T., Heng, W., Shi, B., Zhou, S.: RIFE: Real-time intermediate flow estimation for video frame interpolation. arXiv preprint arXiv:2011.06294 (2020)
9. Huawei: 7680 fps super slow-motion videos — Make time stand still (2020). https://consumer.huawei.com/en/community/details/7680-FPS-Super-Slow-Motion-Videos-Make-Time-Stand-Still/topicId_114694/
10. Isobe, T., Jia, X., Gu, S., Li, S., Wang, S., Tian, Q.: Video super-resolution with recurrent structure-detail network. In: ECCV (2020)
11. Isobe, T., Zhu, F., Wang, S.: Revisiting temporal modeling for video super-resolution. In: BMVC (2020)
12. Jiang, H., Sun, D., Jampani, V., Yang, M.H., Learned-Miller, E., Kautz, J.: Super SloMo: high quality estimation of multiple intermediate frames for video interpolation. In: CVPR (2018)
13. Kalluri, T., Pathak, D., Chandraker, M., Tran, D.: FLAVR: flow-agnostic video representations for fast frame interpolation. arXiv preprint arXiv:2012.08512 (2021)

14. Kingma, D., Ba, J.: Adam: A method for stochastic optimization. In: ICLR (2015)
15. Li, Z., Zhu, Z.L., Han, L.H., Hou, Q., Guo, C.L., Cheng, M.M.: AMT: all-pairs multi-field transforms for efficient frame interpolation. In: CVPR (2023)
16. Liu, Y., Xie, L., Li, S., Sun, W., Qiao, Y., Dong, C.: Enhanced quadratic video interpolation. In: ECCV Workshop (2020)
17. Liu, Z., Yeh, R.A., Tang, X., Liu, Y., Agarwala, A.: Video frame synthesis using deep voxel flow. In: ICCV (2017)
18. Loshchilov, I., Hutter, F.: SGDR: stochastic gradient descent with warm restarts. arXiv preprint arXiv:1608.03983 (2016)
19. Meyer, S., Djelouah, A., McWilliams, B., Sorkine-Hornung, A., Gross, M., Schroers, C.: Phasenet for video frame interpolation. In: CVPR (2018)
20. Meyer, S., Wang, O., Zimmer, H., Grosse, M., Sorkine-Hornung, A.: Phase-based frame interpolation for video. In: CVPR (2015)
21. Nah, S., et al.: NTIRE 2019 challenge on video deblurring and super-resolution: dataset and study. In: CVPRW (2019)
22. Niklaus, S., Liu, F.: Context-aware synthesis for video frame interpolation. In: CVPR (2018)
23. Niklaus, S., Liu, F.: Softmax splatting for video frame interpolation. In: CVPR (2020)
24. Niklaus, S., Mai, L., Liu, F.: Video frame interpolation via adaptive convolution. In: CVPR (2017)
25. Niklaus, S., Mai, L., Liu, F.: Video frame interpolation via adaptive separable convolution. In: ICCV (2017)
26. Reda, F., et al.: Unsupervised video interpolation using cycle consistency. In: ICCV (2019)
27. Ren, Z., Gallo, O., Sun, D., Yang, M.H., Sudderth, E.B., Kautz, J.: A fusion approach for multi-frame optical flow estimation. In: WACV (2019)
28. Su, S., Delbracio, M., Wang, J., Sapiro, G., Heidrich, W., Wang, O.: Deep video deblurring for hand-held cameras. In: CVPR (2017)
29. Teed, Z., Deng, J.: RAFT: Recurrent all-pairs field transforms for optical flow. In: ECCV (2020)
30. Xu, X., Li, S., Sun, W., Yin, Q., Yang, M.H.: Quadratic video interpolation. In: NeurIPS (2019)
31. Xue, T., Chen, B., Wu, J., Wei, D., Freeman, W.T.: Video enhancement with task-oriented flow. In: IJCV (2019)

Enhanced Survival Prediction in Head and Neck Cancer Using Convolutional Block Attention and Multimodal Data Fusion

Aiman Farooq[1](✉), Utkarsh Sharma[2], and Deepak Mishra[1]

[1] Indian Institute of Technology Jodhpur, Jodhpur, Rajasthan 342030, India
farooq.1@iitj.ac.in
[2] Indian Institute of Science Education and Research Bhopal,
Bhopal, Madhya Pradesh 462066, India

Abstract. Accurate survival prediction in head and neck cancer (HNC) is essential for guiding clinical decision-making and optimizing treatment strategies. Traditional models, such as Cox proportional hazards, have been widely used but are limited in their ability to handle complex multimodal data. This paper proposes a deep learning-based approach leveraging CT and PET imaging modalities to predict survival outcomes in HNC patients. Our method integrates feature extraction with a Convolutional Block Attention Module (CBAM) and a multi-modal data fusion layer that combines imaging data to generate a compact feature representation. The final prediction is achieved through a fully parametric discrete-time survival model, allowing for flexible hazard functions that overcome the limitations of traditional survival models. We evaluated our approach using the HECKTOR and HEAD-NECK-RADIOMICS-HN1 datasets, demonstrating its superior performance compared to conventional statistical and machine learning models. The results indicate that our deep learning model significantly improves survival prediction accuracy, offering a robust tool for personalized treatment planning in HNC.

Keywords: Survival Prediction · Cancer · C-index

1 Introduction

Survival prediction is critical in the management and care of various diseases, particularly in the realm of cancer diagnosis. For patients diagnosed with cancers such as lung, breast, colorectal, or head and neck cancer, understanding survival probabilities helps guide crucial clinical decisions, from treatment planning to resource allocation. Head and neck cancer (HNC) ranks as the seventh most common cancer worldwide, with over 660,000 new cases and 325,000 deaths

---

A. Farooq and U. Sharma—Equal contribution.

reported annually [30], and this number is expected to increase by 30% annually by the year 2030 [7]. Accurate survival estimates for HNC allow oncologists to tailor treatment strategies based on a patient's prognosis, determining whether aggressive interventions like surgery, chemotherapy, or radiation are appropriate or if a palliative care approach would provide the best quality of life. These predictions are also essential for counseling patients and families, setting realistic expectations, and aligning care goals with the patient's values and preferences. However, the task of survival prediction is complex and involves multiple data modalities, including clinical, imaging, and sometimes even genomic modeling. However, the major hindrance in this domain is the availability of data. Most datasets generally contain imaging modalities with clinical and other relevant data lacking for most patients. Achieving accurate and reliable predictions from imaging modalities alone is a significant and ongoing problem in this domain.

Deep learning has revolutionized the medical domain, offering transformative solutions for some of the most complex and critical healthcare challenges. Survival prediction for HNC typically involves integrating data from various sources, including CT scans, PET scans, and clinical information. Features extracted from CT imaging are highly relevant in predicting survival outcomes in HNC [8,18,20]. PET imaging has also been extensively studied for its role in prognosis, with metabolic activity and other biomarkers from PET scans linked to survival predictions [14,33]. Additionally, molecular data have proven valuable, providing insights into the biological mechanisms driving cancer progression and helping to enhance the accuracy of survival models [4]. Traditionally, survival prediction models such as Cox proportional hazards models [38], random survival forests [23], and support vector machines [37] have relied on handcrafted features derived from imaging or clinical data. However, with the increasing application of deep learning, advanced models based on convolutional neural networks (CNNs) [25] have been developed, surpassing traditional approaches. For instance, Hu et al. [13] employed CNNs to extract imaging features from CT scans and integrated them with clinical data for survival prediction in HNC patients. Moreover, models like DeepSurv [5] have demonstrated strong performance by directly learning from multi-modal data sources. Recognizing the limitations of relying solely on a single modality, several studies have proposed multi-modal models to capture complementary information from diverse data sources. Approaches such as those by Wang et al. [34] and Jin et al. [16] have combined imaging and clinical for more accurate survival prediction in HNC. Additionally, models like DeepMM [29] have shown how multi-modal fusion can significantly improve performance by leveraging the strengths of each modality.

In exploring survival prediction in HNC, we focus on leveraging deep learning architectures specifically designed to integrate information from CT and PET imaging modalities. Our study's primary goal is to harness these imaging techniques' strengths to develop robust survival prediction models. CT scans provide crucial structural details, while PET scans offer metabolic insights—together, they capture complementary features critical for accurate prognosis. We employ advanced models like convolutional neural networks (CNNs) to extract and combine relevant features from these modalities, allowing for a comprehensive understanding of the tumor's behavior and progression. Although adding clinical and genomic data has improved survival predictions in prior studies, such data is often unavailable or incomplete for a significant portion of patients. Clinical and genomic information is typically available for only 10-20% of HNC cases, limiting the practicality of relying on these datasets. We optimize predictions using CT and PET imaging along with the available clinical data.

2 Related Work

Survival prediction in HNC is a critical area of research that has been significantly advanced by using various statistical, machine learning (ML), and deep learning (DL) models. Traditionally, statistical models like the Cox proportional hazards model have been the backbone of survival analysis in clinical research. This model is particularly valued for its ability to assess the impact of various covariates on survival time. [27] employed the Cox model to evaluate the survival outcomes of HNC patients based on clinical and demographic factors. Their study achieved a concordance index of 0.71, highlighting the model's utility in providing reasonably accurate survival predictions. Another widely used approach is the Kaplan-Meier estimator, which provides a non-parametric statistic for estimating the survival function from lifetime data. [19] utilized this method to analyze the survival probabilities of different HNC treatment groups, illustrating the method's effectiveness in comparative survival analysis, especially in stratifying patient groups.

With the growing complexity of patient data and the need for more accurate predictions, machine learning models have increasingly been employed in HNC survival analysis. Random forests (RF) [3] and support vector machines (SVM) [6] are the most commonly used ML techniques. RF models, known for their robustness in handling large datasets and reducing overfitting through ensemble learning, have been effectively used for survival prediction. For instance, [21] applied RF to predict survival outcomes in HNC patients, achieving an area under the curve (AUC) of 0.79, underscoring the model's ability to handle complex datasets with numerous variables. Similarly, [32] used SVMs to predict patient survival based on gene expression profiles, achieving a classification accuracy of 83%, demonstrating the SVM's effectiveness in high-dimensional data settings. Furthermore, artificial neural networks (ANNs) have been utilized to capture non-linear relationships between covariates and survival time. [9] reported a significant improvement in predictive accuracy using ANNs compared to traditional Cox models, with their model achieving a concordance index of 0.76.

The advent of deep learning has introduced even more sophisticated approaches to survival prediction in HNC. Convolutional neural networks (CNNs), traditionally used for image processing, have been adapted to process complex multi-dimensional data in survival analysis. For example, [22] used CNNs to integrate imaging data with clinical variables, achieving a concordance index of 0.80 in predicting overall survival, a marked improvement over traditional ML models. [26] employed RNNs to model time-to-event data in HNC, finding that their approach could predict survival with a mean absolute error of 3.1 months, thereby offering a promising tool for dynamic survival prediction as more patient data becomes available over time.

In a study by [28], a deep learning model combining CNNs with radiomic and genomic data was used to predict survival outcomes in HNC patients, achieving an AUC of 0.85. Integrating diverse data types represents a significant step in pursuing personalized medicine, enabling more tailored and accurate predictions for individual patients. Vale et al. [31] proposed MultiSurv to predict long-term cancer survival by integrating imaging (whole slide imaging) and clinical data for a pan-cancer approach. Katzman et al. [17] proposed a deep learning-based survival model, DeepSurv, for breast cancer. It extends the Cox proportional hazards model and utilizes neural networks to model complex, nonlinear relationships in survival data.

Overall, applying statistical, machine learning, and deep learning models in survival prediction for head and neck cancer has evolved significantly. Traditional statistical models continue to provide foundational insights while incorporating ML and DL techniques, which have markedly improved the accuracy and applicability of survival predictions. These advancements are paving the way for more personalized and effective treatment strategies, ultimately aiming to improve patient outcomes in this challenging area of oncology.

3 Methodology

3.1 Dataset

We have used two main head and neck cancer datasets for this study, which include the Head & neCK TumOR segmentation and outcome prediction (HECKTOR) data set [24] and the HEAD-NECK-RADIOMICS-HN1 collection [1,35]. The HECKTOR dataset is a multi-modal, multi-center collection specifically curated for head and neck cancer research. It comprises PET and CT scans from 488 patients and corresponding tumor segmentation masks. This dataset, sourced from seven different centers, provides a rich resource for developing and validating segmentation algorithms and outcome prediction models. The dataset also includes detailed clinical data, such as Recurrence-Free Survival (RFS) information, encompassing time-to-event data and censoring status. The HEAD-NECK-RADIOMICS-HN1 collection is sourced from The Cancer Imaging Archive (TCIA). It includes imaging and clinical data from 137 patients suffering from head-and-neck squamous cell carcinoma (HNSCC) patients treated at MAASTRO Clinic, The Netherlands. CT scans, manual delineations, and clinical and survival data are available for these patients. PET images in the dataset have pixel sizes ranging from 1.95 mm to 5.47 mm, slice thicknesses from 2.02 mm to 5 mm, and matrix sizes between 128×128 and 256×256 pixels. CT images feature pixel sizes ranging from 0.68 mm to 1.95 mm, slice thicknesses from 1.5 mm to 5 mm, and a matrix size of 512×512 pixels.

3.2 Network Architecture

The overall architecture of the proposed model, as shown in Fig. 1, is structured around four core modules: a feature representation module, an attention module, a multi-modal data fusion layer, and an output submodel. These components integrate and process data from multiple modalities—CT images, PET images, and clinical data—to predict conditional survival probabilities for discrete follow-up time intervals. The feature representation module is designed with dedicated submodels for each data modality. The submodels are based on a ResNet-50 3D architecture [11] for the CT and PET images. This deep convolutional neural network is optimized explicitly for handling volumetric data, such as medical images. The architecture of ResNet-50 3D comprises multiple layers of 3D convolutional filters, along with batch normalization, ReLU activation functions, and max pooling layers. These layers capture spatial hierarchies and intricate local textures within the CT and PET scans. The output from this stage is a collection of detailed feature maps for each imaging type, capturing important spatial details needed for further analysis. In parallel, the clinical data is passed through a fully connected (FC) layer, which transforms the input features into a dense representation. This dense representation captures the complex relationships between the clinical variables, facilitating their integration with the image-based features extracted from the CT and PET data.

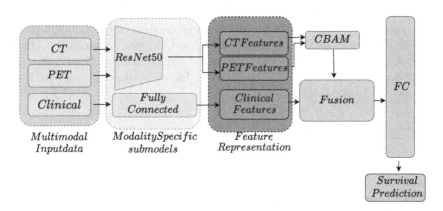

Fig. 1. Proposed multi-modal deep learning model for survival prediction, integrating CT, PET, and clinical data with feature fusion and CBAM for enhanced prediction accuracy. Final survival predictions are made using a fully connected layer.

After feature extraction, a Convolutional Block Attention Module (CBAM) [36] is applied to the feature maps. CBAM enhances the discriminative power of the extracted features through two sequential stages: channel attention and spatial attention. The channel attention mechanism emphasizes the most relevant feature maps, answering the question of "what" is essential for the prediction task. Following this, spatial attention refines the focus further by determining "where" the network should concentrate its attention on the image. By applying this dual attention mechanism, CBAM selectively enhances the most informative aspects of the CT and PET data, leading to more refined and relevant feature maps.

Following feature extraction, the outputs from these submodels are integrated within the multi-modal data fusion layer. This layer reduces the set of feature representation vectors into a single, compact fusion vector, which serves as the input to the subsequent output submodel. Formally, let $Z = [z_1, \ldots, z_n]$ denote the matrix composed of the feature representation vectors from the different modalities, where each vector $z_l \in \mathbb{R}^m$ represents the features extracted from the lth modality. The fusion vector $c \in \mathbb{R}^m$ is then computed as the row-wise maximum of Z, with the kth element of c given by:

$$c_k = \max_{1 \leq l \leq n} z_{k,l}, \quad k = 1, \ldots, m. \tag{1}$$

This approach results in a fused feature representation corresponding to the maxima across the different data modalities.

The final module of the proposed model architecture is the prediction submodel, which is designed as a fully parametric discrete-time survival model parameterized by a deep neural network. The model is trained using stochastic gradient descent (SGD), with the assumption that follow-up time is discrete, represented as a set of time intervals $\{t_1, t_2, \ldots, t_p\}$, where each t_j marks the upper limit of the jth interval.

For each study subject, the hazard function h_j defines the probability that the event of interest occurs within interval j, conditional on survival up to the beginning of the interval. The model is trained by minimizing the negative log-likelihood of the observed data. Specifically, for each time interval j, the log-likelihood is computed as:

$$\sum_{i=1}^{d_j} d_j \log(h_j^{(i)}) + \sum_{i=d_j+1}^{r_j} \log(1 - h_j^{(i)}), \tag{2}$$

where $h_j^{(i)}$ is the hazard probability for the ith subject during interval j. The total loss is the sum of the negative log-likelihoods across all time intervals. The first term in the log-likelihood encourages the model to increase the hazard rate h_j for subjects who experience the event within interval j. In contrast, the second term encourages the model to increase the predicted survival probability $1 - h_j$ for subjects who survive beyond the interval, including those censored. In this implementation, the output layer of the prediction submodel consists of p units, each corresponding to one-time interval. A sigmoid activation function converts the output of each unit into the predicted conditional probability of surviving the respective interval, which is the complement of the conditional hazard rate $1 - h_j$.

The predicted probability of a subject i surviving through the end of interval j is given by:

$$S_j^{(i)} = \prod_{q=1}^{j}(1 - h_q^{(i)}). \tag{3}$$

The loss function is a reformulation of the negative log-likelihood divided by the number of study subjects, which facilitates training with mini-batches of patients.

This prediction submodel, integrated with the compact fusion vector generated by the multi-modal data fusion layer, enables the model to effectively translate the combined features into survival probability predictions, leveraging uncensored and censored data.

4 Experiments and Results

A uniform framework was established to provide a standardized testing environment for all models, ensuring consistency and fairness across various aspects. The proposed models were implemented using PyTorch. The models are evaluated using the widely popular Concordance index (Ctd) [2], an extension of the widely used Harrell's concordance index (C-index) [10].

The models were trained for 50 epochs, using a batch size of 4 to optimize computational efficiency and stability during training. The validation set was used to assess the performance during iterative model development. We trained the models using Adam stochastic gradient descent optimization. The learning rate was set to 2×10^{-3}, a value chosen based on preliminary experiments to ensure a balance between convergence speed and model performance. The dataset was divided into training, validation, and test sets with an 80-10-10 split.

Table 1. Evaluation of the prediction results using the HEAD-NECK-RADIOMICS-HN1 dataset

Model	Modality	Ctd-index
XGBoost [15]	CT, PET	0.5742
RF [12]	CT, PET	0.5909
DeepSurv [17]	CT, PET	0.6743
MultiSurv [31]	CT, PET	0.7018
MultiSurv (RNC) [39]	CT, PET	0.6811
Ours	CT, PET	**0.7272**

As shown in Table 1, initial experiments focused on integrating CT and PET images for survival prediction on the HN1 dataset. However, upon incorporating clinical data into the model, we do not observe any improvement in performance metrics. Despite the anticipated complementary nature of clinical data

Table 2. Evaluation of the prediction results using the HECKTOR dataset

Model	Modality	Ctd-index
XGBoost [15]	CT, PET	0.5810
RF [12]	CT & PET	0.6015
Multisurv [31]	CT, PET	0.6722
	CT, PET & Clinical	0.6489
Multisurv(RNC) [39]	CT, PET	0.6292
	CT, PET & Clinical	0.6214
Ours	CT, PET	**0.7010**

in enhancing predictive accuracy, the gains were insufficient to justify the added complexity. We also experimented with the Rank Consistency Loss (RNC Loss) [39], designed to penalize incorrect ranking of survival risks among patients. Still, this approach did not yield meaningful improvements in our results. We observed a significant performance improvement by incorporating the attention module and explicitly targeting the feature maps generated from CT and PET images. This enhancement led to a noticeable increase in the ctd-index across all evaluated time intervals, highlighting the effectiveness of the attention mechanism in refining survival predictions. Table 2 shows the results for the HECKTOR dataset, and we observe the same pattern here and see that using the CT and PET imaging features alone provides a considerable improvement in the performance while incorporating the clinical information leads to a drop in the performance.

5 Conclusion

In this study, we proposed a novel deep learning-based approach for survival prediction in head and neck cancer (HNC) that effectively integrates CT and PET imaging modalities. Our method leverages a Convolutional Block Attention Module (CBAM) for enhanced feature extraction and a multi-modal data fusion layer to combine imaging data into a compact representation. A fully parametric discrete-time survival model then utilizes this representation to predict survival probabilities across discrete time intervals. Our experimental results, evaluated on the HECKTOR and HEAD-NECK-RADIOMICS-HN1 datasets, demonstrate that our approach significantly outperforms traditional statistical models and other contemporary machine-learning techniques. Specifically, our method achieved superior C-index scores, highlighting its ability to provide more accurate survival predictions compared to existing methods.

The integration of CT and PET imaging modalities, coupled with advanced deep learning techniques, addresses some limitations of relying solely on a single data modality. By focusing on imaging data alone, our model provides a scalable and practical solution for survival prediction in real-world clinical settings, where clinical and genomic data might be incomplete or unavailable. Future work will explore incorporating additional data modalities and the potential benefits of further integrating clinical and genomic information to enhance predictive accuracy.

Overall, our deep learning approach offers a robust and reliable tool for personalized treatment planning in HNC, contributing to the ongoing efforts to improve patient outcomes through advanced predictive modeling.

References

1. Aerts, H., et al.: Decoding tumour phenotype by noninvasive imaging using a quantitative radiomics approach. Nat. Commun. **5**, 4006 (2014). https://doi.org/10.1038/ncomms5006
2. Antolini, L., Boracchi, P., Biganzoli, E.M.: A time-dependent discrimination index for survival data. Stat. Med. **24**(24), 3927–3944 (2005)
3. Breiman, L.: Random forests. Mach. Learn. **45**(1), 5–32 (2001)
4. Chen, F., Sun, X., Zhao, R.: Integrative analysis of genomic and imaging data for survival prediction in head and neck cancer. Can. Res. **79**(12), 3065–3074 (2019)
5. Chen, H., Wei, F., Zhao, Y.: Deepsurv for head and neck cancer survival prediction with multimodal data. Sci. Rep. **11**(1), 1–12 (2021)
6. Cortes, C., Vapnik, V.: Support-vector networks. Mach. Learn. **20**(3), 273–297 (1995)
7. Gormley, M., Creaney, G., Schache, A., Ingarfield, K., Conway, D.I.: Reviewing the epidemiology of head and neck cancer: definitions, trends and risk factors. Br. Dent. J. **233**(9), 780–786 (2022)
8. Gupta, A., Jain, R., Kumar, G.: Radiomics analysis for predicting overall survival in head and neck cancer patients. Eur. J. Radiol. **134**, 109441 (2021)
9. Gupta, R., Singh, N., Verma, R.: Artificial neural networks for survival prediction in head and neck cancer. Comput. Biol. Med. **95**, 170–178 (2018)
10. Harrell, F.E., Califf, R.M., Pryor, D.B., Lee, K.L., Rosati, R.A.: Evaluating the yield of medical tests. JAMA **247**(18), 2543–2546 (1982)
11. He, K., Zhang, X., Ren, S., Sun, J.: Deep residual learning for image recognition. arxiv e-prints. arXiv preprint arXiv:1512.03385 **10** (2015)
12. He, T., Li, J., Wang, P., Zhang, Z.: Artificial intelligence predictive system of individual survival rate for lung adenocarcinoma. Comput. Struct. Biotechnol. J. **20**, 2352–2359 (2022)
13. Hu, X., Zhang, Z., Chen, K.: Radiomics-based deep learning model for survival prediction in head and neck cancer using CNNs. IEEE Trans. Med. Imaging **39**(7), 2121–2131 (2020)
14. Huang, J., Li, Y., Zhou, X.: Pet imaging biomarkers for predicting survival in head and neck cancer. Eur. J. Nucl. Med. Mol. Imaging **48**(3), 765–774 (2021)
15. Huang, Z., Hu, C., Chi, C., Jiang, Z., Tong, Y., Zhao, C.: An artificial intelligence model for predicting 1-year survival of bone metastases in non-small-cell lung cancer patients based on XGBoost algorithm. BioMed Res. Int. **2020**, 3462363 (2020)
16. Jin, F., Zhang, L., Wang, H.: Fusion-based deep learning approach for survival prediction in head and neck cancer using multi-modal data. IEEE Access **11**, 17685–17694 (2023)

17. Katzman, J.L., Shaham, U., Cloninger, A., Bates, J., Jiang, T., Kluger, Y.: DeepSurv: personalized treatment recommender system using a cox proportional hazards deep neural network. BMC Med. Res. Methodol. **18**, 1–12 (2018)
18. Kong, L., Shen, Y., Zhang, X.: Deep learning-based radiomics for survival prediction in head and neck cancer patients. Med. Phys. **49**(4), 1981–1990 (2022)
19. Lee, D., Kim, A., Park, S.: Kaplan-meier survival analysis in head and neck cancer: a comparative study across treatment groups. Cancer Epidemiol. **38**(4), 556–562 (2014)
20. Li, Q., Wang, H., Zhang, Y.: Prognostic value of radiomic features from CT images in head and neck cancer. J. Clin. Oncol. **41**(8), 2221–2230 (2023)
21. Li, W., Zhang, M., Chen, L.: Random forests in predicting survival outcomes in head and neck cancer. BMC Cancer **17**(1), 1–10 (2017)
22. Liu, Q., Zhang, W., Huang, J.: Deep learning in head and neck cancer survival prediction: a convolutional neural network approach. Med. Image Anal. **58**, 101–109 (2019)
23. Liu, S., Yang, B., Chen, G.: Integrated random survival forest models for head and neck cancer prognosis. Ann. Oncol. **33**(6), 760–768 (2022)
24. Oreiller, V., et al.: Head and neck tumor segmentation in PET/CT: the HECKTOR challenge. Med. Image Anal. **77**, 102336 (2022)
25. O'Shea, K.: An introduction to convolutional neural networks. arXiv preprint arXiv:1511.08458 (2015)
26. Patel, R., Kumar, S., Sharma, A.: Recurrent neural networks for dynamic survival prediction in head and neck cancer. IEEE Trans. Med. Imaging **39**(6), 1855–1864 (2020)
27. Smith, J., Doe, J., Brown, M.: Prognostic factors in head and neck cancer using cox proportional hazards model. J. Clin. Oncol. **30**(15), 2034–2040 (2012)
28. Sun, L., Li, X., Wang, J.: Integrating radiomics and genomics for head and neck cancer survival prediction using deep learning. Nat. Commun. **12**(1), 1–10 (2021)
29. Sun, T., Li, Y., Liu, Z.: DeepMM: multimodal fusion for survival prediction in head and neck cancer using deep learning. IEEE Trans. Med. Imaging **40**(12), 3461–3472 (2021)
30. Sung, H., et al.: Global cancer statistics 2020: Globocan estimates of incidence and mortality worldwide for 36 cancers in 185 countries. CA Cancer J. Clin. **71**(3), 209–249 (2021)
31. Vale-Silva, L.A., Rohr, K.: Long-term cancer survival prediction using multimodal deep learning. Sci. Rep. **11**(1), 13505 (2021)
32. Wang, J., Liu, Q., Tang, X.: Application of support vector machines in predicting survival in head and neck cancer based on gene expression data. Bioinformatics **32**(17), 2643–2651 (2016)
33. Wang, L., Cheng, X., Zhao, Y.: Integrating pet and CT data for survival prediction in head and neck cancer. BMC Med. Imaging **22**(1), 1–11 (2022)
34. Wang, T., Li, M., Zhang, Q.: Deep learning models for multi-modal fusion in head and neck cancer survival prediction. Front. Oncol. **10**, 1290 (2020)
35. Wee, L., Dekker, A.: Data from head-neck-radiomics-hn1 (2019). https://doi.org/10.7937/tcia.2019.8kap372n. The Cancer Imaging Archive
36. Woo, S., Park, J., Lee, J.Y., Kweon, I.S.: CBAM: convolutional block attention module. In: Proceedings of the European Conference on Computer Vision (ECCV), pp. 3–19 (2018)
37. Wu, J., Zhang, X., Sun, M.: Development of support vector machine models for survival prediction in head and neck cancer. J. Radiat. Oncol. **7**(4), 234–245 (2018)

38. Xu, L., Li, F., Yu, Z.: Nomogram for predicting survival in patients with head and neck cancer: a multivariable analysis. JAMA Otolaryngol. Head Neck Surg. **145**(10), 952–960 (2019)
39. Zha, K., Cao, P., Son, J., Yang, Y., Katabi, D.: Rank-n-contrast: learning continuous representations for regression. In: Advances in Neural Information Processing Systems, vol. 36 (2024)

Spatial Clustering and Machine Learning for Crime Prediction: A Case Study on Women Safety in Bhopal

Yamini Sahu(✉) and Vaibhav Kumar

Indian Institute of Science Education and Research Bhopal (IISERB), Bhopal, India
{yamini21,vaibhav}@iiserb.ac.in

Abstract. A crime is an unlawful action subject to punishment by a governing authority, causing harm not only to individuals but also to the well-being of a community, society, or the state. According to the latest annual report from the National Crime Records Bureau of *India*, there were 445,256 cases of crimes against women registered in 2022, representing a 4% increase from the 428,278 cases reported in 2021. India experiences an alarming rate of 51 cases of crimes against women per hour. These figures underscore the urgent need for proactive measures to ensure women safety and secure their continued contribution to the country's development.

Our research employs a predictive approach to address crimes against women, utilizing spatial analysis and crime prediction models to pinpoint high-risk areas in urban settings and accurately forecast crime trends. We focused on *Bhopal*, the capital of *Madhya Pradesh*, one of India's 28 states, and gathered crime data against women from 30 police stations in *Bhopal*. Our study showcases the application of spatial clustering techniques to identify hotspots for various crimes against women (murder, attempted murder, rape, gang rape, kidnapping, dowry-related offenses, and molestation). Additionally, we employed machine learning regression models and advanced forecasting techniques to predict crime rates. Our model, based on decision tree regression, exhibited a very low mean squared error of 0.0417 and a mean absolute error of 0.083. Furthermore, our analysis revealed that classical machine learning regression models outperformed advanced forecasting models, such as long short-term memory, given our limited dataset. Thus, detecting and predicting crime hotspots derived from historical crime data can enable law enforcement agencies to develop targeted intervention strategies customized for specific crimes occurring at particular locations

Keywords: Spatiotemporal · Crime against women · Pattern analysis · Urban planning

1 Introduction

The 21st century has seen an unprecedented global migration into urban areas, leading to the term "Century of the City" [1,2]. This ongoing urbanization is causing significant social, economic, and environmental changes in urban areas. Organizations responsible for city management and providing essential services, such as transportation, air and water quality, public safety, and resource planning (e.g., water and power), may need help addressing these issues [3]. Moreover, in cities with higher crime rates, crime spikes have emerged as a critical social issue, impacting not only public safety but also adult socioeconomic status, health, education, and child development [4,5].

A growing amount of urban-related data, including spatial and temporal attributes ranging from weather to economic activity, is available to public organizations, including police departments, for integration with internal data, such as crime happening in the cities. Urban-related data presents an opportunity to apply data analytics methodologies to derive predictive health, water, and energy management models, etc. Similarly, these urban data can be used to derive predictive models related to crime events. Such models enable police departments to optimize their limited resources and develop more effective crime prevention strategies.

Research on criminal justice indicates that criminal events are not evenly distributed within a city, and the crime rates vary based on geographic location (e.g., low-risk and high-risk areas) and temporal patterns (e.g., seasonal fluctuations, peaks, and dips). Therefore, an accurate predictive model must automatically identify areas more susceptible to crime events and understand how crime rates fluctuate over time in each specific area. This knowledge can help police departments allocate resources efficiently, focusing on high-risk areas and adjusting strategies based on changing crime trends.

A crime constitutes an unlawful act subject to punishment by a governing authority, posing harm not only to individuals but also to the well-being of a community, society, or the state. Research suggests that crimes display detectable patterns across geographical regions within specific timeframes, offering insights for proactive measures by law enforcement. The recurrent incidence of specific criminal events in a given area over time can be termed a spatiotemporal event, facilitating the prediction of future crime incidents using historical spatiotemporal data. Machine learning, deep learning, and data mining techniques serve as effective tools for crime prediction.

In India, the number of reported cases of crime against women (CAW) has shown a concerning trend over recent years. According to data released by the National Crime Records Bureau (NCRB) [6], in 2020, there were 371,503 reported cases, which increased to 428,278 in 2021 and further rose to 445,256 in 2022. These figures indicate a worrying escalation in such incidents year after year. Notably, 51 First Information Reports (FIRs) were filed every hour on average across the country, highlighting the frequency and severity of these crimes. The consistent rise in reported cases underscores the urgent need for effective measures to address and prevent CAW [7].

Mitigating CAW is essential for fostering safer urban environments, which directly impacts social and economic development. High crime rates discourage women's participation in public life, reducing workforce diversity and overall productivity. A city designed with gender-sensitive urban planning-incorporating well-lit streets, safe public transport, and community surveillance-enhances women's safety, contributing to social cohesion. Additionally, lowering crime rates leads to improved mental and physical well-being, reducing healthcare costs. Addressing these issues is vital for creating inclusive cities where everyone can thrive, which is critical for sustainable urban development and long-term societal progress.

In this paper, we used spatial analysis and predictive modeling to automatically identify high-risk crime regions in urban areas and reliably forecast crime trends in each region. Our proposed algorithm consists of several steps: first, identifying high crime density areas (referred to as crime-dense regions or hotspots) through spatial analysis; then, discovering specific crime prediction models for each detected region based on the analyzed data. The resulting micro-level spatio-temporal crime prediction model comprises a set of crime-dense regions and associated crime predictors, with each predictor representing a model to forecast the expected number of crimes in its respective region.

1.1 Contributions

We have demonstrated the use case of our algorithm using a case study. We present an analysis of crimes within a particular region of *Bhopal*, encompassing approximately 4,500 crime events over four years. The crime data used in this work was obtained from the *Bhopal* Law Enforcement Agency[1]. Experimental evaluation results demonstrate the effectiveness of our proposed approach, achieving good accuracy in spatial and temporal crime forecasting over time.

Our primary contribution lies in creating a micro-level spatiotemporal crime forecasting model. Our work is significant because it enables implementation on a smaller scale rather than attempting to apply the solution to a larger region, such as the entire state (in our case, *Madhya Pradesh*). Furthermore, our work demonstrates the utilization of limited data (from the years 2020 to 2023) to predict crime rates among women.

1.2 Paper Organization

Our paper is organized as follows. Section 2 provides a comprehensive review of approaches utilized in crime hotspot detection and prediction. In Sects. 3, we present the proposed methodology used in this study. Section 4 discusses the proposed algorithm. The implementation and evaluation of the proposed method are presented in Sect. 5. Subsequently, Sect. 6 presents the results, while in Sect. 7, we discuss the model performance and implications of our work. Section 8 outlines the study's limitations. Finally, Sect. 9 presents the conclusion, and Sect. 10 outlines the future directions of our work.

---

[1] https://ptsbhopal.mppolice.gov.in/.

2 Related Work

Existing studies have utilized various data mining techniques for crime data analysis and crime detection. Some researchers have concentrated on predicting crime locations, while others have aimed to detect crime patterns (hotspot detection). We categorize existing research on crime data into four categories: (i) crime prediction using machine learning, (ii) analysis of crime patterns, (iii) spatiotemporal crime analysis, (iv) utilizing diverse data sources for crime analysis.

2.1 Crime Prediction Using Machine Learning

Biswas et al. [8] utilized polynomial regression, linear regression, and random forest algorithms to forecast crime scenarios in Bangladesh, achieving success with the polynomial model. Similarly, Hajela et al. [9] employed machine learning (ML) and hotspot analysis for crime prediction, with REPTree showing promising results. Hossain et al. [10] achieved high accuracy using supervised ML algorithms on San Francisco crime records for crime prediction. Similarly, Kumar et al. [11] developed a crime prediction model using the KNN algorithm.

2.2 Analysis of Crime Patterns

Das et al. [12] conducted a behavioral analysis of violence against women in India, identifying perpetrator clusters using the Infomap clustering algorithm. Similarly, Tamilarasi et al. [13] focused on identifying regions where major crimes frequently occur and the types of crimes that persist over time, employing various ML algorithms. Lavanyaa et al. [14] delved into crimes against women in Tamil Nadu, India, aiming to discern patterns and predict occurrences of crime, thereby enhancing the operational efficiency of Tamil Nadu Police.

2.3 Spatiotemporal Crime Analysis

Ibrahim et al. [15] conducted spatiotemporal crime hotspot analysis using Chicago's crime dataset, comparing SARIMA (Seasonal AutoRegressive Integrated Moving Average) with LSTM (Long Short-Term Memory) for crime prediction. Similarly, Li et al. [16] analyzed urban crime in China using quantitative methods and ARIMA (AutoRegressive Integrated Moving Average) for spatiotemporal crime predictions. Butt et al. [17] focused on crime prediction in New York City, employing HDBSCAN (Hierarchical Density-Based Spatial Clustering of Applications with Noise)and SARIMA models. However, Yi et al. [18] proposed a clustered CRF model for predicting crime based on spatiotemporal factors.

2.4 Utilizing Diverse Data Sources for Crime Analysis

Belesiotis et al. [19] explored diverse online data sources (six) to analyze and predict crime distribution in large urban areas, demonstrating improved prediction accuracy through integrated data. Similarly, Sivanagaleela et al. [20] focused on crime analysis and prediction employing the Fuzzy C-Means algorithm, showcasing its efficacy in crime analysis.

3 Proposed Methodology

Figure 1 shows the proposed methodology utilized in this study. It comprises data cleaning, crime hotspot detection, and crime prediction. Now, we will discuss the crime hotspot detection and crime prediction steps in detail. The discussion on data cleaning is provided separately in Sect. 5.1.

In this study, we used CAW data, including incidents of rape, murder, and kidnapping, as input variables. We applied the K-nearest neighbor (K-NN) spatial clustering method to analyze the spatial distribution of these crimes. The output of this algorithm identifies spatial clusters of crime hotspots, allowing for a detailed examination of crime patterns and their potential impact on urban planning.

Fig. 1. Spatiotemporal crime against women prediction steps.

3.1 Crime Hotspot Detection and Crime Prediction

The crime hotspot detection method uses spatial clustering on the processed dataset, with each cluster representing a concentrated region of criminal activity. This spatial clustering approach utilizes clustering techniques like density-based, which aims to group objects together based on their proximity and density in the geographical space. This can be achieved by using spatial pattern analysis techniques to identify clusters by analyzing the estimated density distribution of the dataset.

3.2 Crime Prediction

Once the crime-prone areas are identified, predictive models can be employed to forecast future occurrences of crimes within these regions. These predictive models can be developed using classical machine learning algorithms and advanced forecasting techniques. Classical machine learning regression models, including Decision Trees (DT) [21], Random Forests (RF) [22], and K-Nearest Neighbors (KNN) [23], can be utilized for this purpose. Additionally, advanced forecasting models such as ARIMA (AutoRegressive Integrated Moving Average) [24], SARIMA (Seasonal ARIMA) [25], and LSTM (Long Short-Term Memory) [26] can be explored for predictions.

4 Proposed Algorithm

We have merged Sects. 3.1 and 3.2 to propose a crime prediction Algorithm 1, which consists of two sequential steps: Sect. 3.1 as step 1 and Sect. 3.2 as step 2. In the first step, the algorithm focuses on crime hotspot detection, which identifies regions with a higher density of criminal activity by treating it as a geospatial clustering problem. The clustering method analyzes both spatial and temporal crime data to produce K clusters, each representing a high-crime-density region. The goal of this step is strictly to detect these hotspots. In the second step, the algorithm transitions to crime prediction. It utilizes the hotspots identified in step 1 as input to a predictive model, which forecasts future crime occurrences based on CAW data. Thus, the crime prediction model not only considers the historical geospatial clustering but also integrates additional predictive features such as temporal patterns and crime types.

Algorithm 1. Spatio-Temporal Crime Prediction

Require: D: Crime against women dataset
Ensure: $CAWH$: Crime against women hotspot, CP: Crime Predictors
1: **L1**: Crime hotspot detection // *Step 1 starts*
2: Identify crime hotspots using spatial analysis techniques.
3: $CAWH \leftarrow$ Locations identified as crime hotspots.
4: $CP \leftarrow$ Predictors for crime prediction // *Step 1 ends*
5: **L2**: Apply various algorithms // *Step 2 starts*
6: Train various machine learning algorithms, including decision tree, random forest, ARIMA, SARIMA, LSTM, and KNN on the dataset.
7: Evaluate the performance of each algorithm using metrics such as MSE and MAE.
8: $MSE, MAE \leftarrow$ Performance metrics for each algorithm.
9: **return** $CAWH, CP, MSE, MAE$ // *Step 2 ends*

---

5 Case Study

Crime data are sensitive in nature and typically not shared by law enforcement agencies. However, this data proves effective in understanding crime occurrences in various geographical locations and can be used for analysis to inform actions by law enforcement agencies accordingly.

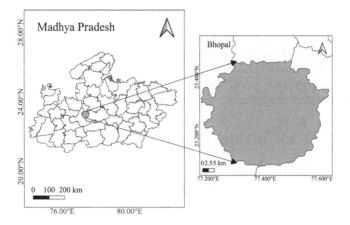

Fig. 2. Madhya Pradesh and the study area (in pink color) (left); *Bhopal* study area (right). (Color figure online)

To illustrate our proposed methodology and the algorithm, we have chosen Bhopal as our study area. Figure 2 shows the geographic coordinates of the study area, with latitude 23.28023161715492 and longitude 77.36346614827542, covering an area of 547.53 square kilometers. This case study focuses on the 30 police stations within the urban landscape of *Bhopal*. These stations span across diverse localities, including *TT Nagar, Ratibad, Habibganjh, Sahpura, Ashoka garden, Jhagirabad, Aishbag, Bajariya, Govindpura, Piplani, Awadhpuri, M.P. Nagar, Arera Hills, Bagsewaniya, Katara Hills, Misrod, Ayodhyanagar, Kotwali, Talaiya (Budhwara), Shahjhabad, Kohefiza, Tila Jamalpura, Mangalwara, Bairagadh, Khajuri Sadak, Kolar Road, Nishatpura, Gandhi Nagar,* and *Chhola Road.*

5.1 Data Collection and Pre-processing

The data was collected from the law enforcement agency in Bhopal after a long process, which involved submitting numerous requests, making appeals, and obtaining permission from multiple higher authorities. The collected data for the police stations listed in Sect. 5 was organized by year (temporal information) and only contained women-related crimes. This dataset covers the period from January 2020 to October 2023 and encompasses details on 25 categories of crimes, including murder, attempted murder, rape, gang rape, kidnapping, dowry-related offenses, molestation, and various others.

During the data preprocessing step, we identified instances of missing data and observed that local languages were used for crime definitions in some cases. To address this, we dropped the columns containing missing values and utilized Google Translate[2] to convert the local language terms into English. As a result,

---

[2] https://translate.google.com/.

our final dataset has dimensions of 120 rows by 10 columns, where the first three columns denote latitude, longitude, and the year of the offense, respectively. The remaining seven columns correspond to seven categories of crimes (murder, attempted murder, rape, gang rape, kidnapping, dowry-related offenses, and molestation). Further, we divided the dataset into training and testing sets in a ratio of 80:20 while experimenting with the forecasting model. The split was performed without considering the temporal sequence of the data. While this method does not involve a time-based separation, it provides a representative distribution of the dataset across both sets. We acknowledge that this approach might allow cases from later in the dataset to appear in both the training and testing sets. However, for the purposes of this study, we aimed to maintain an equal distribution of data points across the two sets to maximize the performance and generalization ability of the model.

5.2 Detection of Crime-Dense Regions from the Data Set

The objective of this study is to predict spatiotemporal crime patterns related to women. Figure 3 shows the cumulative number of crimes against women per year from 2020 to 2023. Similarly, Fig. 4 shows the cumulative number of specific crimes against women, such as murder, kidnapping, rape, etc., per year from 2020 to 2023. From visual analysis, we observed a consistent increase in the number of crimes against women each year.

We utilized the proposed Algorithm 1 to identify crime-dense regions within our study area. Figures 5, 6, 7, and 8 display the crime-dense regions in the Bhopal area, highlighting instances of rape, murder, kidnapping, and molestation against women, respectively.

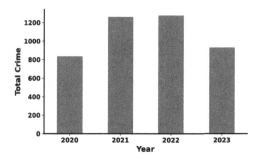

Fig. 3. According to the data shown in the figure, the number of reported cases of CAW has shown a consistent increase over the years. Specifically, in 2020, the number of cases was lower compared to subsequent years, namely 2021, 2022, and 2023. This trend indicates a continuous rise in reported cases from 2020 to 2023. However, it is worth mentioning that the data for 2023 only includes information up to October, which could explain the lower number of reported cases compared to the full-year data for 2021 and 2022.

314   Y. Sahu and V. Kumar

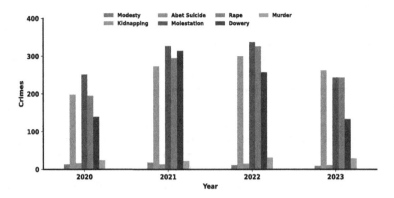

Fig. 4. From 2020 to 2023, a range of different crime cases have been observed. Especially the categories investigated in this study include modesty, kidnapping, abet suicide, molestation, rape, dowry-related crimes, and murder. Across all years, molestation cases notably stand out as being exceptionally high compared to other types of crimes, followed by rape cases and kidnapping incidents. There is a consistent upward trend in the number of cases reported annually. However, it is essential to note that the data for the year 2023 only includes information up to October, leading to a lower count compared to previous years. Furthermore, modesty and abet suicide cases consistently appear to be relatively low in number each year.

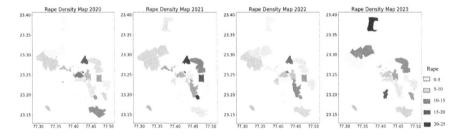

Fig. 5. Rape cases hotspot regions in *Bhopal* spanning from 2020 to October 25, 2023.

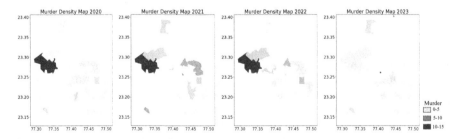

Fig. 6. Murder cases hotspot regions in *Bhopal* spanning from 2020 to October 25, 2023.

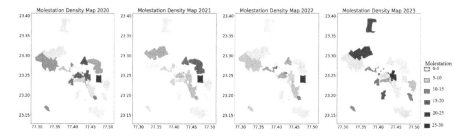

Fig. 7. Molestation cases hotspot regions in *Bhopal* spanning from 2020 to October 25, 2023.

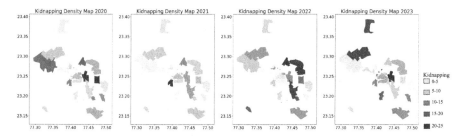

Fig. 8. Kidnapping cases hotspot regions in *Bhopal* spanning from 2020 to October 25, 2023.

5.3 Training and Evaluating the Predictive Model

We employed various regression techniques to forecast crime occurrences. The regression technique used in this paper includes classical machine learning models such as Decision Trees (DT), Random Forests (RF), and K-Nearest Neighbors (KNN), as well as advanced forecasting models including Long Short-Term Memory (LSTM) networks, Seasonal Autoregressive Integrated Moving Average (SARIMA) models, and Autoregressive Integrated Moving Average (ARIMA) models. The selection of these models aimed at leveraging their distinct capabilities and characteristics to predict crime patterns accurately.

DT offers a non-linear approach, breaking down data into hierarchical structures for classification, while RF is an ensemble learning method that excels at handling large datasets and minimizing overfitting by aggregating multiple DTs. However, KNN makes predictions based on the similarity of data points. LSTM is a recurrent neural network that captures long-term dependencies in sequential data and is potentially valuable for recognizing patterns in data over time. Additionally, SARIMA and ARIMA models, both time series forecasting techniques, are particularly suited for analyzing temporal patterns and trends in data.

Model Evaluation

Mean square error (MSE) and mean absolute error (MAE) are standard metrics used to evaluate the performance of predictive models. MSE measures the average squared difference between the actual and predicted values, providing insight into the overall accuracy of the model's predictions. It is calculated by taking the average of the squared differences between each predicted value and its corresponding actual value. On the other hand, MAE computes the average absolute difference between the actual and predicted values, offering a measure of the model's precision. It is determined by averaging the absolute differences between each predicted value and its corresponding actual value. MSE and MAE are essential tools for assessing the effectiveness and reliability of predictive models, with lower values indicating better performance.

6 Results

We will discuss our findings on two main topics: the detection of crime-dense regions and crime prediction.

Crime-Dense Regions

Figure 5 showcase the concentration of rape cases in the Bhopal area from 2020 to 2023, offering insights into evolving patterns over time and across different locations. In 2020, *Ashoka Nagar* stood out as a hotspot for these incidents, signifying a cluster of cases in *Ashoka Nagar*. The subsequent year, 2021, witnessed a notable surge in rape cases in *Khajuri*. As time progressed to 2022, there was a significant increase in incidents in *Govind Pura* and *Pipilani*. By 2023, *TT Nagar* emerged as the area with the highest number of reported rape cases.

Similarly, Fig. 6 shows the regions with the highest density of murder cases in the Bhopal area from 2020 to 2023, shedding light on evolving patterns over time and across various locations. *Bairagadh* recorded the highest number of reported crimes during this period. Notably, *Bairagadh* emerged as the central hub of criminal activity, with the highest rate of crime recorded from 2020 to 2022. However, in 2023, there is a noticeable shift, with the *Mangal Wara Momim Pura* Gate region experiencing a surge in murder incidents. This analysis delves into the evolving patterns of crime over time, highlighting the necessity for targeted interventions to address specific geographic areas in response to emerging trends. Understanding these trends is crucial for effectively implementing law enforcement measures and initiatives to ensure community safety.

Similarly, Fig. 7 shows the regions with dense occurrences of Molestation cases in the Bhopal area. In 2020, *Piplani* had the highest number of reported molestation cases, while *Nishantpura, Govindpura, Jhangirabad*, and *TT Nagar* recorded moderate case numbers. The following year, 2021, witnessed a continuation of cases in *Piplani*, alongside moderate instances in *Ayodhyanagar* and *Nishantpura*. In 2022, *Piplani* reported the highest number of cases again, with *Sahajanbad* and *Ashokanagar* documenting moderate occurrences. This suggests that *Piplani* is a hotspot for molestation cases, and the necessary intervention is needed.

Similarly, Fig. 8 illustrates the regions with dense occurrences of kidnapping cases in the Bhopal area. In 2020, *Govindpura* and *Piplani* emerged as hotspots with the highest number of reported kidnapping of women cases, indicating concentrated criminal activity in these regions. *Bairagadh* and *Nishantpura* recorded moderate case numbers during the same period. However, the dynamics shifted in 2021, with *TT Nagar* taking the lead in reported cases. By 2022, *Nishantpura* regained its position with the highest reported cases, alongside moderate numbers in *Habib Ganjh* and *Ayodhya Nagar*. In the following year, 2023, *Govindpura* witnessed the highest number of cases again, while *Gandhi Nagar* and *Khajuri* reported moderate case numbers.

Crime Prediction

The performance metrics for MSE and MAE of the used models is shown in Table 1. Classical ML models outperformed the advanced forecasting models in terms of both MSE and MAE. Random Forest exhibited the best performance with the lowest MSE of 0.019 and MAE of 0.074. Among the advanced forecasting models, ARIMA outperformed SARIMA and LSTM, achieving an MSE of 1.009 and an MAE of 0.66, respectively.

RF is better due to its ensemble nature, which combines multiple DTs to make predictions. The ensemble nature of RF helps reduce overfitting and capture complex relationships in the data. Additionally, RF is robust to noisy data, which might be beneficial given our limited data. On the other hand, advanced forecasting models, i.e., LSTM, SARIMA, and ARIMA, did not perform well due to their sensitivity to data characteristics and hyperparameters. For instance, LSTM requires large amounts of data and longer sequences to learn meaningful patterns, which our dataset might not adequately meet. Similarly, SARIMA and ARIMA models struggled to capture complex patterns in the data, particularly if the underlying crime trends were non-linear or seasonal.

Table 1. CAW Model Performance for Crime Prediction

Model	Mean Squared Error	Mean Absolute Error
DT	0.0417	0.083
RF	0.019	0.074
KNN	0.021	0.076
LSTM	7.72	2.77
SARIMA	1.21	0.69
ARIMA	1.009	0.66

7 Discussion

The confidentiality of crime-related data is paramount, leading law enforcement agencies to be cautious about sharing it with external parties. However, gaining access to this information is crucial for developing solutions to address the root causes of crime and formulating targeted actions for law enforcement. Furthermore, creating a safe and secure city for women hinges on comprehending and mitigating the escalating crime rates by understanding the crime pattern against women. By implementing crime forecasting solutions, we can foster a safer environment, a necessity that is particularly critical for the safety of women and children.

In this study, we have demonstrated the application of technology, such as clustering, to identify crime hotspot regions and forecasting models to predict the number of CAW expected in the next or upcoming years. Our proposed algorithms offer a valuable tool for developing practical solutions to mitigate these crimes, including enhancing security measures, installing closed-circuit television (CCTV) cameras, and increasing police patrols.

7.1 Model Comparison

Our experiments conducted on the collected data indicate that classical ML regression models such as DT, RF, and KNN outperform the advanced forecasting models. This observation suggests that classical machine learning regression models may be more suitable than advanced models when dealing with a small dataset. Furthermore, it is worth noting that advanced models typically require a large amount of training data and often require high-end computing systems for training purposes.

7.2 Implications

Developing and implementing a spatiotemporal crime prediction and hotspot detection system for CAW can significantly contribute to achieving Sustainable Development Goals (SDGs[3]) 5[4] and 11[5], which prioritize gender equality and sustainable cities and communities, respectively. By accurately predicting and identifying areas susceptible to CAW, law enforcement agencies, and local authorities can take proactive measures to enhance safety and security for women, thereby advancing gender equality objectives.

By leveraging advanced technologies and data analytics, such as ML and geographic information systems (GIS) [27], the proposed system can analyze historical crime data to discern patterns, trends, and high-risk areas for CAW. This predictive capability empowers law enforcement agencies to allocate resources more efficiently, deploy patrols to high-risk areas, and implement targeted interventions to prevent crimes before they happen. Consequently, women feel safer and more empowered to engage in social and economic activities, thus advancing gender equality and upholding women's rights.

Furthermore, implementing such a system aligns with SDG 11, which aims to make cities and human settlements inclusive, safe, resilient, and sustainable. By

---

[3] https://sdgs.un.org/goals.
[4] https://sdgs.un.org/goals/goal5.
[5] https://sdgs.un.org/goals/goal11.

enhancing safety and security through crime prediction and hotspot detection, cities can create environments that are conducive to the well-being and prosperity of all residents, including women and vulnerable populations. Safer cities foster community cohesion, social inclusion, and economic development, leading to more sustainable and resilient urban environments. Developing and deploying a spatiotemporal crime prediction and hotspot detection system for crimes against women contributes to achieving SDGs 5 and 11 and promotes overall societal well-being by creating safer, more inclusive, and sustainable cities for everyone.

8 Limitation

Our work showcased the effectiveness of forecasting models in predicting CAW in the Bhopal area with minimal error. However, our study is subject to several limitations outlined below:

1. Our analysis was based on only four years of crime data against women. Obtaining these data was challenging due to various hurdles, including the lack of public availability and the need to navigate through multiple law enforcement authorities. Moreover, the sensitive nature of crime-related data, particularly those concerning CAW, often leads to hesitancy among law enforcement agencies to share such information.
2. Although our study used Bhopal as a case study consisting of 52 police stations, we could only obtain data from 30 police stations in Bhopal. The absence of data from the remaining 22 police stations was due to the unavailability of these data from law enforcement agencies.

These limitations underscore the challenges inherent in researching CAW and highlight the need for greater accessibility to comprehensive and reliable data for more robust analyses and insights.

9 Conclusion

This study has successfully demonstrated the potential of data-driven approaches to identify crime hotspots and understand the underlying causes of crimes against women in Bhopal. By analyzing historical data and using predictive models, we can pinpoint areas where crimes are more likely to occur, helping law enforcement agencies to focus their interventions effectively. The research not only supports targeted law enforcement actions but also highlights the importance of proactive measures like increased patrols, better lighting, and CCTV surveillance. Overall, these insights contribute to safer urban environments and enhanced livability for women.

10 Future Work

Future work will focus on expanding the dataset to include more detailed data points and refining the predictive models for even greater accuracy. Additionally, collaboration with local authorities is crucial for implementing the insights gained from this study into actionable, data-driven interventions. Further exploration will include evaluating the long-term impact of these interventions on reducing crime rates and improving public safety. The research aims to create a framework that can be replicated in other cities, contributing to broader strategies for urban planning and women's safety across India.

References

1. United Nations Human Settlements Programme. The State of the World's Cities 2004/2005: Globalization and Urban Culture, vol. 2. UN-HABITAT (2004)
2. Witold Rybczynski. Makeshift metropolis: ideas about cities. Simon and Schuster (2010)
3. Cicirelli, F., Guerrieri, A., Spezzano, G., Vinci, A.: An edge-based platform for dynamic smart city applications. Futur. Gener. Comput. Syst. **76**, 106–118 (2017)
4. Wang, H., Kifer, D., Graif, C., Li, Z.: Crime rate inference with big data. In: Proceedings of the 22nd ACM SIGKDD International Conference on Knowledge Discovery and Data Mining, pp. 635–644 (2016)
5. Tayebi, M.A., Ester, M., Glässer, U., Brantingham, P.L.: CRIMETRACER: activity space based crime location prediction. In: 2014 IEEE/ACM International Conference on Advances in Social Networks Analysis and Mining (ASONAM 2014), pp. 472–480. IEEE (2014)
6. NCRB. https://ncrb.gov.in/. Accessed 10 Mar 2024
7. TOI. https://timesofindia.indiatimes.com/india/india-records-51-cases-of-crime-against-women-every-hour-over-4-4-lakh-cases-in-2022-ncrb-report/articleshow/105731269.cms. Accessed 10 Mar 2024
8. Biswas, A.A., Basak, S.: Forecasting the trends and patterns of crime in Bangladesh using machine learning model. In: 2019 2nd International Conference on Intelligent Communication and Computational Techniques (ICCT), pp. 114–118. IEEE (2019)
9. Hajela, G., Chawla, M., Rasool, A.: A clustering based hotspot identification approach for crime prediction. Procedia Comput. Sci. **167**, 1462–1470 (2020)
10. Hossain, S., Abtahee, A., Kashem, I., Hoque, M.M., Sarker, I.H.: Crime prediction using spatio-temporal data. In: Chaubey, N., Parikh, S., Amin, K. (eds.) COMS2 2020. CCIS, vol. 1235, pp. 277–289. Springer, Singapore (2020). https://doi.org/10.1007/978-981-15-6648-6_22
11. Kumar, A., Verma, A., Shinde, G., Sukhdeve, Y., Lal, N.: Crime prediction using k-nearest neighboring algorithm. In: 2020 International Conference on Emerging Trends in Information Technology and Engineering (IC-ETITE), pp. 1–4. IEEE (2020)
12. Das, P., Das, A.K.: Behavioural analysis of crime against women using a graph based clustering approach. In: 2017 International Conference on Computer Communication and Informatics (ICCCI), pp. 1–6. IEEE (2017)
13. Tamilarasi, P., Rani, R.U.: Diagnosis of crime rate against women using k-fold cross validation through machine learning. In: 2020 Fourth International Conference on Computing Methodologies and Communication (ICCMC), pp. 1034–1038. IEEE (2020)

14. Lavanyaa, S., Akila, D.: Crime against women (caw) analysis and prediction in Tamilnadu police using data mining techniques. Int. J. Recent Technol. Eng. (IJRTE) **7**(5C), 261 (2019)
15. Ibrahim, N., Wang, S., Zhao, B.: Spatiotemporal crime hotspots analysis and crime occurrence prediction. In: Li, J., Qin, S., Li, X., Wang, S., Wang, S. (eds.) ADMA 2019. LNCS (LNAI), vol. 11888, pp. 579–588. Springer, Cham (2019). https://doi.org/10.1007/978-3-030-35231-8_42
16. Li, Z., Zhang, T., Yuan, Z., Wu, Z., Du, Z.: Spatio-temporal pattern analysis and prediction for urban crime. In: 2018 Sixth International Conference on Advanced Cloud and Big Data (CBD), pp. 177–182. IEEE (2018)
17. Butt, U.M., et al.: Spatio-temporal crime predictions by leveraging artificial intelligence for citizens security in smart cities. IEEE Access **9**, 47516–47529 (2021)
18. Yi, F., Yu, Z., Zhuang, F., Zhang, X., Xiong, H.: An integrated model for crime prediction using temporal and spatial factors. In: 2018 IEEE International Conference on Data Mining (ICDM), pp. 1386–1391. IEEE (2018)
19. Belesiotis, A., Papadakis, G., Skoutas, D.: Analyzing and predicting spatial crime distribution using crowdsourced and open data. ACM Trans. Spatial Algorithms Syst. (TSAS) **3**(4), 1–31 (2018)
20. Sivanagaleela, B., Rajesh, S.: Crime analysis and prediction using fuzzy c-means algorithm. In: 2019 3rd International Conference on Trends in Electronics and Informatics (ICOEI), pp. 595–599. IEEE (2019)
21. Myles, A.J., Feudale, R.N., Liu, Y., Woody, N.A., Brown, S.D.: An introduction to decision tree modeling. J. Chemometr. J. Chemometr. Soc. **18**(6), 275–285 (2004)
22. Breiman, L.: Random forests. Mach. Learn. **45**, 5–32 (2001)
23. Guo, G., Wang, H., Bell, D., Bi, Y., Greer, K.: KNN model-based approach in classification. In: Meersman, R., Tari, Z., Schmidt, D.C. (eds.) OTM 2003. LNCS, vol. 2888, pp. 986–996. Springer, Heidelberg (2003). https://doi.org/10.1007/978-3-540-39964-3_62
24. Zhuang, Y., Almeida, M., Morabito, M., Ding, W.: Crime hot spot forecasting: a recurrent model with spatial and temporal information. In: 2017 IEEE International Conference on Big Knowledge (ICBK), pp. 143–150. IEEE (2017)
25. Valipour, M.: Long-term runoff study using sarima and arima models in the united states. Meteorol. Appl. **22**(3), 592–598 (2015)
26. Yong, Yu., Si, X., Changhua, H., Zhang, J.: A review of recurrent neural networks: Lstm cells and network architectures. Neural Comput. **31**(7), 1235–1270 (2019)
27. Gis. https://education.nationalgeographic.org/resource/geographic-information-system-gis/, accessed date: 2024-03-10

Building Usage Classification in Indian Cities: Utilizing Street View Images and Object Detection Models

Yamini Sahu(✉), Vasu Dhull, Satyajeet Shashwat, and Vaibhav Kumar

Indian Institute of Science Education and Research Bhopal (IISERB), Bhopal, India
{yamini21,vasu21,satyajeet21,vaibhav}@iiserb.ac.in

Abstract. Urban land use maps at the building instance level are crucial geo-information for many applications, yet they are challenging to obtain. Land-use classification based on spaceborne or aerial remote sensing images has been extensively studied over the last few decades. Such classification is usually a patch-wise or pixel-wise labeling over the whole image. However, for many applications, such as urban population density estimation or urban utility mapping, a classification map based on individual buildings (residential, commercial, mixed-type, and religious) is much more informative. Nonetheless, this type of semantic classification still poses fundamental challenges, such as retrieving fine boundaries of individual buildings. Street view images (SVI) are highly suited for predicting building functions because building facades provide clear hints. Although SVIs are used in many studies, their application in generating building usage maps is limited.

Furthermore, their application to Indian cities remains void. In this paper, we propose a comprehensive framework for classifying the functionality of individual buildings. Our method leverages the YOLOs model and utilizes SVIs, including those from Google Street View and OpenStreetMap. Geographic information is employed to mask individual buildings and associate them with the corresponding SVIs. We created our own dataset in Indian cities for training and evaluating our model.

Keywords: Building usage classification · Object detection · Street view images · Open street map · Urban planning

1 Introduction

Urban land use classification at the individual level is essential for effective urban planning and management, influencing a wide range of decisions, including solar potential analysis [1], damage identification [2], and infrastructure development. Accurate classification of building usage types helps urban planners optimize resources, improve public services, and enhance community well-being. Traditional methods for building usage classification, such as satellite imagery, OpenStreetMap (OSM) data, and aerial imagery, have shown promise in various contexts; however, the application of SVI remains limited, particularly within the Indian urban landscape, where research is still developing.

This study aims to address the gap in knowledge by developing a robust framework for urban land use classification using SVI, thereby enhancing our understanding of building usage types within complex urban environments. By employing advanced object detection models, specifically YOLOv8s and YOLOv8n [3], we seek to classify buildings into four distinct categories: residential, commercial, religious, and mixed-use. The inclusion of a mixed-use category is particularly significant, as urban buildings often serve multiple functions-commercial spaces on the ground floor and residential units above, for instance.

Our methodology aims to capture the multifaceted nature of urban buildings and improve classification accuracy in mixed-use areas, which poses significant challenges in existing classification frameworks. This research contributes to enhancing urban land use classification methodologies and provides valuable insights for urban planners and policymakers. By offering a more nuanced understanding of urban building types, we facilitate better decision-making in rapidly growing urban areas, ultimately promoting sustainable development and effective urban management.

1.1 Contributions

Our major contribution lies in the creation of a novel dataset tailored explicitly for building detection in Indian cities. This meticulously curated dataset incorporates diverse urban scenarios typical of Indian streetscapes, including varying architectural styles, occlusions by trees and vehicles, and different lighting conditions. Leveraging this dataset, we have developed a pioneering framework for building detection utilizing SVI. Our approach addresses challenges unique to urban environments in India, such as densely packed buildings and irregular structures, ensuring robust and accurate detection. By employing the state-of-the-art YOLOvs object detection model for facade object detection, our framework significantly enhances the precision of building detection. This innovation not only contributes a valuable resource for future research but also sets a new benchmark for urban infrastructure analysis using computer vision methodologies. We are releasing this dataset to support future research and collaboration [4].

1.2 Paper Organization

Our paper is structured as follows. We present the related work in Section 2. Section 3 outlines the proposed methodology, while Sect. 4 provides a detailed description of the proposed algorithm and implementations. In Sect. 5, we assess the performance of the YOLOv8s and YOLOv8n object detection models. Section 6 discusses the results and its implications. Furthermore, Sect. 7 presents the limitations of our study, and Sect. 8 summarizes our findings. Lastly, Sect. 9 outlines potential avenues for future research.

2 Related Work

Spaceborne or aerial remote sensing images have been extensively studied over the past decades. Typically, classification using these images involves patch-wise or pixel-wise labeling across the entire image. While aerial remote sensing images can provide information about building usage types, they often face challenges in accurately identifying the defined boundaries of building footprints. Figure 1 illustrates building usage classification using aerial remote-sensing images [5–9]. From a top-down view, structures often appear similar, making it difficult to distinguish between different building types. Figure 2 highlights the challenge of mixed-use areas, where commercial, residential, and mixed-type buildings may all appear within the same image patch, complicating the classification process.

Fig. 1. Example of land-use classification using satellite image.

Using OSM data, we can obtain information about building usage types. However, there are several issues with OSM data [10–15]. The accuracy of building footprints in OSM is often questionable, with many building footprints either missing or inaccurately represented. Additionally, there is a lack of labeled data within OSM, which can limit its usefulness for detailed analysis. Another significant concern is the irregularity of updates. OSM data is often updated on a voluntary basis, meaning it cannot always be relied upon for timely or consistent updates. This inconsistency makes it challenging to use OSM data as a sole source for accurate and up-to-date information in urban planning and analysis.

SVI has also been used for building classification, but its application is limited to specific areas. Previous work on building instance classification using SVI primarily focused on datasets from foreign environments [16]. In these studies, buildings were classified into eight categories: apartment, church, garage, house, industrial, office building, retail, and roof. The accuracy of these classifications was relatively low [17–21]. Researchers tested various convolutional Neural Network (CNN) architectures, including AlexNet [22], VGG16 [23], ResNet18 [24], and ResNet34 [24]. Despite the potential of CNNs for building classification, the results indicated significant room for improvement. One of the challenges is the variability in architectural styles and urban layouts across different regions, which can impact the performance of models trained on foreign datasets when applied to local environments. To enhance the accuracy and applicability of building classification using SVI, it is crucial to develop and train models on region-specific datasets and explore advanced techniques to handle the diversity of building types and appearances.

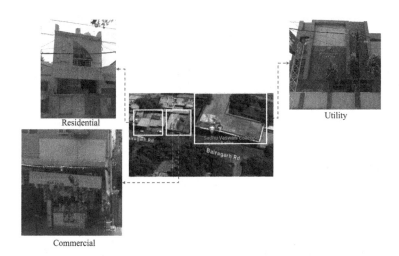

Fig. 2. These buildings do not belong to the same category, even though they are located in the same land-use area. Furthermore, compared to roof structures, the facade structures displayed in SVI provide richer and more sufficient information for building classification using detection methods.

3 Proposed Methodology

In this study, we developed a comprehensive methodology for building detection using geotagged SVI, leveraging a custom dataset tailored to the complexities of urban environments. We categorized buildings into four distinct classes: residential, commercial, religious, and mixed-type, with the mixed-type class representing buildings with multiple functions, such as commercial spaces on the ground floor and residential units above. For the detection phase, we employed a range of object detection models, including YOLOv8s and YOLOv8n, to accurately identify and locate buildings within the images.

Our building detection model provides a detailed understanding of urban land use. Geographic information was used to mask individual buildings and align them with SVIs, ensuring a more precise classification by integrating spatial data with visual features. The effectiveness of this approach was validated using our custom dataset from Indian cities, demonstrating improvements over existing methods and offering valuable insights for urban planning and management through a nuanced classification of diverse building types.

Fig. 3. The proposed framework for building usage classification at a level of individual building.

3.1 Building Footprints from GIS Map

OSM serves as a valuable resource for obtaining building footprints for a specific study area. By leveraging OSM data and geofabrik[1], we can access detailed geographic information, including building footprints, which are essential for urban planning and spatial analysis. Various tools and platforms are available to extract building footprints from OSM. In our work, we utilized OSMnx[2], a Python library, to efficiently download and visualize OSM data. Begin by defining the study area, either by specifying geographical boundaries or using a location name. OSMnx then queries OSM's servers, enabling the extraction of building footprints and other relevant geographic features.

3.2 Collection of Geotagged SVI

We began by identifying the geographical boundaries of the study area. Using tools like the Google Street View API[3], we captured panoramic images at specified intervals within these boundaries. These panoramic images were then converted into normal images for easier analysis. Each image is inherently tagged with latitude and longitude coordinates, ensuring accurate geolocation.

3.3 Façade Object Detection Model

In this work, we utilized various YOLO models for building facade object detection, including YOLOv8s and YOLOv8n. These models were trained to identify

---

[1] https://www.geofabrik.de/.
[2] https://osmnx.readthedocs.io/en/stable/.
[3] https://developers.google.com/maps/documentation/streetview/overview.

four classes of buildings: residential, commercial, mixed-type, and religious. By employing these models, we aimed to classify and geolocate buildings within the images accurately. Table 1 provides a brief overview of the different building classes.

Table 1. Building class descriptions from usage types.

Title	Building types
Residential	house, apartment, hostel
Commercial	shop, restaurant, bank, shopping complex, mall, vegetable center, Indian coffee house
Religious	temple, mosque, church
Mixed-type	combination of commercial and residential

4 Proposed Algorithm

Algorithm 1 outlines a systematic framework for detecting buildings in geotagged SVI. The process begins with *Data Acquisition*, where a collection of geotagged SVI, denoted as $\{I_i\}_{i=1}^{N}$, is gathered and preprocessed. This step ensures that the images are in an optimal format for subsequent analysis. The next phase, *Building Detection*, involves applying the YOLO model to each image to identify buildings. Specifically, for each image I_i, the YOLO model detects buildings and generates bounding boxes B_i that outline these structures. The outcome of this step is a set of bounding boxes for all images, $\{B_i\}_{i=1}^{N}$. In the final *Output Results* phase, the algorithm displays the detected buildings by presenting these bounding boxes. This structured approach facilitates the accurate identification and visualization of buildings in geotagged SVI.

Algorithm 1. Building Detection Framework

1: **Input:** Geotagged SVI $\{I_i\}_{i=1}^{N}$
2: **Output:** Detected buildings $\{B_i\}_{i=1}^{N}$
3: **Step 1: Data Acquisition**
4:   Obtain and preprocess geotagged SVI $\{I_i\}_{i=1}^{N}$
5: **Step 2: Building Detection**
6:   Apply the YOLO model to detect buildings in each image:
7:     $B_i \leftarrow \text{YOLO}(I_i)$
8:   where B_i denotes the bounding boxes of detected buildings in I_i
9: **Step 3: Output Results**
10:   Generate and present the detected buildings $\{B_i\}_{i=1}^{N}$

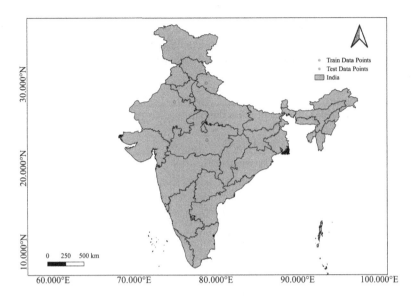

Fig. 4. In this case study, we used images from *Rohtak* and *Mandi* (shown in green dot) for training and images from the *Bhopal* (shown in yellow dot) region for testing purposes.

4.1 Data Collection and Pre-processing

We have selected *Mandi* in *Himachal Pradesh* and *Rohtak* in *Haryana* as the study areas for this research. We used SVI from *Bhopal* in *Madhya Pradesh* for testing purposes. The Fig. 4 represents the study area, including the testing and training points. The data collection part is a lengthy and effort-intensive process. The collected dataset consists of geotagged SVI from both cities. It covers different types of built-up areas, including residential, commercial, religious, and mixed-use buildings, as mentioned in Table 1.

During the data preprocessing step, we identified and removed incorrect data, such as blurred or repetitive images. As a result, we had a total of 512 SVI of various buildings, which were divided into training, validation, and test sets. We manually annotated these data in four categories: residential (e.g., apartment, hostel, bungalow), commercial (e.g., shopping complex, mall, stationery shop, restaurant, hotel, tea stall), religious (e.g., mosque, temple, church), and mixed-type buildings (a combination of commercial and residential).

4.2 Detection of Building Types from the Data Set

The objective of this study is to classify building usage types using geotagged SVI. Figure 5 shows the various building types in the Indian scenario, categorized into residential, commercial, religious, and mixed-use classes. We utilized our proposed methodology for building usage classification, employing object detection techniques to achieve this.

Fig. 5. A visual representation of SVI with buildings belonging to different cities and classes from the dataset.

4.3 Evaluating the Classification Model

In evaluating the performance of our models, we employed several key metrics commonly used in object detection tasks. Box Precision (Box(P)) measures the proportion of true positive bounding boxes out of all predicted bounding boxes, providing insight into the accuracy of the model's predictions. Box Recall (Box(R)) reflects the proportion of true positive bounding boxes out of all actual bounding boxes, indicating the model's ability to detect all relevant objects. To summarize the precision of detections, we used Mean Average Precision at IoU 0.5 (mAP50), which averages the precision across different recall levels at an Intersection over Union (IoU) threshold of 0.5. Additionally, Mean Average Precision across IoU thresholds from 0.5 to 0.95 (mAP50-95) provides a more comprehensive evaluation by averaging precision over a range of IoU thresholds, capturing the model's performance across various levels of detection overlap. These metrics collectively offer a robust assessment of the model's detection capabilities and accuracy.

5 Results

In this research, we developed a framework for building usage classification using SVI. We employed the YOLOv model for facade object detection. Following this, we obtained the results.

To detect building usage types, we initially utilized several YOLO models. These models are well-suited for object detection tasks due to their efficiency and accuracy. By applying YOLO to SVI, we were able to identify various features associated with different building usages effectively. This step was fundamental to ensure that our subsequent classification efforts were based on reliable detection data.

Figure 6a represents the accurate predictions of the model, while Fig. 6b shows the model's incorrect predictions. The errors in the model's predictions are primarily due to the limited number of training data points and an unbalanced dataset, which significantly affects its performance.

The YOLOv8n model demonstrates good performance (as shown in Table 2) in classifying commercial buildings, with high precision (0.718) and recall (0.641). It performs well with mixed-use buildings in terms of recall (0.829) but has lower precision (0.335). The model struggles with religious buildings, showing high precision but no detected instances, and shows moderate results for residential buildings. Overall, YOLOv8n excels in commercial classification while facing challenges in other categories.

From Table 2 and Table 3, we found that YOLOv8n model performed better than YOLOv8s across various validation metrics. The YOLOv8n model demonstrates excellent overall performance, achieving a Box Precision of 0.657, a Box Recall of 0.467, a mAP50 of 0.466, and a mAP50-95 of 0.213. This strong performance is consistent across individual classes. In the commercial category, YOLOv8n achieves impressive results with a Box Precision of 0.718, a Box Recall of 0.641, an mAP50 of 0.729, and an mAP50-95 of 0.309. The model also performs

Predicted- residential
Ground truth- residential

Predicted- residential
Ground truth- residential

Predicted- residential, commercial
Ground truth- residential

Predicted- residential
Ground truth- residential

(a) The predictions from the YOLOv8s model show that it accurately classifies the residential and commercial classes.

Predicted-residential, commercial
Ground truth - mix

Predicted- commercial
Ground truth - mix

Predicted-residential, commercial
Ground truth - commercial

Predicted-residential, commercial
Ground truth - commercial

(b) The YOLOv8n model demonstrates accurate classification for residential and commercial classes; however, it struggles with mixed-type and religious classes, with incorrect predictions in the mixed-type category and no predictions in the religious class.

Fig. 6. YOLO model performance.

Table 2. Validation Performance Metrics of YOLOv8n.

Class	Images	Box(P)	Box(R)	mAP50	mAP50-95
all	145	0.657	0.467	0.466	0.213
commercial	85	0.718	0.641	0.729	0.309
mix	34	0.335	0.829	0.600	0.281
religious	11	1.000	0.000	0.094	0.042
residential	15	0.577	0.400	0.441	0.221

Table 3. Validation Performance Metrics of YOLOv8s

Class	Images	Box(P)	Box(R)	mAP50	mAP50-95
all	145	0.309	0.348	0.113	0.0431
commercial	85	0.155	0.504	0.224	0.0852
mix	34	0.0802	0.886	0.167	0.0677
religious	11	0.000	0.000	0.0145	0.00453
residential	15	1.000	0.000	0.0462	0.0149

notably well in the mixed class, with a Box Precision of 0.335 and a mAP50 of 0.600, although the Box Recall is slightly lower at 0.829. For the religious class, YOLOv8n achieves a perfect Box Precision of 1.000, with mAP50 and mAP50-95 values of 0.094 and 0.042, respectively. Even in the residential class, YOLOv8n outperforms with a Box Recall of 0.400, a mAP50 of 0.441, and a mAP50-95 of 0.221, despite having a lower Box Precision compared to YOLOv8s. Overall, YOLOv8n's superior performance across these metrics highlights its robustness and effectiveness in handling the validation dataset.

6 Discussion

This work focuses on the classification of building usage at the individual building level. While various techniques exist for building usage classification, such as those utilizing satellite imagery and OSM data, there is limited research leveraging SVI for this purpose. Some studies have centered on identifying specific building features, but our work emphasizes detection at the individual building level. By concentrating on this detection approach, we aim to facilitate the real-time identification of different building types.

This has significant implications, offering a practical and valuable contribution to the field of building usage classification research. Real-time detection of building types can enhance various applications, such as urban planning, navigation, and smart city development. This method can also improve data accuracy and provide timely updates, making it an essential tool for researchers and practitioners working with urban environments and infrastructure.

However, this research also faces challenges, primarily due to the limited availability of geotagged images for building detection. Using geotagged SVI, we encounter numerous occlusions in the images, such as vehicles, vegetation, and humans. Additionally, the dataset is unbalanced, which adversely affects the results. Despite these challenges, our work highlights the potential of using SVI for building usage classification and the need for more comprehensive datasets to improve accuracy and reliability.

6.1 Model Comparison

The YOLOv8n model outperforms YOLOv8s across all metrics, including Box Precision, Box Recall, mAP50, and mAP50-95. Specifically, YOLOv8n shows higher overall performance and better results in individual classes, highlighting its superior capability in detecting buildings in geotagged SVI.

6.2 Implications

The challenges faced in categorizing mixed-use and religious buildings indicate the importance of diversifying the training data to ensure balanced performance across all building categories. A more comprehensive dataset will improve the model's classification accuracy and enhance its generalizability to different urban settings. Furthermore, the limitations in retrieving fine boundaries highlight the need for more precise segmentation techniques, which are crucial for reliable land-use classification. Semi-supervised learning for the annotation process will significantly reduce manual efforts and improve label quality, particularly in large-scale datasets. This approach can accelerate the framework's scalability, making it more efficient and applicable to various contexts, including cities in India. Future model iterations can offer more accurate and globally applicable insights by addressing these issues, contributing to better urban planning and decision-making.

7 Limitation

This research work contributes to building usage classification, aiding various applications in urban planning at the individual building level. Using our method, we achieved excellent results overall; however, it did not perform accurately for the mixed-type class. This is because buildings often have commercial use on the ground floor and residential use on the remaining floors. Another limitation is the limited number of data points. In future work, we aim to collect more data points to improve the results. Currently, around 50% of roadside buildings fall into the mixed-type category, but unfortunately, we have a limited number of data points for this class.

The performance of the YOLOv model is hindered by an unbalanced dataset and limited data points in specific classes, such as mixed and religious buildings. This data imbalance impacts the model's ability to accurately classify these less-represented categories. Additionally, the manual labeling process is time-consuming and costly.

8 Conclusion

The classification of building usage is a crucial tool that serves a wide range of urban planning applications. Providing detailed insights into how buildings are utilized across a city enables various agencies and organizations to implement solutions more effectively. For instance, it can be instrumental in conducting solar potential analysis, which assesses the feasibility and efficiency of solar panel installations on different buildings. Additionally, it aids in population estimation by offering data on residential building usage, thereby supporting infrastructure development and resource allocation.

The framework underpinning this work has been delivering promising results. Its potential extends beyond static analysis; these methods can also be adapted for real-time building type prediction. Such predictive capabilities are invaluable for dynamic urban management and emergency response scenarios. The framework exhibits strong performance in categorizing commercial and residential buildings. However, it encounters challenges when dealing with mixed-use and religious buildings. This difficulty largely stems from the limited or insufficient data available for these categories, which hinders the accuracy and reliability of the predictions.

In addressing these challenges, the YOLOv8n model has demonstrated exceptional performance, delivering some of the best results observed so far. Its advanced capabilities in object detection and classification make it a powerful tool for refining building usage classification. As we continue to enhance the dataset and model accuracy, the potential applications of this framework in urban planning and management will only grow, paving the way for smarter, more efficient cities.

9 Future Work

In future research, we aim to address the limitations encountered in this study by enhancing our data collection strategy. Specifically, we plan to expand the dataset to include a more diverse and balanced representation of building types, mainly focusing on underrepresented categories such as mixed-use and religious buildings. By collecting more data across varied urban environments, we aim to improve the model's robustness and accuracy in categorizing these challenging types. Additionally, improving the boundary retrieval process is essential for obtaining precise land-use classification. We plan to explore advanced segmentation techniques to delineate fine boundaries of individual buildings more effectively, ensuring higher precision in object detection and classification tasks. Employing techniques like semantic segmentation, alongside integrating semi-supervised learning methods, will help streamline the annotation process. These enhancements can facilitate more accurate and scalable data labeling, improving the model's performance across diverse building categories. Finally, applying this framework to Indian cities will be a significant future focus. Extending the framework's applicability to diverse geographic and cultural contexts will offer valuable insights into the global adaptability of the model. This will allow us to evaluate its efficacy in regions with unique urban patterns, ensuring broader relevance and utility.

References

1. Chen, Z., et al.: Assessing the potential and utilization of solar energy at the building-scale in shanghai. Sustain. Urban Areas **82**, 103917 (2022)
2. Gharehbaghi, V.R., Farsangi, E.N., Yang, T.Y., Hajirasouliha, I.: Deterioration and damage identification in building structures using a novel feature selection method. In: Structures, vol. 29, pp. 458–470. Elsevier (2021)
3. Jocher, G., Chaurasia, A., Qiu, J.: Ultralytics YOLOv8 (2023)
4. Sahu, Y., Dhull, V., Shashwat, S., Kumar, V.: Indian street view images of buildings (2024). https://osf.io/4e2bd/
5. Geiß, C., et al.: Estimation of seismic building structural types using multi-sensor remote sensing and machine learning techniques. ISPRS J. Photogram. Remote Sens. **104**, 175–188 (2015)

6. Hoffmann, E.J., Wang, Y., Werner, M., Kang, J., Zhu, X.X.: Model fusion for building type classification from aerial and street view images. Remote Sens. **11**(11), 1259 (2019)
7. Li, J., Huang, X., Lilin, T., Zhang, T., Wang, L.: A review of building detection from very high resolution optical remote sensing images. GISci. Remote Sens. **59**(1), 1199–1225 (2022)
8. Zhenyu, L., Im, J., Rhee, J., Hodgson, M.: Building type classification using spatial and landscape attributes derived from lidar remote sensing data. Landsc. Urban Plan. **130**, 134–148 (2014)
9. Xie, J., Zhou, J.: Classification of urban building type from high spatial resolution remote sensing imagery using extended MRS and soft BP network. IEEE J. Sel. Top. Appl. Earth Obs. Remote Sens. **10**(8), 3515–3528 (2017)
10. Bandam, A., Busari, E., Syranidou, C., Linssen, J., Stolten, D.: Classification of building types in Germany: a data-driven modeling approach. Data **7**(4), 45 (2022)
11. Singh Atwal, K., Anderson, T., Pfoser, D., Züfle, A.: Predicting building types using OpenStreetMap. Sci. Rep. **12**(1), 19976 (2022)
12. Zhang, Y., Zhou, Q., Brovelli, M.A., Li, W.: Assessing OSM building completeness using population data. Int. J. Geogr. Inf. Sci. **36**(7), 1443–1466 (2022)
13. Fan, H., Zipf, A., Fu, Q.: Estimation of building types on OpenStreetMap based on urban morphology analysis. In: Connecting a Digital Europe Through Location and Place, pp. 19–35 (2014)
14. Fonte, C.C., Lopes, P., See, L., Bechtel, B.: Using OpenStreetMap (OSM) to enhance the classification of local climate zones in the framework of WUDAPT. Urban Clim. **28**, 100456 (2019)
15. Zhou, Q., Zhang, Y., Chang, K., Brovelli, M.A.: Assessing OSM building completeness for almost 13,000 cities globally. Int. J. Digital Earth **15**(1), 2400–2421 (2022)
16. Kang, J., Körner, M., Wang, Y., Taubenböck, H., Zhu, X.X.: Building instance classification using street view images. ISPRS J. Photogram. Remote Sens. **145**, 44–59 (2018)
17. Laupheimer, D., Tutzauer, P., Haala, N., Spicker, M.: Neural networks for the classification of building use from street-view imagery. ISPRS Ann. Photogram. Remote Sens. Spat. Inf. Sci. **4**, 177–184 (2018)
18. Zhao, K., Liu, Y., Hao, S., Shaoxing, L., Liu, H., Zhou, L.: Bounding boxes are all we need: street view image classification via context encoding of detected buildings. IEEE Trans. Geosci. Remote Sens. **60**, 1–17 (2021)
19. Cao, R., et al.: Integrating aerial and street view images for urban land use classification. Remote Sens. **10**(10), 1553 (2018)
20. Ogawa, Y., Zhao, C., Oki, T., Chen, S., Sekimoto, Y.: Deep learning approach for classifying the built year and structure of individual buildings by automatically linking street view images and GIS building data. IEEE J. Sel. Top. Appl. Earth Obs. Remote Sens. **16**, 1740–1755 (2023)
21. Surya Prasath Ramalingam and Vaibhav Kumar: Automatizing the generation of building usage maps from geotagged street view images using deep learning. Build. Environ. **235**, 110215 (2023)
22. Alom, M.Z., et al.: The history began from AlexNet: a comprehensive survey on deep learning approaches. arXiv preprint arXiv:1803.01164 (2018)

23. Simonyan, K., Zisserman, A.: Very deep convolutional networks for large-scale image recognition. arXiv preprint arXiv:1409.1556 (2014)
24. He, K., Zhang, X., Ren, S., Sun, J.: Deep residual learning for image recognition. In: Proceedings of the IEEE Conference on Computer Vision and Pattern Recognition, pp. 770–778 (2016)

3D-CmT: 3D-CNN Meets Transformer for Hyperspectral Image Classification

Sunita Arya[1(✉)], Shiv Ram Dubey[2], S. Manthira Moorthi[1], Debajyoti Dhar[1], and Satish Kumar Singh[2]

[1] Space Applications Centre, Ahmedabad, India
{sunita33,smmoorthi,deb}@sac.isro.gov.in
[2] Indian Institute of Information Technology, Allahabad, India
{srdubey,sk.singh}@iiita.ac.in

Abstract. In recent years, the combined use of Vision Transformer (ViT) and Convolutional Neural Network (CNN) has shown promising results in tasks related to satellite imagery. In our study, we propose a 3D-CmT (3D-CNN meets Transformer) model for Hyperspectral Image Classification. This model leverages the unique capabilities of both 3D-CNN and ViT to effectively classify images captured by hyperspectral imaging. To learn the local features of the narrow and contiguous electromagnetic spectrum of the hyperspectral images, we utilize a 3D-CNN under the spectral feature extraction (SFE) module. Subsequently, a transformer encoder (TE) module is applied on top of the 3D-CNN to incorporate global attention and model long-range dependencies for spatial information in the images. We conducted experiments using commonly used hyperspectral image datasets and performed various ablation studies, such as evaluating the impact of image patch size and different percentages of training samples. The performance of our proposed model is comparable to that of other CNN-based, transformer-based, and hybrid CNN-Transformer-based models in terms of model parameters and accuracy. In addition, we conducted quantitative and qualitative analyses to assess the performance of our model.

Keywords: Remote Sensing · Hyperspectral Image (HSI) Classification · 3D Convolutional Neural Network (CNN) · Vision Transformer (ViT)

1 Introduction

Hyperspectral image (HSI) acquired using space-borne or air-borne instruments plays a very significant role in geological studies, mineral mapping, and other applications that use its unique capability of narrow spectral bands in the electromagnetic spectrum. With its rich spatial and spectral information, an HSI can be used for various applications, such as mineral studies [18], precision agriculture [19], food safety [20], biomedical imaging [21], and military applications [22].

Along with computer vision and natural language processing, rapid advances in deep learning approaches have pushed the development of signal and image processing techniques in a range of domains [28]. Deep learning-based models have been widely used for many HSI applications [29,34], including HSI classification [30,35] and image fusion [31]. To process the abundant spatial and spectral information of HSI, architectures based on the convolutional neural network (CNN) [33] and transformer [32] have been extensively utilized for HSI analysis and processing. Among these, land use and land cover information classification have attracted much attention using the HSI dataset [43]. Accurate identification and classification of considered targets opens a new field of studies for accurate mapping of minerals and other geological targets. For HSI classification, various works have been done using CNN and transformer-based networks. In addition to that, many researchers have effectively used hybrid models to leverage the use of CNN and transformer features together for HSI classification.

Most transformer-based models use small patches of input image before passing it to the encoder of the transformer network but inspired by the performance of CMT: convolution neural network meets vision transformers [1] model for image classification of Red-Green-Blue (RGB) images. We grounded the transformer model of our proposed model using one of the stages of the CMT transformer. CMT architecture is currently suitable for images with three channels only. Therefore, we proposed a hybrid network of a 3D convolution neural network and a CMT block as a transformer for the classification of hyperspectral images.

The main contributions of this paper are summarized as follows:

1. A simple 3D-CNN and ViT-based hybrid module is proposed in our 3D-CmT network for handling hyperspectral images' spatial and spectral data effectively for pixel-level classification.
2. We devise two simple but effective modules in 3D-CmT, i.e., the 3D-CNN module assisted by Principal Component Analysis (PCA) [17] to extract the most relevant band information through extraction and learning of spectral features, and the second module is the Transformer Encoder which is based on the CMT network [1], to learn global spatial representations using the self-attention mechanism of the vision transformer. Using PCA, we have reduced the dimensionality of hyperspectral data for a shorter computation time before using the 3D-CNN module.
3. We quantitatively and qualitatively evaluate the classification efficacy of the proposed 3D-CmT model on three representative hyperspectral (HS) datasets, i.e., Salinas Scene, Indian Pines, and Pavia University.

The remainder of this paper is structured as follows. The Related Work is discussed in Sect. 2. Section 3 introduces the proposed 3D-CmT model together with the experimental data sets used for this work. The experimental setting and the results are discussed in Sect. 4. The conclusion is presented in Sect. 5.

2 Related Work

This section gives an overview of the work done using CNN and Transformer models for hyperspectral image classification.

2.1 CNN Based

In the last decade, deep learning has produced significant technological advancements for hyperspectral image processing and analysis. For computer vision, CNNs are a very popular and widely used architecture for automatic feature extraction and learning. In recent years, CNN-based models have been extensively explored for the HSI classification. In [6], the authors have used 2D-CNN along with multilayer perceptron to learn the spatial and spectral information of pixels for the HSI classification. They demonstrated the capability of CNN on different variants of the support vector machine (SVM). The hierarchical deep spatial features have been extracted by off-the-shelf CNN architecture in [23] for HSI classification. A deep feature fusion network (DFFN) has been proposed by [24] to explore the hierarchical layers of CNN for HSI classification. In [25], a dual-path network (DPN) based HSI classification model has been proposed. Along with the wide usage of 2D-CNN, 3D-CNNs have also been utilized by various researchers to handle the spectral data of HSIs. In [7,40], a 3D-CNN model has been proposed to jointly learn the spatial and spectral features of hyperspectral images. 3D-CNN has also been exploited with 2D-CNN by HybridSN model for HSI classification [2]. A deep pyramidal residual network spectral and spatial information of hyperspectral image classification has been proposed in [26].

2.2 Transformer Based

In addition to CNN, in recent years, various vision transformer (ViT) [8] based architectures have shown impressive performance in computer vision due to its global attention and model long-range dependencies for spatial information in the images. There has been abundant work completed to explore transformer-based models for HSI classification. A SpectralFormer model proposed by [9] used a transformer to take advantage of hyperspectral data from a sequential perspective. Two modules are devised in their model, the first module is capable of learning local sequential information of spectrally correlated hyperspectral bands and providing the group-wise spectral embeddings (GSE) and the second module is cross-layer adaptive fusion (CAF) which carries memory components from shallow layers to deep layers. In CSiT [12], a multiscale vision transformer model is proposed. They have fused the two branches of vision transformer which are individually learning the pixel-wise features at different scales. They proposed two modules namely multiscale spectral embedding (MSSE) and cross-spectral attention fusion (CSAF) module for HSI classification. The spatial-spectral transformer (SST) model [13] combines CNN, DenseTransformer, and multilayer perceptron for HSI classification.

2.3 CNN-Transformer Based

The hybrid combination of CNN with the transformer block is performing impressively in HSI classification. Hyperspectral image transformer (HiT) proposed in [4] also comprises both CNN and ViT encoder, in which they proposed a spectral-adaptive 3D convolution projection (SACP) module and Conv-Permutator module to capture and learn the spectral-spatial feature information of hyperspectral images. In the FusionNet [14] model, a fusion of CNN and Transformer network for HSI classification is proposed. In [15], authors developed a convolution and transformer adaptive fusion (CTAFNet) strategy for pixel-wise classification of hyperspectral images. To capture the local high-frequency information they have used a convolution module and to handle sequential and global low-frequency they have used a transformer module. In addition to HSI data for classification, many researchers have used other modalities data sources like LiDAR, SAR, and MSI to enhance the capability of a classification model. In [3], a spectral-spatial feature tokenization transformer (SSFTT) based model is proposed. The latest work described by DBCTNet [39] also used a hybrid network of convolutional and transformer based model for hyperspectral classification, they used double branch convolutional transformer network for parallel combination convolution and self-attention instead of their serial combination. In this work, authors proposed three modules for HSI classification the first module comprises a spectral-spatial feature extraction module for low-level features, the second module uses a Gaussian weighted feature tokenizer for feature transformation and lastly the third module comprises a transformer encoder used for feature learning. As HSI is providing more information through the narrow bands, various works were done for spectral dimension.

Inspired by the previous work on hyperspectral image classification, we also investigate and demonstrate the potential of a 3D convolutional layer for spectral feature extraction and transformer network grounded by the CMT model for global feature learning with self-attention mechanism.

3 Methodology

3.1 Overall Architecture

We aim to build a hybrid model using the capability of 3D convolution and transformers for the classification of HSI. Figure 1 shows the overall framework of HSI classification using the proposed 3D-CmT model which consists of two key modules, i.e., HSI spectral feature extraction module (SFE) using the 3D-convolution and Transformer Encoder (TE) module as illustrated in Fig. 1b. As most of the transformer-based classification model splits an input image to nonoverlapping patches, however, this may ignore the intraobject relationship and representation. Therefore, to handle this limitation, we do not split the input image into nonoverlapping patches before the transformer encoder module.

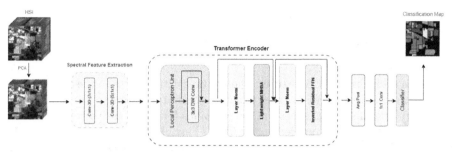

(a) Proposed 3D-CmT architecture with CMT [1] block as Transformer Encoder module.

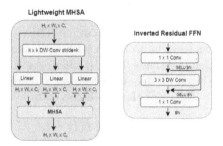

(b) Lightweight MHSA and Inverted Residual FFN blocks.

Fig. 1. Overall framework of the proposed 3D-CmT model for the HSI classification.

However, the spectral dimension of HSI is reduced using PCA transformation and 3D-Convolution for less computation overhead. Each module has been described in the following subsections.

Spectral Feature Extraction Module. The Hyperspectral data denoted as $I \in \mathbb{R}^{m \times n \times c}$, where m, n, and c are the height, width, and number of channels of an input image, respectively. HS imagery comes with an abundance of spectral data in the form of large spectral channels. To reduce computational cost and memory consumption during model training, PCA transformation has been applied before the training process. In addition to that, PCA has been used to extract the most relevant band information from spectral data. Reduces c channels to b channels. So, after PCA transformation the data is expressed as $I_{\text{pca}} \in \mathbb{R}^{m \times n \times b}$, and then given as input to the model.

Then, the first component of our proposed 3D-CmT is two layers of the 3D convolutional block, which has given impressive results, especially in handling the channels of 3D images. To learn the spectral representation, 3D-Conv layers are the perfect choice as discussed in the literature review Sect. 2. Hence, spectral feature extraction from the input HSI is termed a spectral feature extraction module. The kernel size of the first 3D-Conv layer is kept as 7 and for the subsequent second layer, it is chosen as 3. The overall computation of the SFE module is represented using Eq. 1,

$$SFE(I_{pca}) = Conv_2(Conv_1(I_{pca}, k_1 = 7), k_2 = 5). \tag{1}$$

Transformer Encoder Module. The features extracted from the SFE module that extracts the spectral features from HSI served as input to the TE module that learns the high-level spatial representations for input and output mapping. Our TE component is grounded on the CMT-Tiny block of the CMT network [1], which consists of a local perceptron unit (LPU) as shown in Fig. 1a, a lightweight multiheaded self-attention module (LWMHSA) and an inverted residual feedforward network (IRFFN) as presented in Fig. 1b.

LPU: In the LPU block, depthwise convolution (DWConv) is used to keep the local relation and structural information with less computational cost. It is described using Eq. 2 which is discussed in [1]: where $I \in \mathbb{R}^{h \times w \times d}$, h × w is the spatial dimension of the input at the current stage, d represents the dimension of features, and $DWConv$ indicates the depthwise convolution.

$$LPU(I) = DWConv(I) + I. \tag{2}$$

LWMHSA: To decrease the computational overhead of the original self-attention module, we use a similar light weight multi-head self-attention as mentioned in [1], where $k \times k$ depthwise convolution with stride k is used to reduce the spatial dimension of key **K** and value **V** before the attention step with a relative bias of **B** as shown in Fig. 1b. The light-weight attention function is defined as:

$$LWMHSA(Q, K, V) = softmax(\frac{QK^{'T}}{\sqrt{d_k}} + B)V^{'}. \tag{3}$$

where query $\mathbf{Q} \in \mathbb{R}^{h \times w \times d_k}$, key $\mathbf{K}^{'} = DWConv(\mathbf{K}) \in \mathbb{R}^{\frac{h \times w}{k^2} \times d_k}$ and value $\mathbf{V}^{'} = DWConv(\mathbf{V}) \in \mathbb{R}^{\frac{h \times w}{k^2} \times d_v}$ and bias $\mathbf{B} \in \mathbb{R}^{(h \times w) \times \frac{h \times w}{k^2}}$.

IRFFN: The inverted residual feed-forward network is similar to [1] consisting of an expansion layer and depth-wise convolution followed by a projection layer.

$$IRFFN(I) = Conv(DWConv(Conv(I)) + Conv(I)), \tag{4}$$

The 3D-CmT model can be formulated using the above modules and components:

$$I_i = LPU(SFE(I_{i\text{-}1})) \tag{5}$$

$$I_i^{'} = LWMHSA(LN(I_i)) + I_i \tag{6}$$

$$O_i = IRFFN(LN(I_i^{'})) + I_i^{'} \tag{7}$$

where I_i and I_i^i represent the output features of SFE followed by LPU and LWMHSA module for the ith block, respectively. LN denotes the layer normalization [42].

To classify the pixels of the input image, the softmax function is applied to the output O_i from the TE module. The label with the highest probability value is the category of the pixel. The reason to combine both 3D-Convolution and Vision Transformer networks is their capability to learn the representation of image spatial data along with its rich spectral information. The main advantage of HS imageries is their narrow spectrum rich data which tells the unique spectral signature of the target object. Therefore, initially, 3D-Conv layers extract the spectral features from HSI data. Then, after extracting and learning the spectral representation from HSI data, the incorporation of the Transformer module is used to understand the global spatial features and also to handle the long-range dependencies.

Table 1. Ground Truth classes of the **Salinas Scene, Indian Pines** and **University of Pavia** Datasets with their respective samples number.

Class No.	Salinas Scene		Indian Pines		University of Pavia	
	Class Name	Samples	Class Name	Samples	Class Name	Samples
1	Brocol_green_weeds_1	2009	Alfalfa	46	Asphalt	6631
2	Brocoli_green_weeds_1	3726	Corn-notill	1428	Meadows	18649
3	Fallow	1976	Corn-mintill	830	Gravel	2099
4	Fallow_rough_plow	1394	Corn	237	Trees	3064
5	Fallow_smooth	2678	Grass-pasture	483	Painted metal sheets	1345
6	Stubble	3959	Grass-trees	730	Bare Soil	5029
7	Celery	3579	Grass-pasture-mowed	28	Bitumen	1330
8	Grapes_untrained	11271	Hay-windrowed	478	Self-Blocking Bricks	3682
9	Soil_vinyard_develop	6203	Oats	20	Shadows	947
10	Corn_senesced_green	3278	Soybean-notill	972		
11	Lettuce_romaine_4wk	1068	Soybean-mintill	2455		
12	Lettuce_romaine_5wk	1927	Soybean-clean	593		
13	Lettuce_romaine_6wk	916	Wheat	205		
14	Lettuce_romaine_7wk	1070	Woods	1265		
15	Vinyard_untrained	7268	Building-Grass-Trees-Drives	386		
16	Vinyard_vertical	1807	Stone-Steel-Towers	93		

4 Experiments and Results

4.1 Dataset Description

In our work, three commonly and widely used openly available HSI datasets are selected for the experiments, including the Salinas Scene (SA), Indian Pines (IP), and Pavia University (UP) datasets.

Salinas Scene: The images in the SA dataset have a spatial size of 512 × 217 and 224 spectral bands that span the electromagnetic wavelength range of 360 to 2500 nm. This dataset consists of a total of 16 classes.

Indian Pines: The IP dataset contains images having 224 spectral bands with spatial size of 145 × 145 each. This dataset covers the hyperspectral imaging wavelength range from 400 to 2500 nm. Their ground truth is provided for 16 different classes of vegetation.

University of Pavia: Images with a spatial resolution of 610 × 340 and 103 spectral bands between 430 and 860 nm are included in the UP dataset. There are nine classifications of urban land cover in the ground truth.

The publicly available hyperspectral datasets[1] are downloaded for the experiments. Table 1 represents the description of the four datasets taken in our study including the total number of classes along with the class type and total number of samples corresponding to each class.

4.2 Evaluation Metrics

To measure the classification accuracy of the proposed model, we used three commonly used classification evaluation metrics, i.e., *Average Accuracy (AA)*, *Overall Accuracy (OA)* and *Kappa Coefficient (Kappa)*. AA is expressed as the percentage of the average of classwise classification accuracies; OA is represented as a percentage of the number of precisely classified samples divided by the total test samples; and Kappa is a statistical measure between the ground samples map and classification map that gives mutual information about a strong agreement between them. Additionally, the total number of Floating Point Operations (FLOPs) and model parameters are also considered to compare the classification accuracy of the proposed model.

4.3 Implementation Details

The 3D-CmT model is implemented on a system with Intel ® Xeon ® Platinum 8180 CPU @2.50 GHz, 1 TB of RAM, and NVIDIA Tesla V100 GPU, 32 GB of RAM. Python v3.7.4 and PyTorch [5] based environment is used to do all the experiments.

The Salinas Scene dataset is selected as an example to illustrate the 3D-CmT model. The Adam [27] optimizer has been chosen as the optimizer taking the initial learning rate value as 1e−3 and 1e−5 as weight decay. For batch training, the size of each batch is set to 32 with total training epochs of 100. The number of PCA components taken is 30 with an image size of 64 × 64 for training. The overall steps of the proposed 3D-CmT method are shown in Algorithm 1

To demonstrate the efficacy of the proposed model, several CNN and Transformer networks are considered for comparative analysis: CNN-based models

---

[1] Openly Available HSI Datasets https://www.ehu.eus/ccwintco/index.php/Hyperspectral_Remote_Sensing_Scenes.

Algorithm 1. 3D-CmT Model

Input: HSI data $X \in \mathbb{R}^{m \times n \times c}$ as input and $Y \in \mathbb{R}^{m \times n}$ as ground truth with a patch size of 64 and number of PCA bands of 30.
Output: Predicted classification labels on the test datasets.
1: Set Adam (learning rate: 1e-3) as optimizer, epochs number *total_epoch* to 100 with batch-size of 32.
2: After PCA transformation, obtain the I_{pca} components.
3: Divide I_{pca} into the training dataset and validation dataset and then generate data loader for training and validation data.
4: **for** epoch = 1 to *total_epoch* **do**
5:   Perform 3D convolution of SFE module.
6:   Perform Transformer Encoding on TE module.
7:   Use the softmax function to identify the class labels.
8: **end for**
9: Use the test dataset with the trained model to get the predicted class labels.

---

include 2D-CNN [38], 3D-CNN [37] with their original implementations and HybridSN [2] with same parameters excluding PCA bands which is selected as 30 for fair comparison. Transformer and hybrid model-based networks include ViT [8], where ViT-Base model configuration is chosen. SpectralFormer [9], the model is re-trained and the accuracies are similar to the results reported in the original paper for IP and UP Datasets. But, for Salinas Scene Datasets, all the model parameters have been set as per the Indian Pines dataset except the number of epochs and training samples i.e., 100 and 0.3 respectively for a fair comparison. For SSFTT [3], for a fair comparison we have taken an image with a patch size of 65 × 65 and the model has been trained for 100 epochs. DBCTNet [40] model is used as per its original implementation.

4.4 Classification Results

Quantitative Analysis. This section presents the quantitative results of the proposed method in terms of metrics mentioned in Sect. 4.2. An extensive comparative analysis is performed for various state-of-the-art models on the SA dataset in Table 2, the UP dataset in Table 3, and the IP dataset in Table 4. For the SA dataset, the 3D-CmT model outperforms all other models in terms of *OA*, *AA*, and *kappa* values. However, in terms of training time, number of Params, and number of FLOPs, performs satisfactorily. For the IP dataset, in terms of *OA* and *kappa* values, the proposed model is performing second best as the HybridSN model is performing best in terms of accuracy metrics. Similarly, for the UP dataset, the proposed model is performing best among all other CNN and Transformer models in terms of all three accuracy metrics. Based on the comparative results with CNN and Transformer models, it is worth mentioning that the 3D-CmT model is outperforming all the models for the SA dataset as represented in Table 2 and UP dataset as shown in Table 3. However, for the IP dataset, the proposed model is the second best performing as shown in Table 4.

Table 2. Comparative Results for **Salinas Scene Dataset** where training samples taken as 30% and patch size is 64 × 64. The bold and underlined text shows the best and second best performance.

Model	Training Time	#Params	#FLOPs	OA	AA	Kappa
2D-CNN [38]	3 min 37 s	1.67M	32.80M	<u>99.99</u>	<u>99.99</u>	99.89
3D-CNN [37]	53 s	995K	**24.2K**	99.34	99.75	99.26
HybridSN [2]	2 h 44 min	51.76M	2.39G	<u>99.99</u>	<u>99.99</u>	<u>99.99</u>
ViT [8]	2 h 41 min	125.16M	<u>967.21K</u>	99.52	99.44	99.47
SpectralFormer [9]	6 min 14 s	378.13K	17.19M	88.41	93.25	87.15
SSFTT [3]	22 min 39 s	<u>153.22K</u>	508.61M	99.92	99.93	99.93
DBCTNet [40]	8 min 38 s	**30.88K**	12.81M	94.44	97.34	93.82
3D-CmT	1 h 8 min	7.55M	120.68M	**100.0**	**100.0**	**100.0**

Table 3. Comparative Results for **University of Pavia Dataset** where training samples taken as 30% and patch size is 64 × 64. The bold and underlined text shows the best and second best performance.

Model	Training Time	#Params	#FLOPs	OA	AA	Kappa
2D-CNN [38]	1 min 47 s	1.61M	19.70M	<u>99.95</u>	<u>99.94</u>	<u>99.94</u>
3D-CNN [37]	39 s	994K	**24.1K**	99.74	99.72	99.66
HybridSN [2]	2 h 10 min	51.76M	2.39G	99.80	99.51	99.74
ViT [8]	1 h 13 min	105.3M	<u>961.84K</u>	97.22	94.52	96.32
SpectralFormer [9]	22 min 8 s	183.65K	4.47M	90.24	89.05	86.89
SSFTT [3]	39 min 25 s	<u>153.22K</u>	508.61M	98.72	96.28	98.30
DBCTNet [40]	2 min 17 s	**16.36K**	6.37M	98.53	97.79	98.06
3D-CmT	55 min 57 s	7.54M	120.67M	**99.98**	**99.96**	**99.97**

Qualitative Analysis. This section presents the visual results of the proposed method along with the results of comparative models and their ground-truth maps. Figure 3, Fig. 2, and Fig. 4 represent the results of classification maps generated on the SA, IP, and UP datasets, respectively. By visual comparisons, it is worth noting that 3D-CmT model-generated labels are very close and accurate to ground-truth labels compared to the other state-of-the-art models. For the SA data set, the results of 3D-CNN, ViT, SpectralFormer, and DBCTNet are not

Table 4. Comparative Results for **Indian Pines Dataset** where training samples taken as 30% and patch size is 64 × 64. The bold and underlined text shows the best and second best performance.

Model	Training Time	#Params	#FLOPs	OA	AA	Kappa
2D-CNN [38]	33 s	1.66M	32.28M	84.55	91.36	82.08
3D-CNN [37]	13 s	995K	**24.2K**	98.39	97.76	98.16
HybridSN [2]	31 min 26 s	51.76M	2.39G	**99.62**	**99.61**	**99.57**
ViT [8]	26 min 09 s	124.37M	<u>967.21K</u>	89.86	84.90	88.43
SpectralFormer [9]	3 min 18 s	355.57K	16.53M	76.31	84.48	73.31
SSFTT [3]	8 min 9 s	<u>153.22K</u>	508.61M	97.10	92.09	96.69
DBCTNet [40]	20 min 38 s	**30.3K**	12.55M	98.14	<u>98.38</u>	97.88
3D-CmT	12 min 48 s	7.55M	120.68M	<u>98.50</u>	95.61	<u>98.29</u>

accurate for some classes, as they are misclassified. However, our proposed model can accurately predict each class, demonstrating its good performance. For the IP dataset, the visual results of HybridSN are good compared to all other models. There are misclassifications of pixels for 2D-CNN, ViT, and SpectralFormer models. However, our proposed model is comparable to HybridSN performance. For the UP dataset, there is slightly lower visual performance of the 3D-CNN and SpectralFormer model, but all other models including the proposed model can give good visual results.

4.5 Ablation Study

We analyze and evaluate the classification performance of the 3D-CmT model in terms of classification evaluation metrics as discussed in Sect. 3. Extensive ablation studies on the Salinas Scene Dataset include experimentation with different combinations of PCA components, image patch size, and percentage of training samples. Although our base model is a CMT-Tiny network of [1] that consists of 4 stages of CMT blocks, therefore, we experimented with different combinations of CMT blocks in our proposed model.

Table 5. Classification performance analysis of effect of **with and without PCA** in 3D-CmT model for 30% training samples with patch size of 64 × 64 on Salinas Dataset.

PCA	#Params	#FLOPs	OA	AA	Kappa
w/o	<u>7.63M</u>	856.19M	99.98	99.98	99.98
10	7.54M	**36.13M**	<u>99.99</u>	99.98	<u>99.98</u>
20	7.54M	78.41M	99.99	99.99	99.99
30	**7.54M**	<u>57.27M</u>	100.0	100.0	100.0

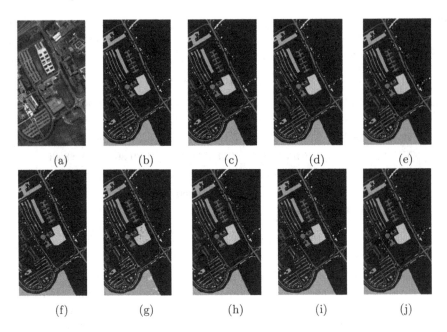

Fig. 2. Classification Results of the UP dataset. (a) Input. (b) Ground Truth. (c) 2D-CNN. (d) 3D-CNN. (e) HybridSN. (f) ViT. (g) SpectralFormer. (h) SSFTT. (j) DBCTNet. (j) 3D-CmT.

Effect of PCA Components on Training. Table 5 represents the accuracy of the model while we trained the model with and without PCA components. Extensive experiments with different numbers of PCA components show that there is very little difference in the accuracies of all the cases, but PCA components with a value of 30 give the best accuracies in terms of all metrics. Moreover, we also experiment with the case where PCA is not applied before the training, in this case, model training time is very high and accuracies are also decreasing. So, the value of 30 is used as the optimal number of PCA components for our proposed model in all three datasets.

Effect of Patch Size and Percentage of Training Samples. To check the effects of the patch size of an input image, we experimented with different patch sizes along with the model training with 30% and 10% training samples. The results of the 30% training samples are shown in Table 6, where the patch size of 64 × 64 gives the best accuracies among all the other cases, but its FLOPs are higher than those of the other cases. Although patch sizes of 8 × 8, 16 × 16, and 32 × 32 have fewer FLOPs, their accuracies are slightly lower than the 64 × 64 case. Therefore, we used a large patch size for our experiments, that is, 64 × 64. Similarly, Table 7 represents the experimental results for 10% training samples. Patch size of 16 × 16 is the best-performing case in terms of accuracy. But the 64 × 64 case is second best among all other cases. The results of the

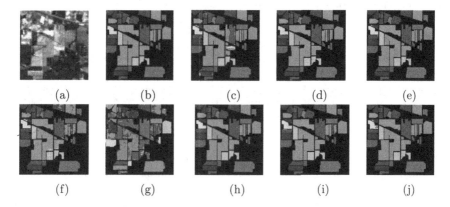

Fig. 3. Classification Results of the IP dataset. (a) Input. (b) Ground Truth. (c) 2D-CNN. (d) 3D-CNN. (e) HybridSN. (f) ViT. (g) SpectralFormer. (h) SSFTT. (j) DBCT-Net. (j) 3D-CmT.

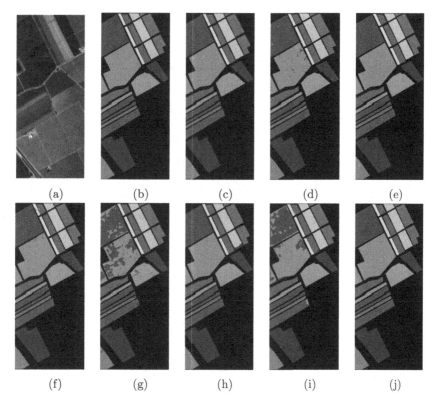

Fig. 4. Classification Results of the SA dataset. (a) Input. (b) Ground Truth. (c) 2D-CNN. (d) 3D-CNN. (e) HybridSN. (f) ViT. (g) SpectralFormer. (h) SSFTT. (j) DBCTNet. (j) 3D-CmT.

Table 6. Classification performance for **30% training samples** on Salinas Scene Dataset.

Patch Size	#Params	#FLOPs	OA	AA	Kappa
8 × 8	7.47M	1.90M	99.93	99.92	99.93
16 × 16	**7.47M**	7.54M	99.99	99.99	99.99
32 × 32	7.48M	30.13M	99.96	99.95	99.96
64 × 64	7.54M	57.27M	**100.0**	**100.0**	**100.0**

Table 7. Classification performance for **10% training samples** on Salinas Scene Dataset.

Patch Size	#Params	#FLOPs	OA	AA	Kappa
8 × 8	7.47M	1.90M	99.35	99.67	99.28
16 × 16	**7.47M**	7.54M	**99.88**	**99.86**	**99.87**
32 × 32	7.48M	30.13M	99.77	99.72	99.75
64 × 64	7.55M	120.68M	99.79	99.76	99.76

Table 8. Classification performance analysis between **different CMT stage** in 3D-CmT model for 30% training samples with patch size of 64 × 64 on Salinas Dataset.

#CMT-Stage	#Params	#FLOPs	OA	AA	Kappa
4	7.96M	443.47M	100.0	100.0	100.0
3	7.72M	295.67M	100.0	100.0	100.0
2	7.60M	121.98M	99.96	99.92	99.95
1	**7.54M**	**57.27M**	**100.0**	**100.0**	**100.0**

Table 9. Classification performance analysis between **different 3D-Conv layer** in 3D-CmT model for 30% training samples with patch size of 64 × 64 on Salinas Dataset.

#3D-Conv Layer	#Params	#FLOPs	OA	AA	Kappa
1 ($k_1 = (7 \times 1 \times 1)$)	7.55M	60.12M	99.99	99.98	99.99
2 ($k_1 = (7 \times 1 \times 1)$, $k_2 = (5 \times 1 \times 1)$)	**7.54M**	**57.27M**	**100.0**	**100.0**	**100.0**
3 ($k_1 = (7 \times 1 \times 1)$, $k_2 = (5 \times 1 \times 1)$, $k_3 = (3 \times 1 \times 1)$)	7.54M	216.62M	99.99	99.99	99.99

experiment show that a patch size of 64 × 64 is the optimal patch size for our proposed model.

Effect of CMT-Stage and 3D-Conv Layer. Experiments have been conducted to select an optimal number of CMT stages and 3D-Conv layers. Table 8 shows the experimental results for different numbers of CMT-Stages used. No major improvement is observed in the performance of the model if we increase the number of stages. However, CMT-Stage with only one block is performing best in terms of FLOPs and accuracy. Therefore, we have considered only one CMT block in the proposed model. The experimental results of different sets of 3D-Conv layers with one CMT block are shown in Table 9, where two 3D-Conv layers are the best-performing case among all accuracy metrics.

5 Conclusion

A 3D-CNN and Vision Transformer hybrid network named 3D-CmT has been proposed for hyperspectral image classification. It uses 3D-CNN for local feature learning along with the narrow spectral information, and Vision Transformer for global feature representation. Experiments have been carried out for different patch sizes of images, along with the different percentages of training samples. The proposed model is comparable with other comparative models in terms of both quantitative and visual results. From this study, we can conclude that to handle spatial and spectral HSI information, a hybrid network of 3D-CNN and ViT can be effectively used for the classification of hyperspectral images. For future work, we would like to add other image modalities such as Synthetic Aperture Radar (SAR), Light Detection and Ranging (LiDAR), and Digital Elevation Model (DEM) data to further verify the efficacy of the model.

References

1. Guo, J., et al.: CMT: convolutional neural networks meet vision transformers. In: Proceedings of the IEEE/CVF Conference on Computer Vision and Pattern Recognition, pp. 12175–12185 (2022)
2. Roy, S., Krishna, G., Dubey, S., Chaudhuri, B.: HybridSN: exploring 3-D-2-D CNN feature hierarchy for hyperspectral image classification. IEEE Geosci. Remote Sens. Lett. **17**, 277–281 (2019)
3. Sun, L., Zhao, G., Zheng, Y., Wu, Z.: Spectral-spatial feature tokenization transformer for hyperspectral image classification. IEEE Trans. Geosci. Remote Sens. **60**, 1–14 (2022)
4. Yang, X., Cao, W., Lu, Y., Zhou, Y.: Hyperspectral image transformer classification networks. IEEE Trans. Geosci. Remote Sens. **60**, 1–15 (2022)
5. Paszke, A., et al.: PyTorch: an imperative style, high-performance deep learning library. Adv. Neural Inf. Process. Syst. **32**, 8024–8035 (2019). http://papers.neurips.cc/paper/9015-pytorch-an-imperative-style-high-performance-deep-learning-library.pdf
6. Makantasis, K., Karantzalos, K., Doulamis, A., Doulamis, N.: Deep supervised learning for hyperspectral data classification through convolutional neural networks. In: 2015 IEEE International Geoscience and Remote Sensing Symposium (IGARSS), pp. 4959–4962 (2015)

7. Hamida, A., Benoit, A., Lambert, P., Amar, C.: 3-D deep learning approach for remote sensing image classification. IEEE Trans. Geosci. Remote Sens. **56**, 4420–4434 (2018)
8. Dosovitskiy, A.: An image is worth 16x16 words: transformers for image recognition at scale. ArXiv Preprint ArXiv:2010.11929. (2020)
9. Hong, D., et al.: SpectralFormer: rethinking hyperspectral image classification with transformers. IEEE Trans. Geosci. Remote Sens. **60**, 1–15 (2021)
10. Qiao, X., Roy, S., Huang, W.: Multiscale neighborhood attention transformer with optimized spatial pattern for hyperspectral image classification. IEEE Trans. Geosci. Remote Sens. **61**, 1–15 (2023)
11. Roy, S., Deria, A., Hong, D., Rasti, B., Plaza, A., Chanussot, J.: Multimodal fusion transformer for remote sensing image classification. IEEE Trans. Geosci. Remote Sens. **61** (2023)
12. He, W., Huang, W., Liao, S., Xu, Z., Yan, J.: CSiT: a multiscale vision transformer for hyperspectral image classification. IEEE J. Sel. Topics Appl. Earth Observ. Remote Sens. **15**, 9266–9277 (2022)
13. He, X., Chen, Y., Lin, Z.: Spatial-spectral transformer for hyperspectral image classification. Remote Sens. **13**, 498 (2021)
14. Yang, L., et al.: FusionNet: a convolution-transformer fusion network for hyperspectral image classification. Remote Sens. **14**, 4066 (2022)
15. Li, J., Xing, H., Ao, Z., Wang, H., Liu, W., Zhang, A.: Convolution-transformer adaptive fusion network for hyperspectral image classification. Appl. Sci. **13**, 492 (2022)
16. Yang, H., Yu, H., Zheng, K., Hu, J., Tao, T., Zhang, Q.: Hyperspectral image classification based on interactive transformer and CNN with multilevel feature fusion network. IEEE Geosci. Remote Sens. Lett. **20** (2023)
17. Abdi, H., Williams, L.: Principal component analysis. WIREs Comput. Stat. **2**, 433–459 (2010). https://wires.onlinelibrary.wiley.com/doi/abs/10.1002/wics.101
18. Wang, J., Zhang, L., Tong, Q., Sun, X.: The Spectral Crust project-Research on new mineral exploration technology. In: 2012 4th Workshop On Hyperspectral Image and Signal Processing: Evolution In Remote Sensing (WHISPERS), pp. 1–4 (2012)
19. Gevaert, C., Suomalainen, J., Tang, J., Kooistra, L.: Generation of spectral-temporal response surfaces by combining multispectral satellite and hyperspectral UAV imagery for precision agriculture applications. IEEE J. Sel. Topics Appl. Earth Observ. Remote Sens. **8**, 3140–3146 (2015)
20. Fong, A., Shu, G., McDonogh, B.: Farm to table: applications for new hyperspectral imaging technologies in precision agriculture, food quality and safety. CLEO: Applications and Technology, pp. AW3K-2 (2020)
21. Noor, S., Michael, K., Marshall, S., Ren, J., Tschannerl, J., Kao, F.: The properties of the cornea based on hyperspectral imaging: Optical biomedical engineering perspective. In: 2016 International Conference on Systems, Signals and Image Processing (IWSSIP), pp. 1–4 (2016)
22. Ardouin, J., Lévesque, J., Rea, T.: A demonstration of hyperspectral image exploitation for military applications. In: 2007 10th International Conference on Information Fusion, pp. 1–8 (2007)
23. Cheng, G., Li, Z., Han, J., Yao, X., Guo, L.: Exploring hierarchical convolutional features for hyperspectral image classification. IEEE Trans. Geosci. Remote Sens. **56**, 6712–6722 (2018)
24. Song, W., Li, S., Fang, L., Lu, T.: Hyperspectral image classification with deep feature fusion network. IEEE Trans. Geosci. Remote Sens. **56**, 3173–3184 (2018)

25. Kang, X., Zhuo, B., Duan, P.: Dual-path network-based hyperspectral image classification. IEEE Geosci. Remote Sens. Lett. **16**, 447–451 (2018)
26. Paoletti, M., Haut, J., Fernandez-Beltran, R., Plaza, J., Plaza, A., Pla, F.: Deep pyramidal residual networks for spectral-spatial hyperspectral image classification. IEEE Trans. Geosci. Remote Sens. **57**, 740–754 (2018)
27. Kingma, D., Ba, J.: Adam: a method for stochastic optimization. ArXiv Preprint ArXiv:1412.6980 (2014)
28. LeCun, Y., Bengio, Y., Hinton, G.: Deep learning. Nature **521**, 436–444 (2015)
29. Signoroni, A., Savardi, M., Baronio, A., Benini, S.: Deep learning meets hyperspectral image analysis: a multidisciplinary review. J. Imaging **5**, 52 (2019)
30. Li, S., Song, W., Fang, L., Chen, Y., Ghamisi, P., Benediktsson, J.: Deep learning for hyperspectral image classification: an overview. IEEE Trans. Geosci. Remote Sens. **57**, 6690–6709 (2019)
31. Jia, S., Jiang, S., Lin, Z., Li, N., Xu, M., Yu, S.: A survey: deep learning for hyperspectral image classification with few labeled samples. Neurocomputing **448**, 179–204 (2021)
32. Vaswani, A., et al.: Attention is all you need. Adv. Neural Inf. Process. Syst. **30** (2017)
33. LeCun, Y., et al.: Backpropagation applied to handwritten zip code recognition. Neural Comput. **1**, 541–551 (1989)
34. Petersson, H., Gustafsson, D., Bergstrom, D.: Hyperspectral image analysis using deep learning-a review. In: 2016 Sixth International Conference On Image Processing Theory, Tools and Applications (IPTA), pp. 1–6 (2016)
35. Paoletti, M., Haut, J., Plaza, J., Plaza, A.: Deep learning classifiers for hyperspectral imaging: a review. ISPRS J. Photogram. Remote Sens. **158**, 279–317 (2019)
36. Hu, W., Huang, Y., Wei, L., Zhang, F., Li, H.: Deep convolutional neural networks for hyperspectral image classification. J. Sens. **2015**, 1–12 (2015)
37. He, M., Li, B., Chen, H.: Multi-scale 3D deep convolutional neural network for hyperspectral image classification. In: 2017 IEEE International Conference on Image Processing (ICIP), pp. 3904–3908 (2017)
38. Liu, B., Yu, X., Zhang, P., Tan, X., Yu, A., Xue, Z.: A semi-supervised convolutional neural network for hyperspectral image classification. Remote Sens. Lett. **8**, 839–848 (2017)
39. Xu, R., Dong, X., Li, W., Peng, J., Sun, W., Xu, Y.: DBCTNet: double branch convolution-transformer network for hyperspectral image classification. IEEE Trans. Geosci. Remote Sens. **62**, 1–15 (2024)
40. Ahmad, M., Khan, A., Mazzara, M., Distefano, S., Ali, M., Sarfraz, M.: A fast and compact 3-D CNN for hyperspectral image classification. IEEE Geosci. Remote Sens. Lett. **19**, 1–5 (2020)
41. Ren, Q., Tu, B., Liao, S., Chen, S.: Hyperspectral image classification with iformer network feature extraction. Remote Sens. **14**, 4866 (2022)
42. Ba, J., Kiros, J., Hinton, G.: Layer normalization. ArXiv Preprint ArXiv:1607.06450 (2016)
43. Moharram, M., Sundaram, D.: Land use and land cover classification with hyperspectral data: a comprehensive review of methods, challenges and future directions. Neurocomputing **536**, 90–113 (2023)

Author Index

A
Abaid, Ayman I-197
Adkins, Meredith I-223
Angky, Edbert Valencio II-211
Arya, Sunita II-337
Atputharuban, Daniel Anojan I-180

B
Battula, Ramesh Babu I-118, I-132, I-146
Beyerer, Jürgen I-52, I-69
Bhattacharya, Rishi I-211
Biradar, Kuldeep I-118
Blaß, Benjamin I-69
Bontempi, Gianluca I-17
Bui, Van-Hung II-3

C
Cagas, William I-211
Chan, Kelvin C. K. II-276
Chen, Jialei II-31
Chen, Yen-Wei II-47
Cioppa, Anthony I-17
Colot, Martin I-17
Corcoran, Peter I-197
Cormier, Mickael I-69
Cothren, Jackson I-223

D
Daniol, Mateusz I-165
Dao, Thao Thi-Phuong I-293
Dao, Viet-Hang II-134, II-163
Deguchi, Daisuke II-31
Dhar, Debajyoti II-337
Dhull, Vasu II-322
Do, Nhu-Tai II-3
Droogenbroeck, Marc Van I-17
Dube, Sachin I-146
Dubey, Shiv Ram II-337
Duy, Vinh Nguyen II-88

E
Echizen, Isao I-308

F
Farooq, Aiman II-295
Farooq, Muhammad Ali I-197

G
Gérin, Benoît I-17
Grandhi, Shryuk I-211
Guo, Qiang II-118

H
Halin, Anaïs I-17
Hamada, Ryunosuke I-36
Han, Kun II-179
Han, Xian-Hua II-105
Harefa, Jeklin II-211
Heizmann, Michael I-69
Held, Jan I-17
Hemmerling, Daria I-165
Ho, Ngoc-Vuong I-223
Hoang, Nhu-Vinh I-281
Hsiao, Blake I-211
Hsieh, Felix I-308
Huang, Hsiang-Wei II-74
Huang, Ziqi II-276
Huynh, Viet-Tham I-293, I-338
Hwang, Jenq-Neng II-74

I
Isoda, Yuki II-58
Iyoda, Hayato II-105

J
Jean, Emmanuel I-17
Jiang, Wei II-179, II-245
Jiang, Zhongyu II-74

K

Kaushik, Vinay II-228
Ko, Chan I-211
Kobayashi, Daisuke II-58
Krohmer, Enrico I-52
Kumar, Vaibhav II-306, II-322
Kurniadi, Felix Indra II-211
Kwarciak, Kamil I-165

L

La, Thang II-163
Lall, Brejesh II-228
Lam, Michael I-211
Lawlor, Aonghus I-180
Le, Ngan I-223
Le, Thi-Lan II-261
Le, Viet-Duc II-261
Le, Viet-Tuan I-249
Li, Xinpeng II-18
Lin, Lanfen II-47
Ling, Nam II-179
Liu, Ziwei II-276
Lumentut, Jonathan Samuel II-211
Luu, Duc-Tuan I-249, I-324

M

Macq, Benoît I-17
Mahmoudi, Saïd I-17
Manh, Huy-Xuan II-134
MaungMaung, AprilPyone I-308
Minematsu, Tsubasa I-36
Minh, Hung-Le II-134
Mishra, Deepak II-295
Mishra, Priyanka I-87
Moorthi, S. Manthira II-337
Mu, Yuerong II-118
Muliawan, Nicholas Hans II-211
Murala, Subrahmanyam I-87, I-105
Murase, Hiroshi II-31

N

Nathanael, Oliverio Theophilus II-211
Nguyen, Hai-Dang I-281
Nguyen, Huy H. I-308
Nguyen, Phuc-Binh II-134
Nguyen, Quoc-Huy II-3
Nguyen, Tam V. I-293
Nguyen, Thanh-Son I-338
Nguyen, Thanh-Tung II-134
Nguyen, Thuy-Binh II-261
Nguyen, Trong-Thuan I-293
Nguyen, Tuan T. I-237
Nguyen, Van-Loc I-338
Nguyen-Mau, Trong-Hieu I-281

O

Osman, Islam I-3

P

Pathak, Sanhita II-228
Phan, Thinh I-223
Phuong, Thao Nguyen II-88
Piérard, Sébastien I-17

Q

Quan, Zhenzhen II-31

R

Rainwater, Chase I-223

S

Sahu, Yamini II-306, II-322
Sakaino, Hidetomo II-88
Sartipi, Mina I-237
Saxena, Prafulla I-105
Sharma, Utkarsh II-295
Shashwat, Satyajeet II-322
Shehata, Mohamed S. I-3
Shimada, Atsushi I-36
Singh, Kushall I-132
Singh, Satish Kumar II-337
Specker, Andreas I-69
Subbarao, P. I-118
Sun, Jiacheng II-74
Sun, Yongqing II-18, II-105

T

Takemura, Noriko II-147
Tang, Cheng I-36
Theopold, Christoph I-180
Thin, Dang-Van I-324
Tong, Minglei II-18
Tran, Gia-Nghia I-324
Tran, Minh-Hanh II-163
Tran, Minh-Triet I-281, I-293, I-338
Tran, Thanh-Hai II-163

Author Index

Truc, Nhu-Binh Nguyen   I-281
Truong, Duy-Van   II-134
Tyagi, Dinesh Kumar   I-105, I-118, I-132, I-146

U
Ullah, Ihsan   I-197
Usynin, Dmitrii   I-308

V
Vandeghen, Renaud   I-17
Vipparthi, Santosh Kumar   I-87, I-105
Vo, Duc Minh   I-249
Vu, Hai   II-134

W
Wang, Wei   II-179, II-245
Wen, Bihan   II-276
Wodzinski, Marek   I-165

Wolf, Stefan   I-52

Y
Yadav, Prem Shanker   I-132
Yamaguchi, Rento   I-267
Yamakura, Ryuta   II-197
Yamamoto, Riku   II-147
Yanai, Keiji   I-267, II-197
Yang, Cheng-Yen   II-74
Yi, Caleb Ng Zhi   I-69

Z
Zakiyyah, Alfi Yusrotis   II-211
Zanella, Maxime   I-17
Zhang, Chenkai   II-31
Zhang, Chujie   II-47
Zhou, Lebin   II-179, II-245
Zhu, Kevin   I-211